Multi-View Geometry Based Visual Perception and Control of Robotic Systems

Jian Chen, Bingxi Jia, and Kaixiang Zhang

Multi-View Geometry Based Visual Perception and Control of Robotic Systems

CRC Press is an imprint of the
Taylor & Francis Group, an **informa** business

MATLAB® is a trademark of The MathWorks, Inc. and is used with permission. The MathWorks does not warrant the accuracy of the text or exercises in this book. This book's use or discussion of MATLAB® software or related products does not constitute endorsement or sponsorship by The MathWorks of a particular pedagogical approach or particular use of the MATLAB® software.

Published in 2018 by CRC Press
Taylor & Francis Group
6000 Broken Sound Parkway NW, Suite 300
Boca Raton, FL 33487-2742

First issued in paperback 2020

© 2018 by Taylor & Francis Group, LLC
CRC Press is an imprint of Taylor & Francis Group, an nforma business

No claim to original U.S. Government works

ISBN-13: 978-0-367-57146-7 (pbk)
ISBN-13: 978-0-8153-6598-3 (hbk)

Library of Congress Cataloging-in-Publication Data

Names: Chen, Jian (Robotics engineer), author.
Title: Multi-view geometry based visual perception and control of robotic systems / Jian Chen.
Description: Boca Raton, FL : CRC Press/Taylor & Francis Group, 2017. | Includes bibliographical references and index.
Identifiers: LCCN 2018008526 | ISBN 9780815365983 (hb : acid-free paper) | ISBN 9780429489211 (ebook)
Subjects: LCSH: Robots–Control systems. | Computer vision.
Classification: LCC TJ211.35 .C44 2017 | DDC 629.8/92637–dc23
LC record available at https://lccn.loc.gov/2018008526

Visit the Taylor & Francis Web site at
http://www.taylorandfrancis.com

and the CRC Press Web site at
http://www.crcpress.com

Contents

Preface

Over the past decade, there has been a rapid development in the vision-based perception and control of robotic systems. Especially, multiple-view geometry is utilized to extract low-dimensional geometric information from abundant and high-dimensional image space, making it convenient to develop general solutions for robot perception and control tasks. This book aims to describe possible frameworks for setting up visual perception and control problems that need to be solved in the context of robotic systems.

The visual perception of robots provides necessary feedback for control systems, such as robot pose information, object motion information, and drivable road information. Since 3D information is lost and image noise exists in the imaging process, the effective pose estimation and motion identification of objects are still challenging. Besides, mobile robots are generally faced with complex scenes making it difficult to robustly detect drivable road space for safe operation. In this book, multiple-view geometry is exploited to describe the scene structure and maps from image space to Euclidean space. Optimization and estimation theories are applied to reconstruct the geometric information of the scene. Then, it is convenient to identify the real-time states of the robot and objects and to detect drivable road region based on geometric information.

The visual control of robots exploits visual information for task description and controls the robots through appropriate control laws using visual feedback. Since depth information is lost in the imaging process of monocular cameras, there exist model uncertainties in the control process. Besides, the limited field of view of the camera and the physical constraints (e.g., nonholonomic constraints) of the robots also have great influences on the stability and robustness of the control process. In this book, multiple-view geometry is used for geometric modeling and scaled pose estimation. Then Lyapunov methods are applied to design stabilizing control laws in the presence of model uncertainties and multiple constraints.

The book is divided into four parts.

- Part I consists of Chapter 1–3. This part is more tutorial than the others and introduces the basic knowledge of the robotics and the multiple-view geometry.

- In Part II, visual perception of robotics is presented. Specifically, Chapter 4 gives an overview of the perception problems. Chapter 5 describes a road construction approach based on iterative optimization and two-view geometry. Considering the complex lighting conditions, an illuminant invariant transform method is presented in Chapter 6 to eliminate the effects of shadows and recover the textures in shadows. After that, a nonlinear estimation strategy is proposed in Chapter 7 to identify the velocity and range of a moving object with a static monocular camera. To eliminate the motion constraint and prior geometric knowledge of the moving object, a static-moving camera system is exploited in Chapter 8 to achieve the velocity and range identification tasks.

- Visual control of robotics is given in Part III. First, Chapter 9 introduces the visual control problems of a robotic system. Then, for the general fully actuated robots with six DOFs, the trajectory tracking control is considered in Chapter 10 for both eye-in-hand and eye-to-hand configurations. In the presence of the system uncertainties, a robust control law is designed in Chapter 11 to asymptotic tracking of a moving object. Considering the field-of-view constraints, i.e., the target should remain visible in the camera view, a trajectory planning method based on navigation function is proposed in Chapter 12 to generate a desired trajectory, moving the camera from the initial pose to a goal pose while ensuring all the feature points remain visible. Besides, for the wheeled mobile robots, adaptive control laws are presented in Chapter 13 to achieve the trajectory tracking and pose regulation tasks, respectively. In Chapter 14, with the existence of unknown camera installing position, a trifocal tensor based controller is designed to address the trajectory tracking and regulation problems. Moreover, to identify the depth information during the control process, a unified controller with online depth estimation is developed in Chapter 15.

- Appendices are provided in Part IV to describe further details of Parts II and III.

The content in Parts II and III of this book (unless noted otherwise) has been derived from the authors' research work during the past several years in the area of visual perception and control. This book is aimed at researchers who are interested in the application of multiple-view geometry and Lyapunov-based techniques to emerging problems in perception and control of robotics.

The authors are grateful to the National Natural Science Foundation of China (Grant 61433013) for supporting our research on control of fuel cell intelligent vehicles. We would like to thank the colleagues and collaborators for their valuable contributions and support, without them this work would not have been possible. We would also like to acknowledge the support from the following members of the Sustainable Energy and Intelligent Vehicles Laboratory in the College of Control Science and Engineering at Zhejiang University: Xinfang Zhang, Qi Wang, Yanyan Gao, Yang Li, and Guoqing Yu.

If the reader finds any errors in the book, please send electronically to

jchen@zju.edu.cn

which will be sincerely appreciated. An up-to-date errata list will be provided at the homepage of the book:

http://person.zju.edu.cn/jchen/714344.html

MATLAB® is a registered trademark of The MathWorks, Inc. For product information, please contact:

The MathWorks, Inc.
3 Apple Hill Drive
Natick, MA 01760-2098 USA
Tel: 508-647-7000
Fax: 508-647-7001
E-mail: info@mathworks.com
Web: www.mathworks.com

Authors

Jian Chen received his BE and ME degrees from Zhejiang University, Hangzhou, China, in 1998 and 2001, respectively, and his PhD degree in electrical engineering from Clemson University, Clemson SC, in 2005. He was a research fellow in the University of Michigan, Ann Arbor, MI from 2006 to 2008, where he worked on fuel cell modeling and control. He joined IdaTech LLC, Bend, OR, in 2008, where he worked on fuel cell back power systems. Later in 2012, he joined Proterra Inc., Greenville, SC, where he was involved in the National Fuel Cell Bus Program. In 2013, he went back to academia and joined the College of Control Science and Engineering, Zhejiang University as a full professor. Moreover, he was recruited by the Chinese Recruitment Program of Global Youth Experts. The research interests of Jian Chen include visual servo techniques, control of fuel cell vehicles, battery management, and nonlinear control.

Bingxi Jia received his BE and PhD degrees both in control science and engineering from the College of Control Science and Engineering, Zhejiang University, Hangzhou, China, in 2012 and 2017, respectively. He joined Zhuying Tech. Inc., Hangzhou, China, in 2017, serving as the senior engineer responsible for the motion planning and control of autonomous vehicles. His research interests include vision-based control, computer vision, and motion planning for intelligent vehicles.

Kaixiang Zhang received his BE degree in automation from Xiamen University, Xiamen, China, in 2014. He is currently pursuing his PhD degree in the College of Control Science and Engineering, Zhejiang University, Hangzhou, China. He is a visiting student in the National Institute for Research in Computer Science and Control, Rennes, France. His current research interests include vision-based control and vision-based estimation.

FOUNDATIONS I

Chapter 1

Robotics

Robotics is concerned with the study of those machines that can replace human beings in the execution of a task, including physical activity, decision-making, and human interaction. This book is devoted to the studies on the robots that accomplish the physical tasks via locomotion. Generally, these types of robots can be divided into robot manipulators and mobile robots. Robot manipulators are also called fixed robots as they are generally mounted in industrial fields and the motion of end effectors are actuated by the motion of joints. The end effectors generally move in 3D space to perform manipulation tasks. Mobile robots are featured with the free mobility in large workspace, including wheeled mobile robots (WMRs), legged mobile robots, unmanned aerial vehicles, and autonomous underwater vehicles. WMRs are widely used in both industrial and daily life scenarios because they are appropriate for typical applications with relatively low mechanical complexity and energy consumption [107].

The key feature of a physical robot is its locomotion via specific mechanical structure. For control development, it should be modeled mathematically to describe its motion characteristics. In the autonomous navigation and manipulation tasks, the motion of robot bases and end effectors can be expressed mathematically by the notion of pose information. In this chapter, the general pose representation of rigid bodies is introduced, and then the robot kinematics is developed.

1.1 Pose Representation

A rigid body in physical space can be described by the position and orientation, which are collectively named as the pose information. Then, the robot locomotion can be described by the pose information, which serves as the output of the robot model. The motion of a rigid body consists of translations and rotations, resulting in the position and orientation descriptions with respect to the reference coordinate system. As shown in Figure 1.1, two coordinate frames $\mathcal{F}$ and $\mathcal{F}'$ exist in 3D Euclidean space. The motion from $\mathcal{F}$ to $\mathcal{F}'$ can be described by a translation and a rotation. Then, the pose of frame $\mathcal{F}'$ with respect to frame $\mathcal{F}$ can be described by the position and the orientation.

1.1.1 Position and Translation

In 3D Euclidean space, the position of the origin of coordinate frame $\mathcal{F}'$ relative to coordinate frame $\mathcal{F}$ can be denoted by the following: 3×1 vector $\begin{bmatrix} x & y & z \end{bmatrix}^T$. The components of this vector are the Cartesian coordinates of $\mathcal{F}'$ in the $\mathcal{F}$ frame, which are the projections of the vector x_f onto the corresponding axes. Besides the Cartesian coordinate system, the position of a rigid body can also be expressed in spherical or cylindrical coordinates. Such representations are generally used for the analysis of specific mechanisms such as the spherical and cylindrical robot joints as well as the omnidirectional cameras. As shown in Figure 1.2, the spherical coordinate system can be viewed as the three-dimensional version of the polar coordinate system, and the position of a point is specified by three numbers: the radial distance r from the point to the coordinate origin, its polar angle θ measured from a fixed zenith direction, and the azimuth angle ϕ of its orthogonal projection on a reference plane that passes through the origin and is orthogonal to the zenith. As shown in Figure 1.3, the position of a point in the cylindrical coordinate system is specified by three numbers: the distance ρ from the point to the chosen reference axis (generally called cylindrical or longitudinal axis), the direction angle ϕ from the axis relative to a chosen reference direction, and the distance z from a chosen reference plane

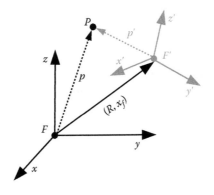

Figure 1.1: Pose representation.

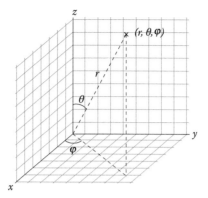

Figure 1.2: Spherical coordinate system.

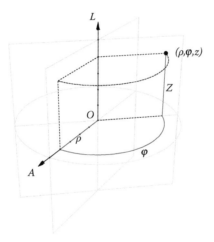

Figure 1.3: Cylindrical coordinate system.

perpendicular to the axis. The latter distance is given as a positive or negative number depending on which side of the reference plane faces the point.

A translation is a displacement in which no point in the rigid body remains in its initial position and all straight lines in the rigid body remain parallel to their initial orientations. The translation of a body in 3D Euclidean space can be represented by the following 3×1 vector:

$$x_f = \begin{bmatrix} x_{fx} \\ x_{fy} \\ x_{fz} \end{bmatrix}. \tag{1.1}$$

Conversely, the position of a body can be represented as a translation that takes the body from a starting position (the origin of frame $\mathcal{F}$) to the current position (the origin of frame $\mathcal{F}'$). As a result, the representation of position can be used to express translation and vice versa.

1.1.2 Orientation and Rotation

The orientation expresses the axis directions of frame $\mathcal{F}'$ with respect to frame $\mathcal{F}$. A rotation is a displacement that the origin of frame $\mathcal{F}'$ coincides with the origin of frame $\mathcal{F}$ and not all the axes of $\mathcal{F}'$ parallel to those of $\mathcal{F}$. As in the case of position and translation, any representation of orientation can be used to create a representation of rotation and vice versa. In the following descriptions in this book, the positive direction of rotational angle is defined by the right-hand rule, as indicated in Figure 1.4.

Rotation matrix

The orientation of frame $\mathcal{F}'$ with respect to frame $\mathcal{F}$ can be denoted by expressing the basis vectors $(\vec{x}',\vec{y}',\vec{z}')$ (the axis directions of $\mathcal{F}'$) in terms of the basis vectors $(\vec{x},\vec{y},\vec{z})$ (the axis directions of $\mathcal{F}'$). The rotation matrix R is used to express the rotation from $\mathcal{F}$ to $\mathcal{F}'$. The components of R are the dot products of the basis vectors of the two coordinate frames:

$$R = \begin{bmatrix} \vec{x}\cdot\vec{x}' & \vec{y}\cdot\vec{x}' & \vec{z}\cdot\vec{x}' \\ \vec{x}\cdot\vec{y}' & \vec{y}\cdot\vec{y}' & \vec{z}\cdot\vec{y}' \\ \vec{x}\cdot\vec{z}' & \vec{y}\cdot\vec{z}' & \vec{z}\cdot\vec{z}' \end{bmatrix}. \tag{1.2}$$

Since the basis vectors are unit vectors, the dot product of any two unit vectors is actually the cosine of the angle between them, the components are commonly referred to as direction cosines.

The rotation matrix R contains nine elements but only three Degrees of Freedom (DOF) to express the rotation and orientation in the 3D Euclidean space.

Figure 1.4: The right-hand rule for rotation.

Therefore, six auxiliary relationships exist among the elements of the matrix. Because the basis vectors of coordinate frame $\mathcal{F}$ are mutually orthonormal, as are the basis vectors of coordinate frame $\mathcal{F}'$, the columns of R formed from the dot products of these vectors are also mutually orthonormal. The rotation matrix is an orthogonal matrix and has the property that its inverse is simply its transpose. This property provides the six auxiliary relationships. Three require the column vectors to have unit length, and three require the column vectors to be orthogonal. Besides, the orthogonality of the rotation matrix can be seen from the expression in (1.2) by considering the frames in reverse order. The orientation of coordinate frame $\mathcal{F}'$ relative to coordinate frame $\mathcal{F}$ is the rotation matrix whose rows are clearly the columns of the matrix R.

Using the above definition, an elementary rotation of frame $\mathcal{F}'$ about the z axis through an angle θ is

$$R_z(\theta) = \begin{bmatrix} \cos\theta & -\sin\theta & 0 \\ \sin\theta & \cos\theta & 0 \\ 0 & 0 & 1 \end{bmatrix}, \tag{1.3}$$

and the same rotation about the y axis is

$$R_y(\theta) = \begin{bmatrix} \cos\theta & 0 & \sin\theta \\ 0 & 1 & 0 \\ -\sin\theta & 0 & \cos\theta \end{bmatrix}, \tag{1.4}$$

and the same rotation about the x axis is

$$R_x(\theta) = \begin{bmatrix} 1 & 0 & 0 \\ 0 & \cos\theta & -\sin\theta \\ 0 & \sin\theta & \cos\theta \end{bmatrix}. \tag{1.5}$$

Three angle representations

As mentioned above, the minimum representation of a rotation in 3D Euclidean space is of three numbers. As a result, a common representation for the orientation of coordinate frame $\mathcal{F}'$ with respect to coordinate frame $\mathcal{F}$ can be denoted as a vector of three angles. Then, the overall rotation can be decomposed into a sequence of rotations by the three angles about specific axes. For example, the rotation matrix R can be expressed by the multiplication of the three matrices $R = R_1 R_2 R_3$ with the rotational order R_1, R_2, R_3. Two commonly used angle representations are the Euler angles and the fixed angles, as described as follows:

■ **Euler angles:** The Euler angles have three components; each represents a rotation about the axes of a moving coordinate frame (i.e., frame $\mathcal{F}'$). As a result, the rotation matrix depends on the order of rotations. In this book, the indication of Euler angles (α, β, γ) expresses the rotation order $Z - Y - X$. Assuming the frames $\mathcal{F}$ and $\mathcal{F}'$ are initially coincident, the

3D rotation can be decomposed into three steps: rotate angle α about the z axis of frame $\mathcal{F}'$, rotate angle β about the y axis of frame $\mathcal{F}'$, and then rotate angle γ about the z axis of frame $\mathcal{F}'$. The rotational matrix can be expressed by the Euler angles as follows:

$$
R = \begin{bmatrix} \cos\alpha\cos\beta & \cos\alpha\sin\beta\sin\gamma - \sin\alpha\cos\gamma & \cos\alpha\sin\beta\cos\gamma + \sin\alpha\sin\gamma \\ \sin\alpha\cos\beta & \sin\alpha\sin\beta\sin\gamma + \cos\alpha\cos\gamma & \sin\alpha\sin\beta\cos\gamma - \cos\alpha\sin\gamma \\ -\sin\beta & \cos\beta\sin\gamma & \cos\beta\cos\gamma \end{bmatrix}
$$

(1.6)

Based on the above expression, the Euler angles can be decomposed by the rotation matrix as follows:

$$
\beta = a\tan2\left(-r_{31}, \sqrt{r_{11}^2 + r_{21}^2}\right)
$$

$$
\alpha = a\tan2\left(\frac{r_{21}}{\cos\beta}, \frac{r_{11}}{\cos\beta}\right)
$$

(1.7)

$$
\gamma = a\tan2\left(\frac{r_{32}}{\cos\beta}, \frac{r_{33}}{\cos\beta}\right).
$$

- **Fixed angles:** Except for the Euler angles, a vector of three angles can also denote the rotational relationship, with each angle representing a rotation about an axis of a fixed frame. As shown in Figure 1.1, $\mathcal{F}$ is the fixed frame and $\mathcal{F}'$ is the moving frame. Take these two frames to be initially coincident, ψ is the yaw rotation about the x axis of $\mathcal{F}$, θ is the pitch rotation about the y axis of $\mathcal{F}$, and ϕ is the roll rotation about the z axis of $\mathcal{F}$. Actually, a set of $X - Y - Z$ fixed angles is equivalent to the same set of $Z - Y - X$ Euler angles ($\alpha = \phi$, $\beta = \theta$, and $\gamma = \psi$).

The representations using three angles are intuitive but suffer from singularity. The singularity occurs when the rotational axis of the second term in the sequence becomes parallel to the rotational axis of the first or the third term. In that case, there are only two effective rotational axes instead of the original three axes, i.e., one DOF is lost. This issue occurs when $\beta = \pm\frac{\pi}{2}$ for Euler angles and $\theta = \pm\frac{\pi}{2}$ for fixed angles.

Angle-vector representation

Different from the three angle representations that decompose the overall rotation into three sequential rotations along specific axes, the angle-vector representation expresses the rotation by a rotational angle θ about a rotational vector v. The rotation matrix is expressed by the angle-vector representation by the Rodrigues' rotation formula as follows:

$$
R = I_{3\times3} + \sin\theta[v]_\times + (1 - \cos\theta)(vv^T - I_{3\times3}),
$$

(1.8)

where $[v]_\times$ is the skew-symmetric matrix defined as

$$[v]_\times = \begin{bmatrix} 0 & -v_z & v_y \\ v_z & 0 & -v_x \\ -v_y & v_x & 0 \end{bmatrix}. \tag{1.9}$$

The rotation angle and vector are encoded in the eigenvalues and eigenvectors of the rotation matrix R. The orthonormal rotation matrix R has one real eigenvalue $\lambda = 1$ and two complex eigenvalues $\lambda = \cos\theta \pm i\sin\theta$. According to the definition,

$$Rv = \lambda v, \tag{1.10}$$

where v is the eigenvector corresponding to the eigenvalue λ. For the case of $\lambda = 1$,

$$Rv = v, \tag{1.11}$$

which implies that the vector is not changed by the rotation. Actually, the rotational axis is the only vector in 3D space satisfying this feature. As a result, the eigenvector corresponding to the eigenvalue $\lambda = 1$ is the rotational vector v.

Unit quaternion representation

The quaternions are actually the extension of complex numbers and were first described by W. R. Hamilton. The concept of quaternions is widely applied to robotics and mechanics in 3D space. A quaternion ε is written as a scalar ε_0 plus a vector ε_v as follows:

$$\begin{aligned} \varepsilon &= \varepsilon_0 + \varepsilon_v \\ &= \varepsilon_0 + \varepsilon_1 \cdot i + \varepsilon_2 \cdot j + \varepsilon_3 \cdot k, \end{aligned} \tag{1.12}$$

where $\varepsilon_0, \varepsilon_1, \varepsilon_2$, and ε_3 are scalars, and i, j, and k are operators satisfying the following rules:

$$\begin{aligned} & ii = jj = kk = -1 \\ & ij = k, jk = i, ki = j \\ & ji = -k, kj = -i, ik = -j. \end{aligned} \tag{1.13}$$

The unit quaternion has the property that $\varepsilon_0^2 + \varepsilon_1^2 + \varepsilon_2^2 + \varepsilon_3^2 = 1$. The unit quaternion can be used to express a rotation of angle θ about the unit rotational vector v as follows:

$$\varepsilon_0 = \cos\frac{\theta}{2}, \varepsilon_v = \sin\frac{\theta}{2} \cdot v. \tag{1.14}$$

1.1.3 Homogeneous Pose Transformation

The preceding sections have addressed representations of position and orientation separately. The motion in 3D Euclidean space can be decomposed into a translation and a rotation. As shown in Figure 1.1, due to the effect of motion, there exists a transformation between the coordinates of point P with respect to frames $\mathcal{F}$ and $\mathcal{F}'$. Denote the coordinate of P with respect to frame $\mathcal{F}$ and $\mathcal{F}'$ as p and p', respectively; the transformation from p' to p is defined as follows:

$$p = R \cdot p' + x_f, \tag{1.15}$$

where R and x_f are the rotation and translation from frame $\mathcal{F}$ to frame $\mathcal{F}'$ expressed in $\mathcal{F}$. To be compact, define the homogeneous coordinates of point P with respect to frames $\mathcal{F}$ and $\mathcal{F}'$ as $\bar{p}$ and $\bar{p}'$, respectively, as follows:

$$\bar{p} = \begin{bmatrix} x & y & z & 1 \end{bmatrix}^T, \bar{p}' = \begin{bmatrix} x' & y' & z' & 1 \end{bmatrix}^T. \tag{1.16}$$

Then, the homogeneous transformation can be rewritten as the following:

$$\bar{p} = \begin{bmatrix} R & x_f \\ 0_{1 \times 3} & 1 \end{bmatrix} \bar{p}' = T\bar{p}', \tag{1.17}$$

where T is the 4×4 homogeneous transformation.

1.2 Motion Representation

In the previous section, the pose representation of rigid bodies in 3D Euclidean space is introduced. This section extends the concepts to the moving objects whose pose is varying, which is a basic description for robots. For robots, the tasks are generally accomplished by following specific paths or trajectories, or by regulating to specific target poses. In this section, the general concepts of path and trajectory are introduced for task description; then, the pose kinematics are described for motion analysis and further control development of robots.

1.2.1 Path and Trajectory

In robotics, a path is a spatial concept in space that leads from an initial pose to a final pose. A trajectory is a path with specified time constraints, i.e., it describes the poses at each specified time instants.

Parameterization

Due to the nature of trajectory, it is straightforward to parameterize the trajectory with time. Then, a trajectory can be denoted as $\mathbf{X}(t)$, where t is the time and $\mathbf{X}$ is the vector representing pose information. For the parameterization of the path, there should be a parameter that varies smoothly and differs everywhere along the path. A commonly used parameter is the arc length accumulating from the start pose. Then, a path can be expressed as $\mathbf{X}(s)$, where s denotes the arc length.

The pose information $\mathbf{X}$ includes position and orientation information. For robots moving in 3D Euclidean space, the pose information includes three DOF position and three DOF orientation. For mobile robots moving on planes, the pose information includes two DOF position and one DOF orientation. For example, a common definition of six DOF pose vector is

$$\mathbf{X} = \begin{bmatrix} x & y & z & \psi & \theta & \phi \end{bmatrix}^T, \tag{1.18}$$

where the coordinates x, y, z denote the position and the angles ψ (yaw angle), θ (pitch angle), ϕ (roll angle) represent the orientation.

Due to the existence of kinematic constraints, the physically feasible trajectories are only a subset of the entire space. For a robot end effector that moves freely with six DOF, the elements in the pose vector vary independently from each other. While for a differential driving wheeled mobile robot, the variation of position is determined by the current orientation; thus, not all trajectories are feasible for this type of robots.

Curve fitting

For the application of control development, the desired path and trajectory generally need to be continuous, i.e., the position, velocity, and acceleration are required to be continuous.

For the case of pose regulation, the initial and final poses are given as $\mathbf{X}_0$ and $\mathbf{X}^*$. The polynomial function is an obvious candidate for curve fitting to parameterize paths or trajectories. For the general case that the position, velocity, and acceleration are given as initial and final conditions, fifth order polynomial function is the minimal structure to satisfy the conditions. For a robot without kinematic constraints (moves freely with six DOF), the elements in the pose vector $\mathbf{X}$ can be designed separately. For example, take one dimension x; the initial conditions $x_0, \dot{x}_0, \ddot{x}_0$ at t_0 and the final conditions $x^*, \dot{x}^*, \ddot{x}^*$ at t^* are given; then, the trajectory can be fitted by the function $x(t) = c_5 t^5 + c_4 t^4 + c_3 t^3 + c_2 t^2 + c_1 t + c_0$, and the coefficients are determined by solving the least square solution of the following equation:

$$\begin{bmatrix} t_0^5 & t_0^4 & t_0^3 & t_0^2 & t_0 & 1 \\ 5t_0^4 & 4t_0^3 & 3t_0^2 & 2t_0 & 1 & 0 \\ 10t_0^2 & 12t_0^2 & 6t_0 & 2 & 0 & 0 \\ t^{*5} & t^{*4} & t^{*3} & t^{*2} & t^* & 1 \\ 5t^{*4} & 4t^{*3} & 3t^{*2} & 2t^* & 1 & 0 \\ 10t^{*2} & 12t^{*2} & 6t^* & 2 & 0 & 0 \end{bmatrix} \cdot \begin{bmatrix} c_5 \\ c_4 \\ c_3 \\ c_2 \\ c_1 \\ c_0 \end{bmatrix} = \begin{bmatrix} x_0 \\ \dot{x}_0 \\ \ddot{x}_0 \\ x^* \\ \dot{x}^* \\ \ddot{x}^* \end{bmatrix} \tag{1.19}$$

For the robots with kinematic constraints, the constraints among the elements in the pose information should be considered in the curve fitting. For example, take the differential driving wheeled mobile robot; it has two position DOFs (x, y) and one orientation DOF θ. It has the feature that the direction of instantaneous velocity coincides with the orientation, i.e., $\theta = atan2(\dot{y}, \dot{x})$. As a result, only the trajectories of x and y need to be designed, and the trajectory of θ can be calculated accordingly.

For the case of path following or trajectory tracking, a set of points are generally given to represent the desired path or trajectory. For control development, a sufficient smooth curve representation is required, i.e., the first and second derivatives exist and are bounded. High order polynomial functions can be used for curve fitting by solving the least square solution of conditions formed by the points. However, the numerical stability decreases with higher order and the fitting accuracy decreases with lower order. Besides, the Runge's phenomenon exists which is a problem of oscillation at the edges of an interval that occurs when using polynomial interpolation with polynomials of high degree over a set of equispaced interpolation points. It shows that going to higher order does not always improve accuracy of curve fitting. Considering the flexibility and accuracy of curve fitting, the spline curves can be used for representation. A spline is a special function defined piecewise by polynomials. It uses low order polynomials in each interval and ensures the continuity between the intervals. Considering the fundamental case of one-dimensional fitting, a set of points $\{x_i\}_{i \in [1,N]}$ at time instants $\{t_i\}_{i \in [1,N]}$ are given. There are various forms of spline functions that can be used for curve fitting. Without generality, the following two cubic spline functions are used:

■ One type of spline function directly uses the values at each data points for interpolation, and the resulting spline curve passes the data points. The spline curve is composed of a series of piecewise cubic polynomials $\{f_i\}_{i \in [1,N-1]}$ with $f_i : [t_i, t_{i+1}] \rightarrow \mathbb{R}$ defined as the following:

$$f_i(t) = a_i(t - t_i)^3 + b_i(t - t_i)^2 + c_i(t - t_i) + x_i. \qquad (1.20)$$

To ensure the smoothness, the following boundary conditions are required:

$$\begin{aligned} f_i(t_{i+1}) &= x_{i+1} \\ \dot{f}_{i-1}(t_i) &= \dot{f}_i(t_i) \\ \ddot{f}_{i-1}(t_i) &= \ddot{f}_i(t_i). \end{aligned} \qquad (1.21)$$

Then, the curve parameters in each interval can be solved by the above boundary conditions.

■ Another type of spline function interpolates the function values based on the values at control points $\{\phi_j\}_{j \in [-1, M-2]}$, which are defined uniformly in the region of time horizon. The distribution of control points is generally sparser than the raw data points ($M < N$) to be smoother and more computationally efficient. The interval between neighboring control points is $\kappa = \frac{t^* - t_0}{M - 3}$. The approximation function is defined in terms of the control points as follows:

$$f(t) = \sum_{k=0}^{3} B_k(s)\phi_{j+l}, \qquad (1.22)$$

where $j = \lfloor \frac{t-t_0}{\kappa} \rfloor - 2$ and $s = \frac{t-t_0}{\kappa} - \lfloor \frac{t-t_0}{\kappa} \rfloor$. The terms $\{B_k(s)\}_{k \in \{0,1,2,3\}}$ denote the spline basis functions defined as follows:

$$B_0(s) = \frac{(1-s)^3}{6}$$

$$B_1(s) = \frac{(3s^3 - 6s^2 + 4)}{6}$$

$$B_2(s) = \frac{-3s^3 + 3s^2 + 3s + 1}{6}$$

$$B_3(s) = \frac{s^3}{6}.$$

(1.23)

1.2.2 Pose Kinematics

For the development of autonomous navigation and manipulation for robots, the task generally described in terms of pose information as introduced above. The control objective is to regulate the robot motion to perform the tasks. In this section, the general six DOF robot is considered which has linear velocities $v = [v_x, v_y, v_z]^T$ and angular velocities $\omega = [\omega_x, \omega_y, \omega_z]^T$ in the frame $\mathcal{F}$ with respect to frame $\mathcal{F}'$. Similarly, denote the linear and angular velocities of frame $\mathcal{F}'$ with respect to $\mathcal{F}$ as $v' = [v'_x, v'_y, v'_z]^T$ and angular velocities $\omega' = [\omega'_x, \omega'_y, \omega'_z]^T$. The positive directions of the linear velocities are the same as the directions of corresponding axes, and the positive directions of the angular velocities are determined by the right-hand rule.

Rotational motion

The direction of the angular velocity vector ω defines the instantaneous axis of rotation, that is, the axis about which the coordinate frame is rotating at a particular instant of time. In general, this axis changes with time. The magnitude of the vector is the rate of rotation about the axis; in this respect, it is similar to the angle-axis representation for rotation introduced above. From mechanics, there is a well-known expression for the derivative of a time-varying rotation matrix

$$\dot{R}(t) = [R\omega]_\times \cdot R(t).$$

(1.24)

Since R is redundant to describe the rotation, it is not appropriate for control development. As mentioned above, the angle-vector representation is widely used in robot applications, which is denoted as $\Theta(t) = u(t)\theta(t)$. The rotation matrix can be expressed in terms of the angle-vector representation as follows:

$$R = I_3 + \sin\theta[u]_\times + 2\sin^2\frac{\theta}{2}[u]_\times^2$$

(1.25)

Taking the derivative of (1.25) and using (1.24), the following expression is obtained:

$$[R\omega] = \sin\theta[\dot{u}]_\times + [u]_\times\dot{\theta} + (1 - \cos\theta)[[u]_\times\dot{u}]_\times, \qquad (1.26)$$

where the following properties are utilized:

$$[u]_\times\zeta = -[\zeta]_\times u \qquad (1.27)$$

$$[u]_\times^2 = uu^T - I_3 \qquad (1.28)$$

$$[u]_\times uu^T = 0 \qquad (1.29)$$

$$[u]_\times[\dot{u}]_\times[u]_\times = 0 \qquad (1.30)$$

$$[[u]_\times\dot{u}]_\times = [u]_\times[\dot{u}]_\times - [\dot{u}]_\times[u]_\times. \qquad (1.31)$$

Taking the derivative of $\Theta(t)$, the following expression is obtained:

$$\dot{\Theta} = \dot{u}\theta + u\dot{\theta}. \qquad (1.32)$$

Multiplying both sides in (1.32) by $(I_3 + [u]_\times^2)$, the following expression is obtained:

$$(I_3 + [u]_\times^2)\dot{\Theta} = u\dot{\theta}, \qquad (1.33)$$

where the following properties are utilized;

$$[u]_\times^2 = uu^T - I_3$$

$$u^T u = 1 \qquad (1.34)$$

$$u^T \dot{u} = 0.$$

Multiplying both sides in (1.32) by $-[u]_\times^2$, the following expression is obtained:

$$-[u]_\times^2\dot{\Theta} = \dot{u}\theta, \qquad (1.35)$$

where the following properties are utilized:

$$u^T u = 1$$
$$u^T \dot{u} = 0. \qquad (1.36)$$

From the expression in (1.26) and expressions in (1.32), (1.33), (1.35), the following expression can be obtained

$$\dot{\Theta} = L_w R\omega \qquad (1.37)$$

where the Jacobian-like matrix $L_w \in \mathbb{R}^{3\times3}$ is defined as

$$L_w = I_3 - \frac{\theta}{2}[u]_\times + \left(1 - \frac{\text{sinc}(\theta)}{\text{sinc}^2(\frac{\theta}{2})}\right)[u]_\times^2 \qquad (1.38)$$

with $\text{sinc}(\theta) = \sin\theta/\theta$.

Translational motion

Denote the rotational and translational matrices from $\mathcal{F}'$ to $\mathcal{F}$ expressed in $\mathcal{F}'$ as R' and x'_f, respectively. The translational vector x_f expressed in $\mathcal{F}$ can be expressed as the following:

$$x_f = -R'^T x'_f. \tag{1.39}$$

Taking the time derivative of (1.39), it can be obtained that

$$\dot{x}_f = -\dot{R}'^T x'_f - R'^T \dot{x}'_f \tag{1.40}$$

Similar to (1.24), the above expression can be rewritten as the following:

$$\dot{x}_f = -[R'^T \omega']_\times \cdot R'^T x'_f - R'^T v'. \tag{1.41}$$

Using the fact that $\omega = R'^T \omega'$ and $v = R'^T v'$:

$$\dot{x}_f = -[\omega]_\times \cdot R'^T x'_f v. \tag{1.42}$$

Then, the time derivative of translational vector x_f can be expressed as the following:

$$\dot{x}_f = -[\omega]_\times \cdot x_f - v \tag{1.43}$$

1.3 Wheeled Mobile Robot Kinematics

In real applications, wheeled mobile robots are widely used for intelligent transportation and manipulation. They are actuated by wheels and generally move on the ground. As a result, the mobile robots have three DOFs, i.e., two position DOFs and one orientation DOF.

1.3.1 Wheel Kinematic Constraints

The first step to a kinematic model of the robot is to express constraints on the motions of individual wheels. The motions of individual wheels would be combined to compute the motion of the robot as a whole. As shown in Figure 1.5, there are four basic wheel types with varying kinematic properties in applications, i.e., fixed wheel, steered wheel, castor wheel, and Swedish wheel.

In the following developments, it is assumed that the plane of the wheel always remains vertical and that there is one single point of contact between the wheel and the ground plane. Besides, there is no sliding at this single point of contact. That is, the wheel undergoes motion only under conditions of pure rolling and rotation about the vertical axis through the contact point. Under these assumptions, we present two constraints for every wheel type. The rolling constraint enforces the concept of rolling contact that the wheel must roll when

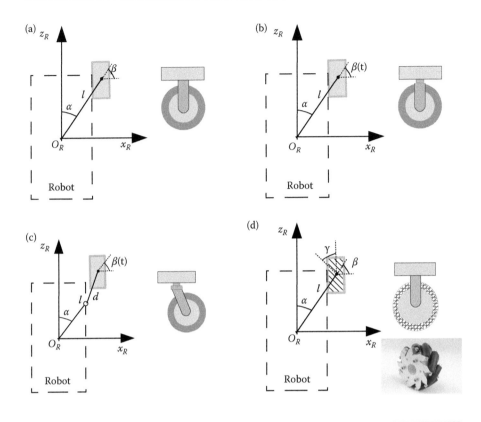

Figure 1.5: Standard basic wheels. (a) Fixed wheel, (b) Steered wheel, (c) Castor wheel, and (d) Swedish wheel.

motion takes place in the appropriate direction. The sliding constraint enforces the concept of no lateral slippage that the wheel must not slide orthogonal to the wheel plane.

As shown in Figure 1.5, the position of the wheel is expressed in polar coordinates by distance l and angle α. The angle of the wheel plane relative to the chassis is denoted by β, which is fixed since the fixed standard wheel is not steerable. The wheel, which has radius r, can spin over time with angular velocity $\omega(t)$. The fixed wheel has no vertical axis of rotation for steering. Its angle to the chassis is thus fixed, and it is limited to motion back and forth along the wheel plane and rotation around its contact point with the ground plane. The steered wheel differs from the fixed standard wheel only in that there is an additional DOF; the wheel may rotate around a vertical axis passing through the center of the wheel and the ground contact point. Castor wheels are able to steer around a vertical axis. However, unlike the steered standard wheel, the vertical axis of rotation in a castor wheel does not pass through the ground contact point. Swedish wheels have no vertical axis of rotation, yet are able to move omnidirectionally like the castor wheel. This is possible by adding a DOF to the fixed standard wheel. Swedish wheels consist of a fixed standard wheel with rollers attached to the wheel perimeter with axes that are antiparallel to the main axis of the fixed wheel component. The rolling and sliding constraints are shown in Tables 1.1 and 1.2, respectively. It should be noted that β is free for the castor wheel and ω_{sw} is free for the Swedish wheel; thus, the sliding constraints of the castor and the Swedish wheels do not affect the motion direction of the robots.

1.3.2   Mobile Robot Kinematic Modeling

For a mobile robot with several wheels, its kinematic constraints can be computed by combining all of the kinematic constraints of each wheel. From the above, the castor and Swedish wheels have no impact on the kinematic constraints of the robot. Therefore, only fixed wheels and steerable standard wheels have impact on robot kinematics, and therefore require consideration when computing the robot's kinematic constraints.

Suppose that the robot has a total of N standard wheels, comprising N_f fixed wheels and N_s steerable wheels. Denote B_f and $B_s(t)$ as the fixed steering angles of fixed wheels and the varying steering angles of the steerable wheels, respectively. Denote W_f and $W_s(t)$ as the angular velocities of fixed and steerable wheels, respectively. The rolling constraint of the robot can be obtained by combining the effects of wheels as follows:

$$J_1(B_s)R(\theta)\dot{X} = J_2W, \tag{1.44}$$

Table 1.1   Wheel Kinematics: Rolling Constraints

Wheel Type	*Rolling Constraint*
Fixed wheel Steered wheel Castor wheel	$[\sin(\alpha+\beta) \quad -\cos(\alpha+\beta) \quad -l\cos\beta]R(\theta)\dot{X} = r\omega$
Swedish wheel	$[\sin(\alpha+\beta+\gamma) \quad -\cos(\alpha+\beta+\gamma)]$ $[-l\cos(\beta+\gamma)]R(\theta)\dot{X} = r\cos\gamma\cdot\omega$

Table 1.2   Wheel Kinematics: Sliding Constraints

Wheel Type	*Sliding Constraint*
Fixed wheel Steered wheel	$[\cos(\alpha+\beta) \quad \sin(\alpha+\beta) \quad l\sin\beta]R(\theta)\dot{X} = 0$
Castor wheel	$[\cos(\alpha+\beta) \quad \sin(\alpha+\beta) \quad d+l\sin\beta]R(\theta)\dot{X} = -d\dot{\beta}$
Swedish wheel	$[\cos(\alpha+\beta+\gamma) \quad \sin(\alpha+\beta+\gamma) \quad l\sin(\beta+\gamma)]$ $R(\theta)\dot{X} = r\omega\sin\gamma + r_{sw}\omega_{sw}$

where $W = [W_f, W_s]^T$ is the angular velocities of all wheels. J_2 is a constant diagonal $N \times N$ matrix whose entries are the corresponding radius of the wheels. J_1 defines the projections for the wheels to their motions along their individual wheel planes:

$$J_1(B_s) = \begin{bmatrix} J_{1f} \\ J_{1s}(B_s) \end{bmatrix}. \tag{1.45}$$

Here, $J_{1f} \in \mathbb{R}^{N_f \times 3}$ and $J_{1s} \in \mathbb{R}^{N_f \times 3}$ are Jacobian-like matrices with each row consisting of the three terms in the equation in Table 1.1.

The sliding constraints can be also obtained as follows:

$$C_1(B_s)R(\theta)\dot{X} = 0, \tag{1.46}$$

where $C_1(B_s) = \begin{bmatrix} C_{1f} & C_{1s}(B_s) \end{bmatrix}^T$. The matrices $C_{1f} \in \mathbb{R}^{N_f \times 3}$ and $C_{1s}(B_s) \in \mathbb{R}^{N_s \times 3}$ are composed of rows consisting of the three terms in the equation in Table 1.2.

1.3.3 Typical Nonholonomic Mobile Robot

The above subsection provides a general tool for kinematic modeling of mobile robots. In the following, the typical differential driving mobile robot is considered as shown in Figure 1.6.

As shown in Figure 1.6, the differential driving mobile robot consists of two independent driving fixed wheels and one castor wheel. The origin of the robot coordinate is located at P. Then, the kinematic model of the robot can be obtained by combining the rolling and sliding constraints as follows:

$$\begin{bmatrix} J_1(B_s) \\ C_1(B_s) \end{bmatrix} R(\theta)\dot{X} = \begin{bmatrix} J_2W \\ 0 \end{bmatrix}. \tag{1.47}$$

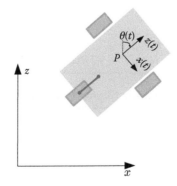

Figure 1.6: Differential driving mobile robot.

Since the castor wheel is not powered and is free to move in any direction, it is ignored in the kinematic model. The two remaining drive wheels are not steerable, and therefore $J_1(B_s)$ and $C_1(B_s)$ simplify to J_{1f} and C_{1f}, respectively. To employ the fixed standard wheel's rolling constraint formula, we must first identify each wheel's values for α and β. Suppose that the robot's local reference frame is aligned such that the robot moves forward along x_r. In this case, for the left wheel, $\alpha = \pi/2, \beta = 0$, for the right wheel, $\alpha = -\pi/2, \beta = \pi$. Then, the kinematic constraints of the robot can be expressed as the following:

$$
\begin{bmatrix} 1 & 0 & l \\ 1 & 0 & -l \\ 0 & 1 & 0 \end{bmatrix} \cdot \begin{bmatrix} \cos\theta & \sin\theta & 0 \\ -\sin\theta & \cos\theta & 0 \\ 0 & 0 & 1 \end{bmatrix} \dot{X} = \begin{bmatrix} r \cdot \omega_l \\ r \cdot \omega_r \\ 0 \end{bmatrix}, \tag{1.48}
$$

Chapter 2

Multiple-View Geometry

In the real world, the lights from the scenario are projected into the camera to generate images, which is the capturing process. Evidently, the images are two-dimensional representations of the three-dimensional world, reflecting the appearance information of the scenario. Based on the appearance information, object recognition and tracking tasks can be properly performed. However, for higher-level tasks such as object and robot localization, path planning, and control, structural information is necessary to be extracted from the appearance information in image space.

Multiple-view geometry provides a versatile tool to describe and analyze the geometric information based on the projected camera views at several poses. For multiple-view cases, the images in different views are related to each other via geometric constraints, which can be generally constructed by the relative pose transformations of camera views and the geometric structure of the scene. For visual perception tasks, geometric constraints can be used to reduce the searching region for the correspondences; then, the scene structure is reconstructed by applying the multiple-view geometry models for the image correspondences. For visual control tasks, since the task and system feedback are both expressed in image space, multiple-view geometry is used to map the high-dimensional and redundant image information to the low-dimensional and compact geometric information.

In this chapter, the projective geometry is introduced first, which provides the representations and transformations of geometric primitives in projective space. Then, the single-view geometry for projective cameras is developed to serve as the basic model for multiple-view geometries. Next, two-view geometric models are developed for two view cases. In visual perception tasks, 3D reconstruction generally needs at least two camera views to determine the environment structure; in visual control tasks, the system control error is generally based on the difference between the two views (current view and target view). As a result, the two-view geometry is most commonly used in vision based robotic tasks. Depending on the captured scene structure, such as planar, nonplanar, and general cases (either planar or nonplanar), corresponding geometric models are proposed for environment modeling. To improve robustness and generality with respect to the tasks and scene structure, three-view geometry is used to describe the geometric constraints among three camera views, including the current view and two auxiliary views. In the last, the online computation of multiple-view geometry constraints is discussed.

2.1 Projective Geometry

2.1.1 Homogeneous Representation of Points and Lines

The image space is the projection of 3D Euclidean space via perspective mapping and has two Degrees of Freedom (DOF). Generally, a point in 2D space can be represented by the pair of coordinates $[x, y]^T \in \mathbb{R}^2$. A line in 2D space is represented by the equation $ax + by + c = 0$; thus, a line can be represented by the vector $[a, b, c]^T$. It should be noted that the correspondence between the line and the vector is not one-to-one, since the lines $ax + by + c = 0$ and $(ka)x + (kb)y + (kc) = 0$ are the same for any nonzero constant k. Thus, the vectors $[a, b, c]^T$ and $k[a, b, c]^T$ represent the same line for any nonzero k. In fact, two such vectors related by an overall scaling are considered as being equivalent. An equivalence class of vectors under this equivalence relationship is known as a homogeneous vector. Any particular vector $[a, b, c]^T$ is a representative of the equivalence class. The set of equivalence classes of vectors in $\mathbb{R}^3$ forms the projective space $\mathbb{P}^2$.

A point $x = [x, y]^T$ lies on the line $l = [a, b, c]^T$ if and only if $ax + by + c = 0$. This may be written in terms of an inner product of vectors representing the point as $[x, y, 1][a, b, c]^T = 0$; that is the point $[x, y]^T \in \mathbb{R}^2$ is represented as a three-vector by adding a final coordinate of 1. Note that for any nonzero constant k and line l, the equation $[kx, ky, k]l = 0$, if and only if $[x, y, 1]l = 0$. It is natural, therefore, to consider the set of vectors $[kx, ky, k]^T$ for varying values of k to be a representation of the point $(x, y)T \in \mathbb{R}^2$. Thus, just as with lines, points are represented by homogeneous vectors. An arbitrary homogeneous vector representative of a point is of the form $X = [x_1, x_2, x_3]^T$, representing the point $[x_1/x_3, x_2/x_3]^T \in \mathbb{R}^2$. Points, then, as homogeneous vectors are also elements of $\mathbb{P}^2$.

2.1.2 Projective Transformation

A projective transformation is an invertible mapping from points in $\mathbb{P}^2$ of one view to points in $\mathbb{P}^2$ of another view. A mapping $h : \mathbb{P}^2 \to \mathbb{P}^2$ is a projectivity if and only if there exists a nonsingular 3×3 matrix H such that for any point in $\mathbb{P}^2$

represented by a vector x, it is true that $h(x) = Hx$. According to the transforming matrix, the projective transformations include the types that are mentioned in Table 2.1.

Isometry

Isometries are transformations of the plane $\mathbb{R}^2$ that preserve Euclidean distance. An isometry is represented as the following:

$$
\begin{bmatrix} x' \\ y' \\ 1 \end{bmatrix} = \begin{bmatrix} \cos\theta & -\sin\theta & x_{fx} \\ \sin\theta & \cos\theta & x_{fy} \\ 0 & 0 & 1 \end{bmatrix} \cdot \begin{bmatrix} x \\ y \\ 1 \end{bmatrix}.
\tag{2.1}
$$

Actually, the isometry is orientation preserving and is a Euclidean transformation. A planar Euclidean transformation has three DOF, one for the rotation and two for the translation. Thus, three parameters must be specified in order to define the transformation. The transformation can be computed from two point correspondences.

Similarity transformation

A similarity transformation (or more simply a similarity) is an isometry composed with an isotropic scaling. In the case of a Euclidean transformation composed with a scaling (i.e., no reflection), the similarity has matrix representation

Table 2.1 **Geometric Properties of Transformations**

Type	Transform Matrix	Invariants
Isometry	$\begin{bmatrix} \cos\theta & -\sin\theta & x_{fx} \\ \sin\theta & \cos\theta & x_{fy} \\ 0 & 0 & 1 \end{bmatrix}$	Length, angle and area
Similarity transformation	$\begin{bmatrix} s\cos\theta & -s\sin\theta & x_{fx} \\ s\sin\theta & s\cos\theta & x_{fy} \\ 0 & 0 & 1 \end{bmatrix}$	Angle, ratios of lengths and areas
Affine transformation	$\begin{bmatrix} a_{11} & a_{12} & x_{fx} \\ a_{21} & a_{22} & x_{fy} \\ 0 & 0 & 1 \end{bmatrix}$	Parallel lines, ratio of lengths of parallel line segments, and ratio of areas
Projective transformation	$\begin{bmatrix} h_{11} & h_{12} & h_{13} \\ h_{21} & h_{22} & h_{23} \\ h_{31} & h_{32} & h_{33} \end{bmatrix}$	Concurrency, collinearity, and cross ratio.

$$
\begin{bmatrix} x' \\ y' \\ 1 \end{bmatrix} = \begin{bmatrix} s\cos\theta & -s\sin\theta & x_{fx} \\ s\sin\theta & s\cos\theta & x_{fy} \\ 0 & 0 & 1 \end{bmatrix} \cdot \begin{bmatrix} x \\ y \\ 1 \end{bmatrix}, \tag{2.2}
$$

where the scalar s represents the isotropic scaling. A similarity transformation is also known as an equiform transformation, because it preserves "shape" (form). A planar similarity transformation has four DOF, the scaling accounting for one more DOF than a Euclidean transformation. A similarity can be computed from two point correspondences.

Affine transformation

An affine transformation (or more simply an affinity) is a nonsingular linear transformation followed by a translation. It has the matrix representation as follows:

$$
\begin{bmatrix} x' \\ y' \\ 1 \end{bmatrix} = \begin{bmatrix} a_{11} & a_{12} & x_{fx} \\ a_{21} & a_{22} & x_{fy} \\ 0 & 0 & 1 \end{bmatrix} \cdot \begin{bmatrix} x \\ y \\ 1 \end{bmatrix}. \tag{2.3}
$$

A planar affine transformation has six DOF corresponding to the six matrix elements. The transformation can be computed from three point correspondences. An affinity is orientation preserving or orientation reversing, which is based on whether A is positive or negative, respectively.

Projective transformation

A projective transformation is a general nonsingular linear transformation of homogeneous coordinates. This generalizes an affine transformation, which is the composition of a general nonsingular linear transformation of inhomogeneous coordinates and a translation. The projective transformation is defined as follows:

$$
\begin{bmatrix} x' \\ y' \\ 1 \end{bmatrix} = \begin{bmatrix} h_{11} & h_{12} & h_{13} \\ h_{21} & h_{22} & h_{23} \\ h_{31} & h_{32} & h_{33} \end{bmatrix} \cdot \begin{bmatrix} x \\ y \\ 1 \end{bmatrix}. \tag{2.4}
$$

The matrix has nine elements with only their ratio significant, so the transformation is specified by eight parameters. A projective transformation between two planes can be computed from four point correspondences, with no three collinear on either plane.

2.2 Single-View Geometry

The single-view geometry is the fundamental model to describe the geometric relationship between 2D image space and 3D Euclidean space. According to the different camera types, typical single-view geometry models include the pinhole model (for perspective cameras) and the spherical model (for fish-eye cameras and omnidirectional cameras). For the applications of robotics and autonomous vehicles, perspective cameras are mainly used for environment perception and robot control. In this section, the pinhole model is introduced.

2.2.1 Pinhole Camera Model

As shown in Figure 2.1, the camera C maps the 3D point O onto image plane as point p. Define the coordinate of point O with respect to C as $\bar{m} \triangleq \begin{bmatrix} x & y & z \end{bmatrix}^T$, with its normalized coordinate defined as $m \triangleq \begin{bmatrix} x/z & y/z & 1 \end{bmatrix}^T$; define the coordinate of image point p as $p \triangleq \begin{bmatrix} u & v & 1 \end{bmatrix}^T$. The single-view geometry describes the mapping from the normalized coordinate m to image coordinate p:

$$p = A \cdot m, \tag{2.5}$$

where A is the intrinsic camera matrix defined as:

$$A \triangleq \begin{bmatrix} \alpha_u & -\alpha_u \cot \theta & u_0 \\ 0 & \dfrac{\alpha_v}{\sin \theta} & v_0 \\ 0 & 0 & 1 \end{bmatrix}, \tag{2.6}$$

where α_u and α_v are the focal lengths of camera C in terms of pixels along the horizontal (u) and vertical (v) directions, respectively. $p_0 \triangleq \begin{bmatrix} u_0 & v_0 & 1 \end{bmatrix}^T$ is the intersecting point of the optical axis and the image plane, which is also called the principal point. θ represents the angle between the horizontal and vertical axes. Generally, the camera intrinsic parameters (i.e., α_u, α_v, u_0, v_0, and θ) can be calibrated off-line.

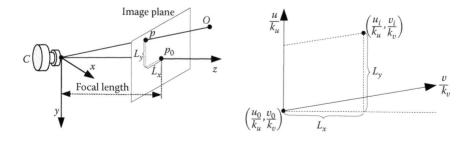

Figure 2.1: Single-view geometry.

2.2.2 Camera Lens Distortion

For general perspective cameras, there are two types of distortions:

■ Radial distortion: It is symmetric; ideal image points are distorted along radial directions from the distortion center. This is caused by imperfect lens shape.

■ Decentering distortion: This is usually caused by improper lens assembly; ideal image points are distorted in both radial and tangential directions.

The distortion can be expressed as power series in radial distance $r = \sqrt{(u-u_0)^2 + (v-v_0)^2}$ as follows:

$$\delta_u = (u-u_0)(k_1 r^2 + k_2 r^4 + k_3 r^6 + \cdots)$$
$$+ (p_1(r^2 + 2(u-u_0)^2) + 2p_2(u-u_0)(v-v_0))(1+p_3 r^2 + \cdots) \tag{2.7}$$

$$\delta_v = (v-v_0)(k_1 r^2 + k_2 r^4 + k_3 r^6 + \cdots)$$
$$+ (p_1(r^2 + 2(v-v_0)^2) + 2p_2(u-u_0)(v-v_0))(1+p_3 r^2 + \cdots), \tag{2.8}$$

where k_i are coefficients of radial distortion and p_j are coefficients of decentering distortion. It is likely that the distortion function is totally dominated by the radial components and especially dominated by the first term. It has also been found that any more elaborated modeling not only would not help (negligible when compared with sensor quantization) but also would cause numerical instability.

2.3 Two-View Geometry

The two-view geometry describes the relationship between two cameras in both 2D image space and 3D Euclidean space, representing the characteristics of two camera systems or the motion process of single cameras. Generally, real scenes consist of planar and nonplanar objects. As shown in Figure 2.2, there exist a point O_π on the plane π and a set of nonplanar points $\{O_i\}$. The two cameras $\mathcal{C}$ and $\mathcal{C}'$ are placed at different poses with the transformation matrices (R, x_f)m, where $R \in \mathbb{SO}(3)$ and $x_f \triangleq \begin{bmatrix} x_{fx} & x_{fy} & x_{fz} \end{bmatrix}^T \in \mathbb{R}^3$ represent the rotation and translation matrices from $\mathcal{C}'$ to $\mathcal{C}$ expressed in $\mathcal{C}'$. Define the camera intrinsic matrices of cameras $\mathcal{C}$ and $\mathcal{C}'$ as A and A', respectively. In the case that the image planes of the two cameras do not coincide, the line between the origins of $\mathcal{C}$ and $\mathcal{C}'$ intersects with the image planes of cameras $\mathcal{C}$ and $\mathcal{C}'$ at points e and e', which are generally called epipoles.

Traditional two-view geometric models include the homography and epipolar geometry. The homography relies on the relative pose information (among the object and the cameras) and describes the relationship between the projections in two views of planar objects. Epipopar geometry does not rely on the scene

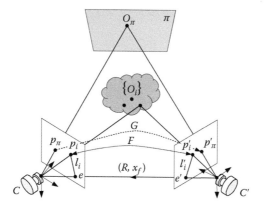

Figure 2.2: Two-view geometric model.

structure and describes the relationship between the projections in two views of 3D points. However, the estimation of epipolar geometry is degenerated. In this section, the homography and epipolar geometry are introduced for planar and nonplanar scenes. Besides, a general two-view geometry is constructed for general scenes (both planar and nonplanar) with respect to a reference plane.

2.3.1   *Homography for Planar Scenes*

Homography model development

Consider the point O_π on plane π; define the coordinates of point O_π with respect to C and C' as $\bar{m}_\pi \triangleq \begin{bmatrix} x_\pi & y_\pi & z_\pi \end{bmatrix}^T \in \mathbb{R}^3$ and $\bar{m}'_\pi \triangleq \begin{bmatrix} x'_\pi & y'_\pi & z'_\pi \end{bmatrix}^T \in \mathbb{R}^3$, respectively. Then, the transformation relationship between $\bar{m}_\pi$ and $\bar{m}'_\pi$ is defined as:

$$\bar{m}'_\pi = R\bar{m}_\pi + x_f. \tag{2.9}$$

Define n as the normal vector of plane π with respect to C and d as the distance from the origin of C to plane π along the vector n. The following relationship holds:

$$d = n^T \bar{m}_\pi. \tag{2.10}$$

Using (2.9) and (2.10) can be rewritten as:

$$\bar{m}'_\pi = H\bar{m}_\pi, \tag{2.11}$$

where H is the Euclidean homography defined as:

$$H \triangleq R + x_f \frac{n^T}{d}. \tag{2.12}$$

Define the normalized coordinates of $\bar{m}_\pi$ and $\bar{m}'_\pi$ as m_π and m'_π, respectively, as follows:

$$m_\pi \triangleq \frac{\bar{m}_\pi}{z_\pi}, m'_\pi \triangleq \frac{\bar{m}'_\pi}{z'_\pi}. \tag{2.13}$$

Substituting (2.13) into (2.12), the following relationship holds:

$$m'_\pi = \frac{z_\pi}{z'_\pi} H m_\pi. \tag{2.14}$$

Define the image points corresponding to point O_π in the views of C and C' as $p_\pi \triangleq \begin{bmatrix} u_\pi & u_\pi & 1 \end{bmatrix}^T \in \mathbb{P}^3$ and $p'_\pi \triangleq \begin{bmatrix} u'_\pi & v'_\pi & 1 \end{bmatrix}^T \in \mathbb{P}^3$, respectively. Based on the single-view geometry, it can be obtained that:

$$p'_\pi = \frac{z_\pi}{z'_\pi} G p_\pi, \tag{2.15}$$

where

$$G \triangleq A' H A^{-1} \tag{2.16}$$

is defined as the Projective Homography matrix.

In real applications, the corresponding points in two views are used to estimate the projective homography matrix based on the model (2.15). At least four pairs of feature points can determine the homography matrix. For vision-based control, the homography matrix can be decomposed to obtain the pose information $(R, x_f/d, n)$.

Homography decomposition for pose estimation

To facilitate the control development, the homography matrix needs to be decomposed to obtain the relative pose information. Standard algorithms for homography decomposition obtain numerical solutions using the singular value decomposition (SVD) of the matrix. It is shown that in the general case there are two possible solutions to the homography decomposition, and *a priori* knowledge of the normal vector n can be used to determine the unique one. Then, the rotation matrix R and the scaled translation matrix x_f/d are obtained for further control development.

Generally, the decomposition process is based on the SVD of the homography matrix. Two of the commonly used methods are proposed by Faugeras [67] and Zhang [226] as follows:

■ Faugeras singular value decomposition-based decomposition:
After the SVD decomposition of the homography matrix H, it can be obtained that:

$$H = U \Lambda V^T, \tag{2.17}$$

where U and V are orthogonal matrices, and Λ is a diagonal matrix, which contains the singular values of matrix H. Then, Λ can also be considered as a homography matrix defined as

$$\Lambda = R_\Lambda + x_{h,\Lambda} n_\Lambda^T, \tag{2.18}$$

where the raw pose information R, x_h, n can be computed as

$$
\begin{aligned}
R &= U R_\Lambda V^T \\
x_h &= U x_{h,\Lambda} \\
n &= V n_\Lambda.
\end{aligned}
\tag{2.19}
$$

Since the matrix Λ is diagonal, then it is simple to compute the components of the rotation matrix, translation and normal vectors. First, x_h can be easily eliminated from the three vector equations coming out from (2.18) (one for each column of this matrix equation). Then, imposing that R_Λ is an orthogonal matrix, we can linearly solve for the components of n, from a new set of equations relating only these components with the three singular values (see [67] for the detailed development). As a result of the decomposition algorithm, we can get up to eight different solutions for the triplets: $\{R, t, n\}$.

■ Zhang SVD-based decomposition [226]
Notice that a similar method to obtain this decomposition is proposed in [11]. The closed-form expressions for the translation vector, normal vector and rotation matrices are obtained from SVD decomposition of the homography matrix.

Compute the SVD decomposition of $H^T H$

$$H^T H = V \Lambda^2 V^T, \tag{2.20}$$

where the eigenvalues and corresponding eigenvectors are as follows:

$$\Lambda = diag(\lambda_1, \lambda_2, \lambda_3), \ V = \begin{bmatrix} v_1 & v_2 & v_3 \end{bmatrix} \tag{2.21}$$

with the eigenvalues ordered as

$$\lambda_1 \geq \lambda_2 = 1 \geq \lambda_3, \tag{2.22}$$

Denote x'_h as the scaled translational vector from $\mathcal{C}$ to $\mathcal{C}'$ expressed in $\mathcal{C}$, i.e., $x'_h = R^T x_h$. From the fact that $x'_h \times n$ is an eigenvector associated to the unitary eigenvalue of matrix $H^T H$ and that all the eigenvectors must be orthogonal, the following relations hold:

$$\|x'_h\| = \lambda_1 - \lambda_3, n^T x'_h = \lambda_1 \lambda_3 - 1 \tag{2.23}$$

and

$$
\begin{aligned}
v_1 &\propto v'_1 = \zeta_+ x'_h + n \\
v_2 &\propto v'_2 = x'_h \times n \\
v_3 &\propto v'_3 = \zeta_- x'_h + n
\end{aligned}
\tag{2.24}
$$

with v_i being unitary vectors, while v_i are not. The scalar functions $\zeta_{+,-}$ are scalar functions of the eigenvalues given by:

$$\zeta_{+,-} = \frac{1}{2\lambda_1\lambda_3}\left(-1\pm\sqrt{1+4\frac{\lambda_1\lambda_3}{(\lambda_1-\lambda_3)^2}}\right). \tag{2.25}$$

The norms of $v'_{1,3}$ are computed from the eigenvalues:

$$\begin{aligned}\|v'_1\| &= \zeta_+^2(\lambda_1-\lambda_3)^2 + \zeta_+(\lambda_1\lambda_3-1)+1 \\ \|v'_3\| &= \zeta_-^2(\lambda_1-\lambda_3)^2 + \zeta_-(\lambda_1\lambda_3-1)+1.\end{aligned} \tag{2.26}$$

Then, $v'_{1,3}$ can be computed as

$$v'_1 = \|v'_1\|v_1, v'_3 = \|v'_3\|v_3. \tag{2.27}$$

Then, the following expressions are used to compute the solutions for the couple translation vector and normal vector:

$$\begin{aligned}x'_h &= \pm\frac{v'_1-v'_3}{\zeta_+-\zeta_-}, n = \pm\frac{\zeta_+v'_3-\zeta_-v'_1}{\zeta_+-\zeta_-} \\ x'_h &= \pm\frac{v'_1+v'_3}{\zeta_+-\zeta_-}, n = \pm\frac{\zeta_+v'_3+\zeta_-v'_1}{\zeta_+-\zeta_-}.\end{aligned} \tag{2.28}$$

The rotational matrix can be obtained as

$$R = H(I+x'_h n^T)^{-1} \tag{2.29}$$

Since multiple solutions are obtained from the above methods, the following constraints can be used to eliminate impossible solutions. Besides, motion continuity can be also used to determine the unique one in the moving process.

◼ Reference-plane noncrossing constraint
This is the first physical constraint that allows to reduce the number of solutions from 8 to 4. This constraint imposes that: *Both frames, C and C', must be in the same side of the object plane.* This means that the camera cannot go in the direction of the plane normal further than the distance to the plane. Otherwise, the camera crosses the plane and the situation can be interpreted as the camera seeing a transparent object from both sides. The constraint can be rewritten as:

$$1+n^T R^T x_f > 0. \tag{2.30}$$

◼ Image point visibility
This additional constraint allows to reduce from 4 to 2 feasible solutions. The following additional information is required: *For all the reference points being visible; they must be in front of the camera.* The constraint can be rewritten as:

$$m_i n > 0, m'(Rn) > 0. \tag{2.31}$$

2.3.2 *Epipolar Geometry for Nonplanar Scenes*

Epipolar geometry model development

Epipolar geometry doesn't rely on the scene structure and only relies on the intrinsic parameters and the relative pose information between two views. As shown in Figure 2.2, consider an arbitrary point O_i in the scene; define the coordinates of O_i with respect to $\mathcal{C}$ and $\mathcal{C}'$ as $\bar{m}_i \triangleq \begin{bmatrix} x_i & y_i & z_i \end{bmatrix}^T \in \mathbb{R}^3$ and $\bar{m}'_i \triangleq \begin{bmatrix} x'_i & y'_i & z'_i \end{bmatrix}^T \in \mathbb{R}^3$, respectively; define the image points of O_i in cameras $\mathcal{C}$ and $\mathcal{C}'$ as $p_i \triangleq \begin{bmatrix} u_i & v_i & 1 \end{bmatrix}^T \in \mathbb{P}^3$ and $p'_i \triangleq \begin{bmatrix} u'_i & v'_i & 1 \end{bmatrix}^T \in \mathbb{P}^3$, respectively. The relationship between image points p_i and p'_i can be described by the fundamental matrix F as follows:

$$p'^T_i F p_i = 0, \tag{2.32}$$

where F is of 3×3, with rank 2.

For the image points p_i and p'_i, denote the lines $l'_i = F p_i$ and $l_i = F p'_i$ as their epipolar lines in the other images. As shown in Figure 2.2, the lines between the camera origins intersects with the image planes at points e and e', which are called epipoles. The relationship between the epipoles and the fundamental matrix is given by:

$$Fe = 0, F^T e' = 0. \tag{2.33}$$

For the physical configuration in Figure 2.2, the fundamental matrix is determined by the camera intrinsic parameters and the relative pose information as follows:

$$F = A'^{-T} [x_f]_\times R A^{-1}. \tag{2.34}$$

In real applications, the corresponding points in two views can be used to estimate the fundamental matrix F based on the epipolar geometry in (2.32). Then, the epipoles e, e' are calculated and used for vision-based control. For the general case of six DOF, at least seven pairs of corresponding points determine the fundamental matrix F. Besides the fundamental matrix, the essential matrix E is defined with respect to normalized coordinates as follows:

$$E = A'^T FA. \tag{2.35}$$

Then, the essential matrix E is related to the relative information as follows:

$$E = [x_f]_\times R. \tag{2.36}$$

The epipolar constraint is expressed with respect to E as follows:

$$m'_i E m_i = 0. \tag{2.37}$$

However, the estimation of epipolar geometry degenerates at some cases (i.e., the epipolar geometry cannot be determined uniquely even with enough corresponding points), as follows:

- The points and the camera origins are on the quadratic surface;

- The points are co-planar;

- The translation between the two cameras is zero.

The conditions above limits the application scenarios of epipolar geometry, especially for vision-based control systems in planar scenes or when there only exists rotational motion of the camera. Besides, the epipolar geometry can only determine the point-to-line relationship between two views (e.g., the point p_i in C corresponds to the line l'_i in C'), while the homography determines the point-to-point relationship.

Pose estimation from epipolar geometry

To obtain the pose information from the epipolar geometry, the essential matrix E is decomposed as follows: Perform the SVD decomposition of the essential matrix E as $E = U diag([1,1,0])$, there are two possible factorizations $E = SR$ as follows:

$$S = UZU^T, \ R = UWV^T \quad \text{or} \quad UW^TV^T, \tag{2.38}$$

where

$$W = \begin{bmatrix} 0 & -1 & 0 \\ 1 & 0 & 0 \\ 0 & 0 & 1 \end{bmatrix}, \ Z = \begin{bmatrix} 0 & 1 & 0 \\ -1 & 0 & 0 \\ 0 & 0 & 0 \end{bmatrix}. \tag{2.39}$$

Then, according to the property that $Sx_f = 0$, the translational vector $x_h = x_f/d$ can be obtained as $x_h = U \cdot \begin{bmatrix} 0 & 0 & 1 \end{bmatrix}^T$. As a result, four solutions are obtained by decomposing as follows:

$$R = UWV^T, \ x_h = \pm U \cdot \begin{bmatrix} 0 & 0 & 1 \end{bmatrix}^T$$
$$R = UW^TV^T, \ x_h = \pm U \cdot \begin{bmatrix} 0 & 0 & 1 \end{bmatrix}^T. \tag{2.40}$$

The difference between the translational vectors is obvious that the direction of the two views is reversed. The difference between the rotational vectors is the rotation through π about the line joining the two camera centers. Then, the unique solution can be obtained by testing with a single point to determine if it is in front of both cameras.

2.3.3 General Scenes

As described above, the homography is well defined and can be uniquely determined in various camera poses, but it only describes the characteristics of planar objects in two views. The epipolar geometry describes general nonplanar scenes, but it degenerates for planar scenes or purely rotational motion. In the previous approaches, multiple homographies are exploited to describe nonplanar scenes discretely, with the homography planes parallel with each other. However, the accuracy of the discrete model is limited and multiplies the model complexity. In the following, a general two-view geometric model is developed to describe the scene continuously.

As shown in Figure 2.3, there exists a plane π and a nonplanar point O_i in the scene. Define the coordinates of point O_i with respct to C and C' as $\bar{m}_i \triangleq \begin{bmatrix} x_i & y_i & z_i \end{bmatrix}^T \in \mathbb{R}^3$ and $\bar{m}_i' \triangleq \begin{bmatrix} x_i' & y_i' & z_i' \end{bmatrix}^T \in \mathbb{R}^3$, respectively; define the image points of O_i in the cameras C and C' as $p_i \triangleq \begin{bmatrix} u_i & v_i & 1 \end{bmatrix}^T \in \mathbb{P}^2$ and $p_i' \triangleq \begin{bmatrix} u_i' & v_i' & 1 \end{bmatrix}^T \in \mathbb{P}^2$, respectively. Construct a ray from the origin of camera C to point O_i which intersects with plane π at point $O_{\pi i}$. Define the coordinates of point $O_{\pi i}$ with respect to C and C' as $\bar{m}_{\pi i}$ and $\bar{m}_{\pi i}'$, respectively; define the image point of $O_{\pi i}$ in camera C' as $p_{\pi i}'$. Construct a plane π' through point O_i parallel to plane π, with the distance denoted as D_i. The positive direction of D_i is indicated in Figure 2.3. Similar to (2.11), the coordinates of point O_i with respect to C and C' can be related as:

$$\bar{m}_i' = H'\bar{m}_i \tag{2.41}$$

where H' is defined as:

$$H' \triangleq R + x_f \frac{n^T}{d - D_i}. \tag{2.42}$$

Using (2.12), the expression (2.42) is rewritten as:

$$H' = H + x_f \frac{D_i n^T}{d(d - D_i)}. \tag{2.43}$$

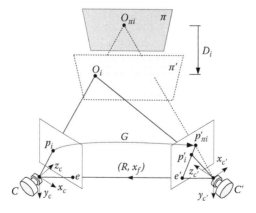

Figure 2.3: Two-view geometric model with respect to a reference plane.

Substitute (2.43) into (2.41), it can be obtained that:

$$\bar{m}'_i = H\bar{m}_i + x_f \frac{D_i n^T}{d(d - D_i)} \bar{m}_i. \tag{2.44}$$

By using $n^T \bar{m}_i = d - D_i$, the following expression is obtained:

$$\bar{m}'_i = H\bar{m}_i + \frac{D_i}{d} x_f. \tag{2.45}$$

Denote the normalized coordinates of $\bar{m}_i$ and $\bar{m}'_i$ as m_i and m'_i, respectively, as follows:

$$m_i \triangleq \frac{\bar{m}_i}{z_i}, \ m'_i \triangleq \frac{\bar{m}'_i}{z'_i}. \tag{2.46}$$

Substitute (2.46) into (2.45), the relationship between the normalized coordinates is obtained as follows:

$$
\begin{aligned}
m'_i &= \frac{z_i}{z'_i} H m_i + \frac{D_i}{d z'_i} x_f \\
&= \frac{z_i}{z'_i} \left(H m_i + \frac{D_i}{z_i} \cdot \frac{x_f}{d} \right).
\end{aligned}
\tag{2.47}
$$

Using the single-view geometry $p_i = A m_i, p'_i = A' m'_i$, it is obtained that:

$$p'_i = \frac{z_i}{z'_i} \left(G p_i + \beta_i A' \cdot \frac{x_f}{d} \right), \tag{2.48}$$

where

$$\beta_i \triangleq \frac{D_i}{z_i} \tag{2.49}$$

is the projective parallax with respect to the homography G.

Using the two-view geometry (4.1), points in the scene can be described by the projective homography of the reference plane and the projective parallax with respect to the reference plane. This model is suitable for general road scenes, in which the points in regions of interest are near a reference plane.

2.4 Three-View Geometry

2.4.1 General Trifocal Tensor Model

As shown in Figure 2.4, there exists three cameras $\mathcal{C}$, $\mathcal{C}'$ and $\mathcal{C}''$ in the scene. Denote the rotation and translation from $\mathcal{C}'$ to $\mathcal{C}$ expressed in $\mathcal{C}'$ as R' and x'_f, respectively, denote the rotation and translation from $\mathcal{C}''$ to $\mathcal{C}$ expressed in $\mathcal{C}''$ as R'' and x''_f, respectively. Consider point O in the scene; denote its image points in the three views $\mathcal{C}$, $\mathcal{C}'$, and $\mathcal{C}''$ as $p \triangleq \begin{bmatrix} u & v & 1 \end{bmatrix}^T$, $p' \triangleq \begin{bmatrix} u' & v' & 1 \end{bmatrix}^T$, and

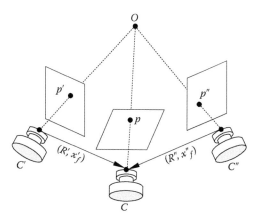

Figure 2.4: Three-view geometry.

$p'' \triangleq \begin{bmatrix} u'' & v'' & 1 \end{bmatrix}^T$, respectively. Correspondingly, denote the normalized coordinates of the point O with respect the three cameras as $m \triangleq \begin{bmatrix} m_x & m_y & 1 \end{bmatrix}^T$, $m' \triangleq \begin{bmatrix} m'_x & m'_y & 1 \end{bmatrix}^T$, and $m'' \triangleq \begin{bmatrix} m''_x & m''_y & 1 \end{bmatrix}^T$, respectively. Then, single-view geometry in (2.5) can be used to relate the image coordinates and the normalized coordinates.

The trifocal tensor consists of three matrices (T_1, T_2, T_3) and has 26 independent variables. It describes the geometric constraints among three views, including the relationships of point-point-point, point-point-line, point-line-point, point-line-line, and line-line-line correspondences. For the applications of vision-based control systems, feature points are extracted and matched in the relevant views, and the constraint equation is defined as:

$$[m']_\times \left(\sum_i m''_i T_i \right) [m]_\times = 0_{3 \times 3}, \tag{2.50}$$

where m''_i is the i-th element in m''. In this book, the function $T(C, C', C'')$ is used to describe the configuration shown in Figure 2.4. Besides, the trifocal tensor elements $\{T_i\}_{i \in 1,2,3}$ is related to the relative pose information among the three views as follows:

$$T_i = r'_i \cdot x_f''^T - x'_f \cdot r_i''^T, \tag{2.51}$$

where r'_i and r''_i are the i-th column vectors of the rotation matrices R' and R'', respectively.

Compared to homography and epipolar geometry, the trifocal tensor describes the relationship among three views and provides more information, and the redundant information improves the system robustness. Besides, the trifocal tensor doesn't rely on the scene structure and is suitable for vision-based control applications in general scenes.

2.4.2 Pose Estimation with Planar Constraint

As shown in Figure 2.5, the global coordinate system is located at C^*. Denote $(x(t), z(t))$ and $(x_0(t), z_0(t))$ as the positions of C_0 and C in the coordinate of C^*, respectively; denote θ_0 and θ as the right-handed rotation angles about y-axis that align the rotations of C_0 and C with C^*, respectively. In the following, $R \triangleq \begin{bmatrix} r_1 & r_2 & r_3 \end{bmatrix} \in SO(3)$ and $x_f \triangleq \begin{bmatrix} x_{fx} & 0 & x_{fz} \end{bmatrix}^T \in \mathbb{R}^3$ denote the rotation and translation from C to C^* expressed in C, respectively, which are given by:

$$R = \begin{bmatrix} \cos\theta & 0 & \sin\theta \\ 0 & 1 & 0 \\ -\sin\theta & 0 & \cos\theta \end{bmatrix}, \; x_f = \begin{bmatrix} -x\cos\theta - z\sin\theta \\ 0 \\ x\sin\theta - z\cos\theta \end{bmatrix}. \tag{2.52}$$

Similarly, $R_0 \triangleq \begin{bmatrix} r_{01} & r_{02} & r_{03} \end{bmatrix} \in SO(3)$ and $x_{f0} \triangleq \begin{bmatrix} x_{f0x} & 0 & x_{f0z} \end{bmatrix}^T \in \mathbb{R}^3$ denote the rotation and translation from C_0 to C^* expressed in C_0, respectively, which are given by:

$$R_0 = \begin{bmatrix} \cos\theta_0 & 0 & \sin\theta_0 \\ 0 & 1 & 0 \\ -\sin\theta_0 & 0 & \cos\theta_0 \end{bmatrix}, \; x_{f0} = \begin{bmatrix} -x_0\cos\theta_0 - z_0\sin\theta_0 \\ 0 \\ x_0\sin\theta_0 - z_0\cos\theta_0 \end{bmatrix}. \tag{2.53}$$

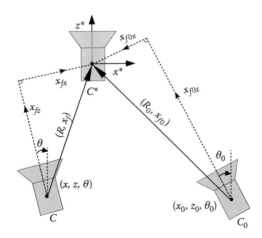

Figure 2.5: Three-view geometry with planar constraint.

Denote T_{ijk} as the element on the j-th row and k-th column of T_i; the trifocal tensor elements can be expressed by the pose parameters up to a scale as follows:

$$T_{111} = x_{f0x}\cos\theta - x_{fx}\cos\theta_0, \quad T_{131} = -x_{f0x}\sin\theta - x_{fz}\cos\theta_0,$$
$$T_{113} = x_{f0z}\cos\theta + x_{fx}\sin\theta_0, \quad T_{133} = -x_{f0z}\sin\theta + x_{fz}\sin\theta_0,$$
$$T_{212} = -x_{fx}, T_{221} = x_{f0x}, T_{223} = x_{f0z}, T_{232} = -x_{fz}, \quad (2.54)$$
$$T_{311} = x_{f0x}\sin\theta - x_{fx}\sin\theta_0, \quad T_{331} = x_{f0x}\cos\theta - x_{fz}\sin\theta_0,$$
$$T_{313} = x_{f0z}\sin\theta - x_{fx}\cos\theta_0, \quad T_{333} = x_{f0z}\cos\theta - x_{fz}\cos\theta_0.$$

The other elements of the trifocal tensor are zero due to the planar motion constraint.

Denote $d \triangleq \sqrt{x_{f0x}^2 + x_{f0z}^2}$ as the distance between C_0 and C^*. Assume that the origin of C^* doesn't coincide with that of C_0, i.e., $d \neq 0$ and $T_{221}^2 + T_{223}^2 \neq 0$. Then, the trifocal tensor elements can be normalized as follows:

$$\bar{T}_{ijk} \triangleq \alpha \cdot \frac{T_{ijk}}{\sqrt{T_{221}^2 + T_{223}^2}}, \quad (2.55)$$

where α is a sign factor and can be determined by $\text{sign}(T_{221}) \cdot \text{sign}(x_{f0x})$ or $\text{sign}(T_{223}) \cdot \text{sign}(x_{f0z})$. The signs of x_{f0x} and x_{f0z} do not change in the control process, and can be obtained in the teaching process using *a priori* knowledge.

Using the trifocal tensor among the three views, the translation vector can be estimated up to a scale d as:

$$x_h \triangleq \begin{bmatrix} x_{hx} & 0 & x_{hz} \end{bmatrix}^T = \frac{x_f}{d}, \quad (2.56)$$

where x_{hx}, x_{hz} are given by:

$$x_{hx} = -\bar{T}_{212}, \quad x_{hz} = -\bar{T}_{232}. \quad (2.57)$$

The orientation angle θ can be determined by the following expressions:

$$\sin\theta = \bar{T}_{221}(\bar{T}_{333} - \bar{T}_{131}) - \bar{T}_{223}(\bar{T}_{331} + \bar{T}_{133}),$$
$$\cos\theta = \bar{T}_{221}(\bar{T}_{111} - \bar{T}_{313}) + \bar{T}_{223}(\bar{T}_{113} + \bar{T}_{311}). \quad (2.58)$$

Then, the scaled position coordinates $x_m \triangleq \frac{x}{d}, z_m \triangleq \frac{z}{d}$ can be calculated as follows:

$$\begin{bmatrix} x_m \\ z_m \end{bmatrix} = -\begin{bmatrix} \cos\theta & -\sin\theta \\ \sin\theta & \cos\theta \end{bmatrix} \cdot \begin{bmatrix} x_{hx} \\ x_{hz} \end{bmatrix}. \quad (2.59)$$

2.5 Computation of Multiple-View Geometry

In the applications of vision-based control, multiple-view geometry is generally estimated by the images captured by the camera at different poses. Image feature extraction and matching methods are used to obtain the corresponding feature points in relevant views.

2.5.1 Calibration of Single-View Geometry

Accurate camera calibration is a necessary prerequisite for the extraction of precise and reliable 3D metric information from images. Various algorithms for camera calibration have been reported over the years and the research on conventional perspective cameras is relatively mature. Recent approaches concentrate on the study on the calibration of unified camera models [172] and camera networks [207]. Readers may refer to the overall reviews in [74,89,173]. Three typical methods are summarized as follows:

- Tsai's calibration model [205] assumes that some parameters of the camera are provided by the manufacturer to reduce the initial guess of the estimation. It requires n features points ($n > 8$) per image and solves the calibration problem with a set of n linear equations based on the radial alignment constraint. A second order radial distortion model is used while no decentering distortion terms are considered. The two-step method can cope with either a single image or multiple images of a 3D or planar calibration grid, but grid point coordinates must be known.

- The technique developed by Heikkila and Silven [88] first extracts initial estimates of the camera parameters using a closed-form solution (direct linear transformation) and then a nonlinear least-squares estimation employing the Levenberg–Marquardt algorithm is applied to refine the input/output and compute the distortion parameters. The model uses two coefficients for both radial and decentering distortion, and the method works with single or multiple images and with 2D or 3D calibration grids.

- Zhang's calibration method [224] requires a planar checkerboard grid to be placed at different orientations (more than two) in front of the camera. The developed algorithm uses the extracted corner points of the checkerboard pattern to compute a projective transformation between the image points of the n different images, up to a scale factor. Afterwards, the camera interior and exterior parameters are recovered using a closed-form solution, while the third- and fifth-order radial distortion terms are recovered within a linear least-squares solution. A final nonlinear minimization of the reprojection error, solved using a Levenberg–Marquardt method, refines all the recovered parameters. Zhang's approach is quite similar to that of Triggs (1998), which requires at least five views of a planar scene.

The use of a calibration pattern or set of markers is one of the most reliable ways to estimate a camera's intrinsic parameters. If a small calibration rig is used, e.g., for indoor robotics applications or for mobile robots that carry their own calibration target, it is best if the calibration object can span as much of the workspace as possible, as planar targets often fail to accurately predict the components of the pose that lie far away from the plane. A good way to determine whether the calibration has been successfully performed is to estimate the covariance in the parameters and then project 3D points from various points in the workspace into the image in order to estimate their 2D positional uncertainty.

When a finite workspace is being used and accurate machining and motion control platforms are available, a good way to perform calibration is to move a planar calibration target in a controlled fashion through the workspace volume. This approach is sometimes called the N-planes calibration approach and has the advantage that each camera pixel can be mapped to a unique 3D ray in space, which takes care of both linear effects modeled by the calibration matrix A and nonlinear effects such as radial distortion. In this case, the pattern's pose has to be recovered in conjunction with the intrinsics.

2.5.2 Computation of Two-View Geometry

2.5.2.1 Computation of Homography

Considering the configuration in Figure 2.2, denote the matched feature points in images $\mathbf{I}$ and $\mathbf{I}'$ as $\{p_j\}$ and $\{p'_j\}$, respectively. For convenience, the single-view geometry is used to normalize the image coordinates as $\{m_j\}$ and $\{m'_j\}$. Denote the Euclidean homography H as:

$$H = \begin{bmatrix} h_{11} & h_{12} & h_{13} \\ h_{21} & h_{22} & h_{23} \\ h_{31} & h_{32} & h_{33} \end{bmatrix}. \tag{2.60}$$

Denote $h_{9\times1}$ as the column vector containing the homography elements, defined as:

$$h \triangleq \begin{bmatrix} h_{11} & h_{12} & h_{13} & h_{21} & h_{22} & h_{23} & h_{31} & h_{32} & h_{33} \end{bmatrix}^T. \tag{2.61}$$

For one pair of corresponding feature points $m = \begin{bmatrix} m_x & m_y & 1 \end{bmatrix}^T$ and $m' = \begin{bmatrix} m'_x & m'_y & 1 \end{bmatrix}^T$, the following equations are obtained using (2.14):

$$m'_x = \frac{h_{11}m_x + h_{12}m_y + h_{13}}{h_{31}m_x + h_{32}m_y + h_{33}}$$
$$m'_y = \frac{h_{21}m_x + h_{22}m_y + h_{23}}{h_{31}m_x + h_{32}m_y + h_{33}}. \tag{2.62}$$

Then, above equations can be rewritten into the following expression:

$$\Pi_{2\times9}h_{9\times1} = 0_{2\times1} \tag{2.63}$$

where $\Pi_{2\times9}$ is the coefficient matrix defined as follows:

$$\Pi_{2\times9} = \begin{bmatrix} m_x & m_y & 1 & 0 & 0 & 0 & -m_x m'_x & -m'_x m_y & -m'_x \\ 0 & 0 & 0 & m_x & m_y & 1 & -m_x m'_y & -m_y m'_y & -m'_y \end{bmatrix}. \tag{2.64}$$

Then, corresponding N point pairs can be used to construct the following linear equation:

$$\Pi_{2N \times 9} h_{9 \times 1} = 0_{2N \times 1} \tag{2.65}$$

2.5.2.2 Computation of Epipolar Geometry

Considering the configuration in Figure 2.2, denote the matched feature points in images **I** and **I'** as $\{p_j\}$ and $\{p'_j\}$, respectively. For convenience, the single-view geometry is used to normalize the image coordinates as $\{m_j\}$ and $\{m'_j\}$. Denote the essential matrix E as:

$$E = \begin{bmatrix} e_{11} & e_{12} & e_{13} \\ e_{21} & e_{22} & e_{23} \\ e_{31} & e_{32} & e_{33} \end{bmatrix}. \tag{2.66}$$

Denote $e_{9 \times 1}$ as the column vector containing the elements in the essential matrix, which is given by:

$$e_{9 \times 1} = \begin{bmatrix} e_{11} & e_{12} & e_{13} & e_{21} & e_{22} & e_{23} & e_{31} & e_{32} & e_{33} \end{bmatrix}^T. \tag{2.67}$$

For one pair of corresponding feature points $m = \begin{bmatrix} m_x & m_y & 1 \end{bmatrix}^T$ and $m' = \begin{bmatrix} m'_x & m'_y & 1 \end{bmatrix}^T$, the following equation are obtained:

$$m'_x m_x e_{11} + m'_x m_y e_{12} + m'_x e_{13} + m'_y m_x e_{21} + m'_y m_y e_{22} + m'_y e_{23}$$
$$+ m_x e_{31} + m_y e_{32} + e_{33} = 0. \tag{2.68}$$

Then, corresponding N point pairs can be used to construct the following linear equation:

$$\Pi_{1 \times 9} e_{9 \times 1} = 0. \tag{2.69}$$

2.5.3 Computation of Three-View Geometry

In multiple-view geometry, two-view geometry has been widely used in relevant fields. Thus, there are several open source online estimation methods for two-view geometry, such as the functions *findHomography*, *findFundamentalMat* in OpenCV. However, the trifocal tensor has not been used for vision-based control before 2010 and its only estimation was not well implemented.

Considering the configuration in Figure 2.4, in the applications of vision-based control, the trifocal tensor is generally estimated by the three images captured by the camera at three different poses. In real applications, image feature point extraction and matching methods (e.g., Scale-Invariant Feature Transform (SIFT), Speeded-Up Robust Features, etc.) are used to obtain the corresponding feature points in three views. Then, the matched feature points $(\{p_j\}, \{p'_j\}, \{p''_j\})$ are used to estimate the trifocal tensor, where $j \in \{1, ..., N\}$. For convenience, the single-view geometry is used to normalize the image coordinates as $\{m_j\}, \{m'_j\}, \{m''_j\}$.

2.5.3.1 Direct Linear Transform

Considering one point triple m, m', m'', expand the tensor equation in (2.50) to obtain the following expression:

$$\Pi_{4\times27}t_{27\times1} = 0_{4\times1} \tag{2.70}$$

where $t_{27\times1}$ is the column vector containing the trifocal tensor elements, defined as:

$$t \triangleq \begin{bmatrix} T_1^{11} & T_1^{12} & T_1^{13} & T_1^{21} & T_1^{22} & T_1^{23} & T_1^{31} & T_1^{32} & T_1^{33} & \cdots & T_3^{33} \end{bmatrix}^T \tag{2.71}$$

In (2.71), T_i^{kl} is the element on the k-th row and l-th column in T_i. In (2.70), Π is a coefficient matrix of 4×27, which is defined as:

$$\Pi = \begin{bmatrix} m_x'' \pi_1 & m_y'' \pi_1 & \pi_1 \\ m_x'' \pi_2 & m_y'' \pi_2 & \pi_2 \\ m_x'' \pi_3 & m_y'' \pi_3 & \pi_3 \\ m_x'' \pi_4 & m_y'' \pi_4 & \pi_4 \end{bmatrix} \tag{2.72}$$

where $\pi_1, \pi_2, \pi_3, \pi_4$ are 1×9 row vectors, defined as:

$$\begin{aligned}
\pi_1 &= \begin{bmatrix} -1 & 0 & m_x & 0 & 0 & 0 & m_x' & 0 & -m_x m_x' \end{bmatrix} \\
\pi_2 &= \begin{bmatrix} 0 & -1 & m_y & 0 & 0 & 0 & m_x' & -m_y m_x' \end{bmatrix} \\
\pi_3 &= \begin{bmatrix} 0 & 0 & 0 & -1 & 0 & m_x & m_y' & 0 & -m_x m_y' \end{bmatrix} \\
\pi_4 &= \begin{bmatrix} 0 & 0 & 0 & 0 & -1 & m_y & 0 & m_y' & -m_y m_y' \end{bmatrix}
\end{aligned} \tag{2.73}$$

Based on (2.70), corresponding N point triples are used to construct the following linear equation:

$$\Pi_{4N\times27}t_{27\times1} = 0_{4N\times1} \tag{2.74}$$

At least seven triples are needed to solve the least squares solution of t.

2.5.3.2 Matrix Factorization

Define the following vectors:

$$U = \begin{bmatrix} -1 \\ 0 \\ m_x \end{bmatrix}, \quad V = \begin{bmatrix} 0 \\ -1 \\ m_y \end{bmatrix}, \quad U' = \begin{bmatrix} 1 \\ 0 \\ -m_x' \end{bmatrix}, \quad V' = \begin{bmatrix} 0 \\ 1 \\ -m_y' \end{bmatrix}. \tag{2.75}$$

Then using tensor multiplication, $\pi_1, \pi_2, \pi_3, \pi_4$ is expressed as:

$$\pi_1 = U'^T \otimes U^T, \quad \pi_2 = U'^T \otimes V^T, \quad \pi_3 = V'^T \otimes U^T, \quad \pi_4 = V'^T \otimes V^T \tag{2.76}$$

Thus $\Pi_{4\times 27}$ can be decomposed as:

$$\Pi_{4\times 27} = (I_4 \otimes m''^T) \begin{bmatrix} I_6 \otimes U'^T \\ I_6 \otimes V'^T \end{bmatrix} \begin{bmatrix} I_9 \otimes U^T \\ I_9 \otimes V^T \end{bmatrix} \tag{2.77}$$

Popularize (2.77) to N triples, it can be obtained that

$$\Pi_{4N\times 27} = P_{4N\times 12N} \cdot Q_{12N\times 18N} \cdot L_{18N\times 27} \tag{2.78}$$

where

$$P_{4N\times 12N} = diag(I_4 \otimes m_1''^T, \ldots, I_4 \otimes m_N''^T) \tag{2.79}$$

$$Q_{12N\times 18N} = diag\left(\begin{bmatrix} I_6 \otimes U_1'^T \\ I_6 \otimes V_1'^T \end{bmatrix}, \ldots, \begin{bmatrix} I_6 \otimes U_N'^T \\ I_6 \otimes V_N'^T \end{bmatrix}\right) \tag{2.80}$$

$$L_{18N\times 27} = \begin{bmatrix} \begin{bmatrix} I_9 \otimes U_1^T \\ I_9 \otimes V_1^T \end{bmatrix} \\ \vdots \\ \begin{bmatrix} I_9 \otimes U_N^T \\ I_9 \otimes V_N^T \end{bmatrix} \end{bmatrix} \tag{2.81}$$

Denote $h = L_{18N\times 27}t, l = Q_{12N\times 18N}h$, it can be obtained that

$$\begin{cases} I_{18N\times 27} \cdot t - h = 0 \\ Q_{12N\times 18N} \cdot h - l = 0 \\ P_{4N\times 12N} \cdot l = 0 \end{cases} \tag{2.82}$$

Perform SVD decomposition to $Q_{12N\times 18N} \cdot L_{18N\times 27}$:

$$Q_{12N\times 18N} \cdot L_{18N\times 27} = \bar{U}_{12N\times 12N}^T \begin{bmatrix} D_{27\times 27} \\ 0_{(12N-27)\times 27} \end{bmatrix} \bar{V}_{27\times 27} \tag{2.83}$$

Based on the above developments, it can be obtained that:

$$Q_{12N\times 18N}L_{18N\times 27}t = l \tag{2.84}$$

which can be rewritten as:

$$\begin{bmatrix} D_{27\times 27} \\ 0_{(12N-27)\times 27} \end{bmatrix} \bar{V}_{27\times 27}t = \bar{U}_{12N\times 12N}l \tag{2.85}$$

As a result, it can be obtained that:

$$\begin{cases} \bar{U}_1 l - D_{27\times 27}\bar{V}_{27\times 27}t = 0 \\ \bar{U}_2 l = 0 \end{cases} \tag{2.86}$$

where

$$\begin{bmatrix} \bar{U}_{1,27\times 12N} \\ \bar{U}_{2,(12N-27)\times 12N} \end{bmatrix} = \bar{U}_{12N\times 12N} \tag{2.87}$$

Since $\bar{U}_2 l = 0, \bar{U}_2 \bar{U}_1^T = 0_{(12N-27) \times 27}$ then $l = \bar{U}_1 c, c \in \mathbb{R}^{27}$. It follows that:

$$\begin{cases} Q_{12N \times 18N} L_{18N \times 27} t = l \\ P_{4N \times 12N} l = 0 \end{cases} \Leftrightarrow \begin{cases} c - D_{27 \times 27} V_{27 \times 27} t = 0 \\ P_{4N \times 12N} \bar{U}_1^T c = 0 \end{cases} \tag{2.88}$$

As a result,

$$\begin{bmatrix} D_{27 \times 27} V_{27 \times 27} & -I_{27 \times 27} \\ 0 & P_{4N \times 12N} \bar{U}_1^T \end{bmatrix} \cdot \begin{bmatrix} t \\ c \end{bmatrix} = 0 \tag{2.89}$$

Finally the trifocal tensor elements are obtained as follows:

$$t = (D_{27 \times 27} V_{27 \times 27})^{-1} v_1 \tag{2.90}$$

where v_1 is the eigen vector of $P_{4N \times 12N} \bar{U}_1^T$ and can be obtained from SVD decomposition.

2.5.4 Robust Approaches

This section has introduced several methods for the computation of multiple-view geometry using image measurements. In real scenes, the captured images suffer from image noise and discretization errors, and the extraction and matching of feature points are sensitive to light variation, image noise, and image view, etc. [81]. As a result, the above estimation methods should be combined with robust approaches.

Point normalization

For the computation of the above multiple-view geometry, the following normalization steps are required for better conditioning:

1. The points are translated so that their centroid is at the origin.

2. The points are then scaled so that the average distance from the origin is equal to $\sqrt{2}$.

3. This transformation is applied to each of the two images independently.

The above transform process can be described by a similarity transformation with translation and scaling. Given three images $\mathbf{I}, \mathbf{I}', \mathbf{I}''$, denote the matrices for normalization as M, M', and M'', respectively.

Using the normalized points, the computation of multiple-view geometry is processed as follows:

1. Extract and match the corresponding feature points $\{p_i\}$, $\{p_i'\}$, and $\{p_i''\}$ in the relevant images $\mathbf{I}, \mathbf{I}'$, and $\mathbf{I}''$;

2. Transform the feature points for normalization as $\bar{p}_i = M \cdot p_i, \bar{p}_i' = M' \cdot p_i'$ and $\bar{p}_i'' = M'' \cdot p_i''$;

3. Compute the multiple-view geometry (homography $\bar{H}$, essential matrix $\bar{E}$, and trifocal tensor $\bar{T}$) using the normalized points;

4. De-normalize the calculated multiple-view geometry, $H = M'^{-1}\bar{H}$ for homography, $E = M'^T\bar{M}$ for essential matrix.

Error analysis

In real applications, the measured feature points are not accurate and the computations of multiple-view geometry suffer from errors. In applications, redundant corresponding points (more than the minimum case) are used to deal with the effects of image noise and point mismatches. For robust estimation, the redundant points are generally used in groups to eliminate the noisy or mismatched feature points. As a result, the estimation error need to be analyzed quantitatively to show the reliability of the corresponding feature points. Generally, the following error measurements are used:

■ Algebraic distance: In the computation of multiple-view geometry models, linear equations are constructed using image measurements to express the geometric constraints. The estimation of multiple-view geometry is simplified into the minimization of the residual vector of the linear equations. Taking the computation of homography for example, the linear equation is expressed as $\Pi_{2\times9}h_{9\times1} = 0_{2\times1}$, then the algebraic distance is defined by $\|\Pi_{2\times9}\tilde{h}_{9\times1}\|$, where $\hat{h}$ is the vector of the estimated homography matrix.

■ Geometric distance: The geometric distance is computed by the difference between the raw points and the transformed points using the geometric model. Taking the computation of homography for example, denote the raw points as $\{p_i\}$ and $\{p'_i\}$, and the estimated homography matrix as $\tilde{H}$, then the algebraic distance for one point is computed by $\|p'_i - \hat{H} \cdot p_i\|$.

■ Image reprojection error: Different from the above distances that rely on the difference among feature points, the image reprojection error is directly defined with respect to image intensities in the image space.

Random sample consensus method

To be general, the Random sample consensus (RANSAC) method is described as follows. The objective of the RANSAC method is to robustly fit a model to a data set S which contains outliers. The algorithm is performed as follows:

1. Randomly select a sample of s data points from S and instantiate the model from this subset.

2. Determine the set of data points Si which are within a distance threshold t of the model. The set Si is the consensus set of the sample and defines the inliers of S.

3. If the size of *Si* (the number of inliers) is greater than some threshold *T*, re-estimate the model using all the points in *Si* and terminate.

4. If the size of *Si* is less than *T*, select a new subset and repeat the above.

5. After *N* trials the largest consensus set *Si* is selected, and the model is re-estimated using all the points in the subset *Si*.

In the RANSAC process, the distance threshold *t* describes the probability of a point to be an inlier and is usually chosen empirically in practice. It is often computationally infeasible and unnecessary to try every possible sample. Instead the number of samples *N* is chosen sufficiently high to ensure with a probability, *p*, that at least one of the random samples of *s* points is free from outliers. Usually *p* is chosen at 0.99. Suppose *w* is the probability that any selected data point is an inlier, and thus $\varepsilon = 1 - w$ is the probability that it is an outlier. Then at least *N* selections (each of *s* points) are required, where $(1 - ws)N = 1 - p$, so that $N = \log(1 - p) / \log(1 - (1 - \varepsilon)^s)$.

Chapter 3

Vision-Based Robotic Systems

In the everyday life of human beings, vision is the most important sensing modality and provides rich information about the environment. The integration of vision sensors into robotic systems has brought great flexibility and efficiency for applications. In vision-based robotic systems, robot states and environment structure are measured by visual feedback. Compared with other sensing modalities, such as Global Positioning System (GPS), ultrasonic sensors, Light Detection And Ranging (LiDARs), Radars, etc., vision sensors are more descriptive and can be widely used in various environments.

Just like the behavior of human beings, robot tasks can be described by visual information and the controller regulates the robot to accomplish the task using the estimated states based on visual feedback. For example, if the mobile robot is ordered to move toward some specific place, it is easier and more flexible to tell it what the place looks like instead of where the place is. For conventional robotic systems, the exact location (Euclidean coordinate) of the desired place should be determined first, and then the robot is localized in real time to provide feedback to the control system. For vision-based robotic systems, the image to be seen in the desired place is provided, and the captured image at the view of the robot is used to provide visual feedback to the visual controller. The task is finished when the current image coincides with the desired image.

In real applications, the robot workspace is limited and the robot should share spaces with other obstacles. As a result, the robot should also avoid obstacle collision in the control process. This issue is especially important for mobile robots with larger workspace than robot manipulators that generally works in a limited space. However, the definition of obstacle is rather geometrical. Even though the descriptive information is provided by the vision sensors, it is not easy to accurately detect the obstacles in workspace. For the application of obstacle avoidance, the drivable road regions (the spaces that can be driven through safely by the robot) are necessary for safety, and the location and motion information of obstacles provides further information for efficient obstacle avoidance.

3.1 System Overview

3.1.1 System Architecture

The vision-based robotic system coordinates vision system and control system to accomplish specific tasks for robots. As shown in Figure 3.1, given the robotic task, the robot system consists of visual perception and visual control modules. Visual perception provides necessary feedback information for the planning and control of the robot system.

The visual perception module includes the estimation of robot states and the perception of environment information. In the moving process of the robot, system states are estimated by comparing the images captured at current pose and the desired pose. Generally, system states can be expressed in several spaces, such as the Euclidean space, image space, and the scaled geometric space. For environment perception, the images captured at various views can be used to

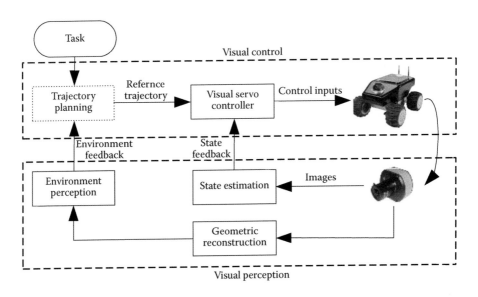

Figure 3.1: Vision-based robot perception and control systems.

reconstruct the geometric information of the environment. Then, the drivable road region for mobile robots is detected and reconstructed geometrically based on geometric and appearance information. The motion information of the obstacles (including the pose and velocity information) is estimated based on the reconstructed geometric information.

The visual control module includes the trajectory planning and visual servo controller to regulate the robot for task accomplishment. Generally, the task is given by one or several target images, implying what the robot ought to see at target poses. The visual servo controller relies on the state estimation based on visual feedback. Different from conventional robot controllers that rely on position measurements, visual servo controllers should deal with the model uncertainties and constraints induced by the vision system. Considering the obstacles and limited field-of-view (FOV) of vision sensors, appropriate reference trajectory are planned by the trajectory planning algorithm taking into account the environment perception results and robot states.

3.1.2 Physical Configurations

To incorporate vision sensors into the robot control system, there are mainly two types of configurations, i.e., eye-in-hand system and eye-to-hand system, as shown in Figures 3.2 and 3.3, respectively.

In an eye-in-hand system, the camera is installed at the end effector of the manipulator or on the mobile robot, and move with the motion of the robot. As shown in Figure 3.2b, the camera C is installed rigidly at the end of the manipulator, with a rigid transformation with respect to the end effector $\mathcal{F}$. For the application of robotic manipulators, the end effector is regulated to the desired

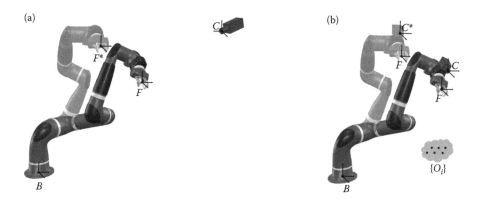

Figure 3.2: Classical configurations for robotic manipulators. (a) Eye-to-hand system and (b) eye-in-hand system.

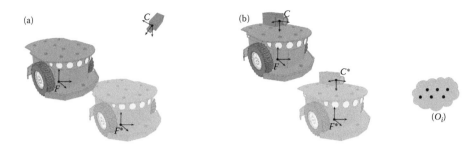

Figure 3.3: Classical configurations for mobile robots. (a) Eye-to-hand system and (b) Eye-in-hand system.

pose $\mathcal{F}^*$ to accomplish the task. The task can be also expressed with respect to the camera frame, i.e., the control objective is to regulate the camera $\mathcal{C}$ to $\mathcal{C}^*$. For the case of mobile robots in Figure 3.3b, the navigation task of the mobile robot can also be expressed with respect to the camera frame. Image(s) are captured by the camera to express the desired pose(s), so that the task is accomplished by regulating the mobile robot based on the feedback of the camera. In an eye-to-hand system, the camera is fixed in the workspace to monitor the motion of the robot. The task can also be expressed by the image captured when the robot is at the desired pose. System error vector is generally computed by the displacements in images of the end effector (in Figure 3.2a) or the mobile robot (in Figure 3.3a).

It should be noted that these two configurations both have advantages and disadvantages. The eye-in-hand system is easier to configure since the camera can be simply installed on the robot and does not need to be installed at specific positions in the workspace in the eye-to-hand system. Besides, the system feedback is computed by the images, but the captured targets in eye-to-hand systems are generally further than those in eye-in-hand systems, reducing the control precision. However, due to the limited FOV of cameras, the workspace of eye-in-hand systems is limited since conventional approaches require the correspondence of the specific images. Comparably, the workspace of the eye-to-hand system is relatively bigger and is also determined by the FOV of cameras. However, eye-to-hand configuration also has FOV constraints, i.e., the end effector or the mobile robot should be in the FOV of the camera to construct effective feedback. For the coordinate control of multiple robots, the eye-to-hand configuration is more suitable since there is a fixed global frame naturally. In practice, the physical configuration is generally determined by the requirements of the task.

3.2 Research Essentials

As shown in Figure 3.4, considering the complexity of the environment and the high dimensionality of visual information, unified modeling and analysis is

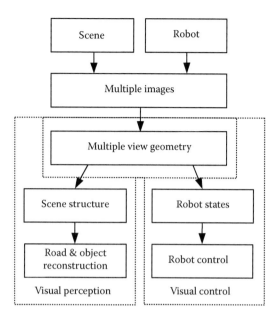

Figure 3.4: Multiple-view geometry based robot perception and control systems.

conducted for the environment based on multiple-view geometry. Then, the scene structure and self states are analyzed from a geometry point of view, abbreviating geometric information for the visual perception and control tasks of robots.

The visual perception tasks are considered in Part II of this book, including the drivable road detection and reconstruction for mobile robots and the moving object motion estimation for general robots. In the control process of mobile robots, the detection of drivable road region is necessary to guarantee the driving safety. However, due to the complexity of the environments, the robust detection of road region from images is challenging caused by varying road appearance, image noise, illumination conditions, etc. In this book, two-view geometry will be used to reconstruct the 3D information of the environment, based on which the range and 3D information of road regions can be obtained. For the motion estimation of moving objects, multiple-view geometry is used to reconstruct the relative pose information of the observed objects, and the motion information is estimated by designing adequate observers.

The visual control tasks are considered in Part III of this book, including the control tasks for general robots and nonholonomic mobile robots. In the field of visual servo control, there exist two main challenges: the model uncertainties induced by the robot-vision system, the constraints induced by the limited FOV of cameras and the motion properties of robots. In this book, multiple-view geometry is exploited to reconstruct the relative geometric information among the views in the tasks and scaled pose information can be generally obtained which includes an unknown scale factor. To deal with model uncertainties in visual control systems, advanced control strategies are designed to guarantee the stability and convergence of the closed control systems, such as adaptive control and robust control. To deal with the FOV constraint caused by the camera, appropriate path can be planned to ensure the visibility of the target in the control process and visual servo controllers are designed to track the desired path to accomplish the task.

VISUAL PERCEPTION OF ROBOTICS

II

Chapter 4

Introduction to Visual Perception

4.1 Road Reconstruction and Detection for Mobile Robots

4.1.1 Previous Works

For the application of mobile robots, environment perception is crucial which mainly includes the detection of vehicles [191], pedestrians [97], traffic signs [182], drivable road regions [92,184], etc. Drivable road detection is an essential functionality and has been widely studied in the past few years. Previous works generally use range sensors (e.g., ultra sonic sensors, laser scanners, etc.), vision sensors, and the fusion of multi-modal sensors [73]. Comparably, vision sensors are cheaper and provide richer environment information. However, developing a reliable vision-based road detection system is a hard task, which requires flexibility and robustness to work with different physical configurations, light conditions, road types, and background objects. As shown in Figure 4.1, the main difficulty is that the road appearance changes both spatially and temporally, i.e., the appearance of the road in one image varies in different areas due to intrinsic or environmental characteristics, and the road type varies with time going by as the vehicle keeps going. In terms of the physical configuration of vision-based road detection systems, monocular and binocular camera systems are commonly used.

Appearance-based approaches

Monocular camera systems are generally easier to configure and of low cost. Standard approaches are based on appearance information, such as color

Figure 4.1: Various road images including structured roads (first row) and unstructured roads (second row).

[3,87,198], texture [118,216], etc. Color-based methods generally classify road pixels by color information based on road appearance models that are constructed from sample regions [3] or trained from ground truth data [126]. However, since the appearance of the road is affected by environmental factors (e.g., shadows), the success of road detection relies on the generality of the road model, which is not adaptable for new scenes. In general, the road region is surrounded by typical textures, such as lane markings for marked roads [61,135] and edges for unmarked roads [160]. However, these types of models are not flexible and representative enough to describe the real world. Besides, great efforts are needed to manually design such models, especially when there exist shadows that greatly influence the appearance of the road surface.

To reduce the dependence on the prior knowledge of road models, supervised machine learning methods can be applied to road modeling if enough ground truth data is provided. Features of image regions are used as the inputs to the model of which the outputs are the class probability of being road or non-road. In [187], the scene image is divided into grids. Features (e.g., average, entropy, energy, and variance from different color channels) are extracted from each grid as the inputs to an artificial neural network. However, the resolution of segmentation is low, and the features are not representative enough for complex scenes. In addition, convolutional neural networks (CNNs) have been established as a powerful class of models for image recognition problems without the need of manually designing image features. In [157], deep CNNs are exploited for model training of road regions, and the image pixels are classified into road and non-road ones. A Network-in-Network architecture is used in [157], taking advantage of a large contextual window for better road detection. In [162], a hierarchical approach is proposed to label semantic objects and regions in scenes. The multiple class classification problem is divided into a hierarchical series of simple machine learning subproblems, which is suitable for the unified scene understanding including roads, pedestrians, vehicles, etc. However, in outdoor scenes, the road appearance is greatly affected by the complex lighting conditions. Therefore, the models trained with enough lights are not descriptive enough for the recognition of road regions in shadows, causing ambiguity in monocular road detection without 3D information.

As mentioned before, the road appearance is strongly affected by shadows for outdoor uses, limiting the reliability of the system. The existence of shadows changes the intensity of pixels and weakens the textures, causing ambiguities for road detection. Detecting and removing shadows properly is still challenging. Previous works are based on statistical learning [84,122], image decomposition [183], shadow's physical models [200], etc. However, for the application of road detection and stereo matching, these methods are generally too complex in terms of implementation and computation. Other kinds of approaches are preferred, which are based on direct image transforms. These methods try to find some illuminant invariant spaces to reduce the effect of shadows, and generally have a complexity of $O(N)$. In [2,4], a one-channel illuminant invariant image space is proposed and has been widely used in road detection systems recently. In [90,91], the log-chromaticity image transform is proposed and has been used for robust stereo matching with varying illuminations. In [45], chromaticity coordinates are used to compute illuminant invariant images based on which road regions are segmented. In [149], a one-channel image space is computed by the combination of the logarithm of three image channels and is used for visual localization. However, there is a trade-off between recovering accuracy and computational efficiency, and in these methods much color information or textures are lost.

Stereo vision-based approaches

Comparably, binocular camera systems provide more information except for the appearances, and the typical approaches rely on the disparity maps extracted from stereo matching, which are more robust to changes in appearance. In [121], a road detection algorithm is proposed based on "u-disparity" and "v-disparity" images, which provide a side view and top view of the scene to classify the road region, so that the road region can be segmented conveniently at discontinuities. In [215], both planar and nonplanar roads are classified based on u,v-disparity maps using intrinsic road attributes under stereo geometry. In [163], the image is segmented into cells with each cell labeled with the largest elevation in it. Then, they are classified as road or background using the elevation distribution. However, the performance of these methods rely on the quality of stereo matching, and it is generally a hard task to generate accurate and dense disparity maps in real time, especially for textureless and occluded areas. In practice, typical stereo vision systems are costly because they should be carefully configured, i.e., the two cameras are parallel and displaced horizontally with respect to each other, so that corresponding image points are on the same row to reduce the searching region. To improve measuring accuracy, a wider baseline is generally needed, while one has to search over a larger space to find corresponding points and there exist more ambiguities, limiting the flexibility of the system.

It is obvious that both appearance and stereo vision-based methods have advantages and disadvantages. Generally, appearance information is more descriptive but has difficulties in modeling especially in complex scenarios. Stereo vision provides the scene geometry independent from the scene appearance but is not discriminative enough in the far region due to the limited image resolution. In previous approaches, image appearance features and geometric features are fused for robust detection. In [114], road features in intensity images are used for validation and correction of extracted road boundaries from disparity images. In [146], road regions obtained from disparity images are used as prior knowledge for super pixel classification in intensity images.

Geometry-based approaches

In perspective images, a distinguishing feature is that the image of an object that stretches off to infinity can have finite extent. An infinite line is projected as a line terminating at a vanishing point in the image. As a result, parallel lines in the world, e.g., road curbs, are projected as converging lines in the image, and their image intersection is the vanishing point for the direction of the road. In structured roads (such as city lanes and highways shown in the first row of Figure 4.1), the road curbs are generally indicated by lane markings which are a strong feature for road detection. In unstructured roads (such as pavements and country roads shown in the second row of Figure 4.1), the lane markings are not available but there generally exist parallel textures left by vehicles. Once the vanishing point is estimated in the image, the road boundaries are determined coarsely by selecting the dominant lines through the vanishing point [118]. Since the scene may be complex in the real world, there exist many other textures that serve as disturbances in the road vanishing point detection. Several image filtering (e.g., Gabor, Laplacian of Gaussian, etc.) and voting methods [118,119] are proposed to select the dominant road features. However, these methods only detect the main road region (straight part) and are not applicable to highly curved roads. Besides, the estimation of vanishing points is not sensitive to heavy traffic and shadows.

Assuming the road is planar, its images captured at two poses can be related by a homography [86], which has been successfully used for vision-based control [31]. To relax the dependency on the appearance information, another kind of methods is based on homography [132] to classify road points from non-road points. Previous works are based on the warping error of two images [5] (the subtraction between the current image and the previous image warped by homography), the transforming error of feature points [44], or both of them [170], so that image points on the desired plane are classified and can be used as prior knowledge for ground region segmentation. However, homography-based methods only work for planar roads, while most outdoor road surfaces are not strictly planar. In [60], multiple homographies with different depth parameters are used to identify objects of different heights on the road. While the algorithm works in a discrete manner relying on the layers of height and the computational complexity is multiplied. Homography-based methods provide an elegant way to determine ground surfaces with a few prior appearances. However, outdoor road regions are typically weakly or repetitively textured, so that the robust extraction of feature points is hard and homography-based warping error suffers from ambiguities.

4.1.2 A Typical Vehicle Vision System

4.1.2.1 System Configuration

In the following sections, the concerned problem is the drivable road reconstruction with an on-vehicle two-camera system. As shown in Figure 4.2, two cameras are installed rigidly on the vehicle. Denote the coordinate systems of the vehicle and the cameras as $\mathcal{F}, \mathcal{C}$, and $\mathcal{C}'$, respectively. The origin of $\mathcal{F}$ is located at the

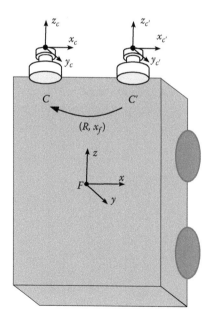

Figure 4.2: On-vehicle two-camera system configuration.

center of the vehicle, and the xz axis plane of $\mathcal{F}$ is parallel to the grounding plane of the vehicle, with the x axis parallel to the wheel axis and the z axis perpendicular to the wheel axis. For simplicity, camera $\mathcal{C}$ is installed so that the axes of $\mathcal{C}$ are parallel to those of $\mathcal{F}$. Camera $\mathcal{C}'$ can be installed arbitrarily as long as it has corresponding views with $\mathcal{C}$. $R \in \mathbb{SO}(3)$ and $x_f \in \mathbb{R}^3$ denote the rotation and translation from $\mathcal{C}'$ to $\mathcal{C}$ expressed in $\mathcal{C}'$, which can be calibrated beforehand.

Generally, the road surface is assumed to be geometrically smooth, and the drivable road region is defined as the region that is connected to the current vehicle grounding plane without large discontinuities of 3D structure (e.g., height information). Then, the road reconstruction task aims to segment the road region from the images and reconstruct the 3D structure of the road surface.

Since the proposed approach relies on the reconstructed 3D information of the scene, compared to a monocular camera that performs structure from motion using two successively captured images, the 3D reconstruction from two simultaneous images captured by the two-camera system is more reliable and robust for dynamic environments. As a result, the two-camera system is preferred.

4.1.2.2   Two-View Geometry Model

As shown in Figure 4.3, there exist a point O_i and a plane π in the scene. The ray from the origin of $\mathcal{C}$ to O_i intersects with the plane π at the point $O_{\pi i}$; the normal of the plane π is denoted as n in the coordinate system of $\mathcal{C}$, and the distance from $\mathcal{C}$ to plane π is denoted as d. The rotation and translation from $\mathcal{C}'$ to $\mathcal{C}$ are defined as R and x_f, respectively. The coordinates of O_i with respect to $\mathcal{C}$ and $\mathcal{C}'$ are denoted as $\bar{m}_i \triangleq \begin{bmatrix} x_i & y_i & z_i \end{bmatrix}^T \in \mathbb{R}^3$ and $\bar{m}'_i \triangleq \begin{bmatrix} x'_i & y'_i & z'_i \end{bmatrix}^T \in \mathbb{R}^3$, respectively, and the image points corresponding to O_i captured by $\mathcal{C}$ and $\mathcal{C}'$ are denoted as $p_i \triangleq \begin{bmatrix} u_i & v_i & 1 \end{bmatrix}^T \in \mathbb{P}^2$ and $p'_i \triangleq \begin{bmatrix} u'_i & v'_i & 1 \end{bmatrix}^T \in \mathbb{P}^2$, respectively. Similarly, the coordinates of $O_{\pi i}$ with respect to $\mathcal{C}$ and $\mathcal{C}'$ are denoted as $\bar{m}_{\pi i}$ and $\bar{m}'_{\pi i}$ respectively, and the image point corresponding to point $O_{\pi i}$ captured by

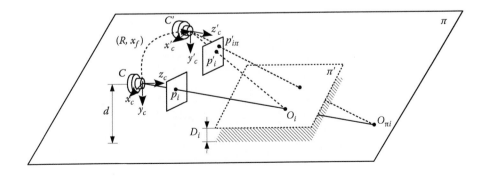

Figure 4.3: Geometric model with respect to the reference plane π.

C' is denoted as $p'_{\pi i}$. According to the pinhole model, the points in image space and 3D space can be related by intrinsic camera matrices, which are denoted as A and A' for cameras C and C', respectively. Generally, the intrinsic matrix A can be expressed with several parameters [86]: α_u and α_v represent the focal lengths of the camera C expressed in terms of pixel units in x and y directions, respectively. θ is the angle between the two image axes, and $p_0 \triangleq \begin{bmatrix} u_0 & v_0 & 1 \end{bmatrix}^T$ is denoted as the coordinate of the principal point, which is the intersection between the optical axis and the image plane. Similarly, the intrinsic parameters of camera C' are defined as $\alpha'_u, \alpha'_v, u'_0, v'_0$ and θ'. In practice, the above parameters in A and A' can be calibrated using various methods with small errors [86].

Based on the developments in Appendix A, the image points p_i, p'_i corresponding to the point O_i can be related by a projective homography G and a projective parallax β_i with respect to the plane π:

$$p'_i = \frac{z_i}{z'_i}\left(Gp_i + \beta_i A' \cdot \frac{x_f}{d}\right) \tag{4.1}$$

where $G \triangleq A'\left(R + x_f \frac{n^T}{d}\right)A^{-1}$ is the projective homography and $\beta_i \triangleq \frac{D_i}{z_i}$ is the projective parallax relative to the homography.

4.1.2.3 Image Warping Model

Denote the coordinates of image points on one image row in $\mathbf{I}$ as $\mathbf{P} \triangleq \begin{bmatrix} p_1 \cdots p_N \end{bmatrix}$, where $p_i = \begin{bmatrix} u_i & v_i & 1 \end{bmatrix}^T$ is the homogeneous coordinate of the pixel on the i-th column, and N is the count of image columns. Similarly, the coordinates of corresponding points in $\mathbf{I}'$ are denoted as $\mathbf{P}' \triangleq \begin{bmatrix} p'_1 \cdots p'_N \end{bmatrix}$, where $p'_i = \begin{bmatrix} u'_i & v'_i & 1 \end{bmatrix}^T$ is the coordinate of the corresponding point of p_i. Using the geometric model in (4.1), the warping function $\begin{bmatrix} u'_i & v'_i \end{bmatrix}^T = w(\beta_i, p_i)$ from image $\mathbf{I}$ to image $\mathbf{I}'$ can be expressed as:

$$\begin{bmatrix} u'_i & v'_i \end{bmatrix}^T = w(\beta_i, p_i)$$
$$= \begin{bmatrix} \dfrac{g_{11}u_i + g_{12}v_i + g_{13} + \beta_i(\alpha'_u(x_{fx} - x_{fy}\cot\theta') + u'_0 x_{fz})/d}{g_{31}u_i + g_{32}v_i + g_{33} + \beta_i x_{fz}/d} \\ \dfrac{g_{21}u_i + g_{22}v_i + g_{23} + \beta_i(\alpha'_v x_{fy}/\sin\theta' + v'_0 x_{fz})/d}{g_{31}u_i + g_{32}v_i + g_{33} + \beta_i x_{fz}/d} \end{bmatrix} \tag{4.2}$$

where g_{ij} is the element on the i-th row and j-th column of G, and β_i is the projective parallax at point p_i.

4.1.2.4 Vehicle-Road Geometric Model

As shown in Figure 4.4a, for the application of road reconstruction, a good choice for the reference plane π is the grounding plane of the vehicle, which is defined by the contact points of four wheels with the ground. The coordinate of point O_i

(a)

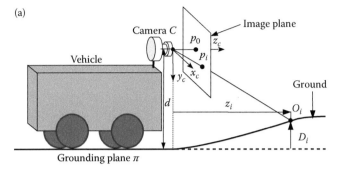

(b)

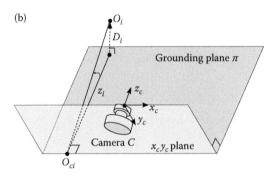

Figure 4.4: Configuration of the road reconstruction system. (a) Side view of the road detection system and (b) top view of the road detection system.

with respect to C can be expressed as $\bar{m}_i = \begin{bmatrix} x_i & d - D_i & z_i \end{bmatrix}^T$ and its normalized coordinate is $m_i = \begin{bmatrix} x_i/z_i & (d - D_i)/z_i & 1 \end{bmatrix}^T$. According to the pinhole model for perspective cameras, $p_i = A \cdot m_i$, then it can be obtained that

$$D_i = \frac{d\alpha_v \beta_i}{\alpha_v \beta_i + (v_i - v_0)\sin\theta},$$

$$x_i = z_i \left(\frac{u_i - u_0}{\alpha_u} + \frac{v_i - v_0}{\alpha_v}\cos\theta \right), \tag{4.3}$$

$$z_i = \frac{d\alpha_v}{\alpha_v \beta_i + (v_i - v_0)\sin\theta}.$$

From the above, for the given two-camera system, a point O_i in the scene can be described by the projective parallax β_i with respect to the grounding plane. Then, its coordinates and height with respect to the grounding plane can be directly calculated using the vehicle-road geometric model, based on which road regions can be detected conveniently.

4.1.2.5 More General Configurations

The above vehicle geometry considers the commonest case that the axes of camera $\mathcal{C}$ are parallel to those of the vehicle $\mathcal{F}$, while $\mathcal{C}'$ can be installed arbitrarily as long as it has corresponding view with $\mathcal{C}$. For more general configurations, i.e., $\mathcal{C}$ is also installed arbitrarily and the rotation from $\mathcal{C}$ to $\mathcal{F}$ is denoted as R_c. The grounding plane is also selected as the reference plane. $\mathcal{C}_r$ is denoted as a rotation of camera $\mathcal{C}$ and is parallel to $\mathcal{F}$, and the coordinate of O_i with respect to $\mathcal{C}_r$ is denoted as $\bar{m}_i = \begin{bmatrix} x_i & d - D_i & z_i \end{bmatrix}^T$, which is same to the standard configuration in Figure 4.4. Then, the coordinate of O_i with respect to $\mathcal{C}$ can be expressed as $\bar{m}_{ci} \triangleq \begin{bmatrix} x_{ci} & y_{ci} & z_{ci} \end{bmatrix}^T = R_c \bar{m}_i$, and the parallax information is expressed as $\beta_i = D_i / z_{ci}$. Similarly, the vehicle-road geometry can be solved based on the camera model.

From the above, for the two-camera system with arbitrary configuration, a point O_i in the scene can be described by the projective parallax β_i with respect to the grounding plane. Then, its coordinates and height with respect to the grounding plane can be directly calculated using the vehicle geometry, based on which road regions can be detected conveniently.

4.2 Motion Estimation of Moving Objects

Using visual sensors to identify the unknown interested information of an object is an important problem with many applications. Based on different targets, the problem can be divided into structure from motion (SfM), motion from structure (MfS), structure and motion (SaM), etc. The SfM problem is aimed to estimate the 3-D Euclidean coordinates of feature points on an object with known relative motion between the camera and the object. Conversely, the MfS problem utilizes the known geometry of the object to estimate the relative motion. The SaM is an extended problem, whose objective is to identify both the Euclidean coordinates of feature points and the relative motion information [37].

Various methods have been developed to solve the SfM, MfS, and SaM problems, such as batch methods [85,164,208] and nonlinear observer methods [33,34,38–40,48–50,57,82,112,161,196]. The batch methods usually exploit projective techniques, linear optimization, and matrix factorization to reconstruct the 3-D Euclidean information from a sequence of images. Generally, the convergence of the batch methods cannot be guaranteed and it is hard to achieve real-time tasks by using these methods. In comparison to batch methods, the nonlinear observer methods usually ensure the estimate convergence with analytical proofs of stability and can be used for real-time tasks [48].

Some observers have been designed for the SfM and MfS problems. In [34], the sliding mode method is utilized for the range identification of an object by designing a discontinuous observer, which achieves uniformly ultimately bounded convergence. A continuous observer guaranteeing asymptotic range identification is presented in [57]. Based on the immersion and invariance methodology, reduced order observers with asymptotic convergence are given in [112,161]. Specifically, both the velocity and the linear acceleration of the camera are required to facilitate the observer design. In [82], a range estimator is developed based on nonlinear contraction theory and synchronization. To improve the transient response of the observer, an active estimation framework is proposed in [196]. All the works mentioned above use a moving camera to estimate the range of a static object. In [50], the structure of an object moving on a ground plane is estimated by a moving camera based on the unknown input observer approach. To solve the MfS problem, a continuous estimator is designed in [39] to identify the velocities of a moving object with a fixed camera. Specifically, homography-based techniques are used in the development of the observer design, and a single length between two feature points on the object is assumed to be known *a priori*.

Compared to the SfM and MfS problems, the SaM problem tends to use less knowledge to estimate more unknown states. In [48], with the knowledge of camera linear velocity and acceleration in one dimension, a reduced order observer is developed for the structure estimation of a static object. A homography-based adaptive observer is presented in [33] to use a fixed camera to estimate the range of a moving object. Both in [33] and [48], the unknown velocities are identified by the estimator designed in [39]. Using a moving camera to identify the structure and motion of a moving object is studied in [38,49]. In addition, a method that combines static and moving objects to estimate the moving camera's linear velocity and the object structure is proposed in [40].

All the above nonlinear observer methods utilize a single camera to achieve tasks. However, to the best of our knowledge, existing nonlinear observer methods that use a moving camera or a static camera to identify the structure and motion of a moving object suffer from different limitations. For example, a single length between two feature points on the object and a rotation matrix are required to be known in [33,38], and the object is assumed to be moving with constant velocities in [49]. To relax or eliminate the restriction for the moving object, a feasible idea is to combine a static camera and a moving camera to solve this SaM issue. In practice, multi-camera systems exist in many scenes and can capture richer image information of the unknown object than the single camera. Especially, using a static-moving camera system to identify the structure and motion of a moving object has realistic applications. The problem comes from scenarios such as to determine the speed and range of cars moving on the roads, which are observed by both the camera attached on a helicopter and the traffic surveillance camera attached on a building.

4.3 Scaled Pose Estimation of Mobile Robots

4.3.1 Pose Reconstruction Based on Multiple-View Geometry

For feedback construction, the relevant image views are exploited to reconstruct geometric information. As described in the previous sections, two-view and three-view geometries are generally used for scaled pose estimation in mobile robots.

■ Trifocal tensor based method: Denote $d' \triangleq \sqrt{x_{f0x}^2 + x_{f0z}^2}$ as the distance between C_0 and C^*. Using the trifocal tensor among the three views, the translation vector can be estimated up to a scale d' as:

$$x_h \triangleq \begin{bmatrix} x_{hx} & 0 & x_{hz} \end{bmatrix}^T = \frac{x_f}{d'}, \tag{4.4}$$

where x_{hx}, x_{hz} are given by:

$$x_{hx} = -\bar{T}_{212}, \ x_{hz} = -\bar{T}_{232}. \tag{4.5}$$

The orientation angle θ can be determined by the following expressions:

$$\begin{aligned} \sin\theta &= \bar{T}_{221}(\bar{T}_{333} - \bar{T}_{131}) - \bar{T}_{223}(\bar{T}_{331} + \bar{T}_{133}), \\ \cos\theta &= \bar{T}_{221}(\bar{T}_{111} - \bar{T}_{313}) + \bar{T}_{223}(\bar{T}_{113} + \bar{T}_{311}). \end{aligned} \tag{4.6}$$

Then, the scaled position coordinates $x_m \triangleq \frac{x}{d}, z_m \triangleq \frac{z}{d}$ can be calculated as follows:

$$\begin{bmatrix} x_m \\ z_m \end{bmatrix} = - \begin{bmatrix} \cos\theta & -\sin\theta \\ \sin\theta & \cos\theta \end{bmatrix} \cdot \begin{bmatrix} x_{hx} \\ x_{hz} \end{bmatrix}. \tag{4.7}$$

■ Homography-based method: This type of methods rely on the decomposition of homography matrix between the current and final frames, and d'' is denoted as the distance from the origin of view C^* to the plane along the normal vector, which is the scaled factor in the estimated pose information. Based on the development in Chapter 2, the rotational matrix R and the translation vector can be estimated up to a scale d'' as:

$$x_h'' \triangleq \begin{bmatrix} x_{hx} & 0 & x_{hz} \end{bmatrix}^T = \frac{x_f}{d''}. \tag{4.8}$$

■ Epipole-based method: To avoid the short baseline problem, a simple strategy similar to the "virtual target" [144] is used; the virtual views C_0' and $C^{*'}$ are obtained by performing a translation along the z-axis of the robot from C_0 and C^*, respectively. In Figure 4.5c, $\varepsilon_1, \varepsilon_2, \ldots, \varepsilon_6$ are denoted as the horizontal coordinates of the epipoles, and d''' is denoted as the distance from C_0' to $C^{*'}$, which is the scale factor in the estimated pose information. Assuming the images are normalized to unit

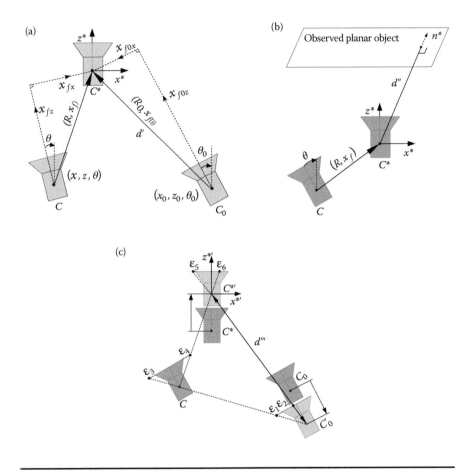

Figure 4.5: Different multiple-view geometry-based pose reconstruction methods. (a) Trifocal tensor based, (b) Homography-based, and (c) Epipole-based.

focal length, the orientation information θ and scaled position information $x'''_m \triangleq \frac{x}{d'''}, z_m \triangleq \frac{z}{d'''}$ can be calculated with respect to $C^{*'}$ after some geometry operations as follows:

$$x'_m = -\frac{\varepsilon_6(\varepsilon_2 - \varepsilon_1)\sqrt{(\varepsilon_3^2 + 1)(\varepsilon_4^2 + 1)}}{(\varepsilon_4 - \varepsilon_3)\sqrt{(\varepsilon_1^2 + 1)(\varepsilon_2^2 + 1)(\varepsilon_6^2 + 1)}}$$

$$z'_m = -\frac{(\varepsilon_2 - \varepsilon_1)\sqrt{(\varepsilon_3^2 + 1)(\varepsilon_4^2 + 1)}}{(\varepsilon_4 - \varepsilon_3)\sqrt{(\varepsilon_1^2 + 1)(\varepsilon_2^2 + 1)(\varepsilon_6^2 + 1)}} \qquad (4.9)$$

$$\theta = \arctan \varepsilon_4 - \arctan \varepsilon_6.$$

It is obvious that singularity exists when $\varepsilon_3 = \varepsilon_4$, which means that C_0', C, and $C^{*\prime}$ are collinear.

4.3.2 Dealing with Field of View Constraints

Generally, reconstructed pose information are expressed with respect to the final view, which relies on the trifocal tensor among the start, current, and final images. The corresponding points and lines in the three images are usually used to estimate the trifocal tensor. For systems using perspective cameras, the three views are easy to have no correspondence due to the limited field of view. This limitation can be partially solved by using omnidirectional camera systems [14], which extend the field of view naturally. However, existing interest point detection and description algorithms suffer weak repeatability, invariance and precision with large rotational or translational displacements [81]. This means that the feature points are not consistent and subject to noise. What's more, this problem gets worse for omnidirectional cameras because of the image distortion. In pose estimation systems, this problem is generally handled using multi-sensor fusion, such as multi-camera systems [6] and the integration of vision system, inertial measurement unit (IMU), and odometry [201]. A simple strategy is that when the image measurements are not available, the system is switched to other sensors.

In this section, key frame strategy is proposed to overcome this limitation, which exploits the nature of trifocal tensor, i.e., the trifocal tensor is invariant to scene structure and the transformations among three views can be normalized to a common scale. As shown in Figure 4.6, the main idea is that some key frames are selected along the desired trajectory, so that neighboring key frames have enough corresponding feature points. Local pose information of the current frame is estimated with respect to two most similar key frames, and then it is transformed iteratively until the final frame to get the global pose information. Since the desired pose information and the current pose information can

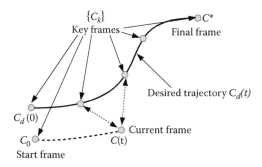

Figure 4.6: Key frame strategy.

be estimated in the same manner, this section only considers the current pose estimation without loss of generality.

The concepts of "key frame" have been proposed in video summarization [1] and visual odometry [181], which are different from the proposed strategy. In video summarization, key frames are defined as the representative frames that abstract the video information; in visual odometry, key frames are used for a better triangulation of 3D points from images; while in this chapter, the key frame strategy is used to generate continuous scaled pose information in large workspace. In this section, the procedure of key frame strategy is further described.

4.3.2.1 Key Frame Selection

The rule for key frame selection is that neighboring key frames have enough corresponding feature points. In Algorithm 1, key frames are selected according to the matching ratio of feature points (function CORRESPONDINGRATIO in Algorithm 1). $\mathcal{C}_{k,j}$ denotes the j-th key frame, and the constant $\tau \in (0,1)$ denotes the threshold for key frame selection. The goal of Algorithm 1 is to ensure that the matching ratio between neighboring key frames is close to the threshold τ.

Algorithm 1 Key frame selection

Input:
    Desired trajectory, $\{\mathcal{C}_d\}$;
    Corresponding ratio threshold, τ;
Output:
    Key frames, $\{\mathcal{C}_k\}$;
  1: Add $\mathcal{C}_0, \mathcal{C}_d(0)$ into $\{\mathcal{C}_k\}$;
  2: $j \leftarrow 2$;
  3: **for** each $\mathcal{C}_t$ in $\{\mathcal{C}_d\}$ **do**
  4:     **if** CORRESPONDINGRATIO$(\mathcal{C}_t, \mathcal{C}_{k,j}) < \tau$ **then**
  5:         Add $\mathcal{C}_t$ into $\{\mathcal{C}_k\}$;
  6:         $j \leftarrow j+1$;
  7:     **end if**
  8: **end for**
  9: Add $\mathcal{C}^*$ into $\{\mathcal{C}_k\}$ if $\mathcal{C}^*$ is not in $\{\mathcal{C}_k\}$;
10: **function** CORRESPONDINGRATIO$(\mathcal{C}_t, \mathcal{C}_p)$
11:     Extract feature points from images of $\mathcal{C}_t, \mathcal{C}_p$;
12:     $count_p \leftarrow$ count of feature points in $\mathcal{C}_p$;
13:     Match feature points from $\mathcal{C}_t, \mathcal{C}_p$;
14:     $count_m \leftarrow$ count of matched feature points;
15: **return** $count_m / count_p$
16: **end function**

If a larger τ is set, neighboring key frames are required to have more corresponding points, so that the estimation of trifocal tensor is more accurate and the workspace is larger; meanwhile, the scale factor d is smaller, causing the pose estimation system to be more sensitive, i.e., a small change of x_f will cause a big change of x_h and vice versa. Thus, the threshold τ should be selected as a trade-off.

Remark 4.1 The key frame strategy works with other similarity measures except for the matching ratio of feature points; the discussion on different criteria is not in the scope of this section. In practice, the count of feature points varies with different feature extraction algorithms and different images, while some of them are not stable enough. A common strategy is to select a certain number of strong feature points according to the magnitude of feature descriptors [11].

4.3.2.2   Pose Estimation

In the control system, pose information is expressed with respect to the final frame, and the procedure for pose estimation using key frames is described in Algorithm 2. The main idea is that first local pose information is estimated using

---

Algorithm 2 Pose estimation

Input:
  Current frame, $\mathcal{C}$;
  Key frames, $\{\mathcal{C}_k\}$;
Output:
  Pose information with respect to $\mathcal{C}^*$, (x_{hx}, x_{hz}, θ);
  1: Select two most similar key frames $\mathcal{C}_{k,i-1}, \mathcal{C}_{k,i}$;
  2: Estimate the trifocal tensor $T(\mathcal{C}_{k,i-1}, \mathcal{C}, \mathcal{C}_{k,i})$;
  3: Compute the scaled coordinate with respect to $\mathcal{C}_{k,i}$ as $X_m(x_{m,i}(t), z_{m,i}(t), \theta_i(t))$;
  4: $k \leftarrow$ count of $\{\mathcal{C}_k\}$;
  5: **for** each $j \in [i, k-1]$ **do**
  6:     $X_m \leftarrow$ POSETRANSFORM(X_m, j);
  7: **end for**
  8: Compute (x_{hx}, x_{hz}, θ) from X_m;
  9: **function** POSETRANSFORM$(X_{m,j}, j)$
 10:     $T \leftarrow T(\mathcal{C}_{k,j-1}, \mathcal{C}_{k,j+1}, \mathcal{C}_{k,j})$;
 11:     Compute $(R_{j+1}, x_{f,j+1}/d_{j-1})$ from T;
 12:     Compute d_{j-1}/d_{j+1} using (4.10);
 13:     Compute $X_{m,j+1}$ using (4.11);
 14: **return** $X_{m,j+1}$
 15: **end function**

---

two most similar key frames, which are selected according to the function COR-RESPONDINGRATIO, and then it is transformed to the next key frame iteratively until the final frame to obtain global pose information. This process is applicable for current and desired frames, and then global pose information is available to construct the control system.

As shown in Figure 4.7, the function POSETRANSFORM defined in Algorithm 2 transforms the coordinate of $\mathcal{C}$ from $\mathcal{C}_{k,j}$ to $\mathcal{C}_{k,j+1}$. In Figure 4.7, $R_{j-1}, x_{f,j-1}$, and d_{j-1} denote the rotation, translation, and distance from $\mathcal{C}_{k,j-1}$ to $\mathcal{C}_{k,j}$, respectively, and $R_{j+1}, x_{f,j+1}$ and d_{j+1} denote the rotation, translation, and distance from $\mathcal{C}_{k,j+1}$ to $\mathcal{C}_{k,j}$, respectively. Given the normalized trifocal tensor $\bar{T}$ calculated by $T(\mathcal{C}_{k,j-1}, \mathcal{C}_{k,j+1}, \mathcal{C}_{k,j})$, the rotational matrix R_{j+1} and scaled translational matrix $x_{f,j+1}/d_{j-1}$ can be calculated using (2.53), (4.5) and (4.6). And the depth ratio d_{j-1}/d_{j+1} is given by:

$$\frac{d_{j-1}}{d_{j+1}} = \sqrt{\frac{1}{\bar{T}_{212}^2 + \bar{T}_{232}^2}}. \tag{4.10}$$

Based on previous steps, the scaled position of $\mathcal{C}$ with respect to $\mathcal{C}_{k,j+1}$ is given by:

$$\begin{bmatrix} x_{m,j+1}(t) \\ 0 \\ z_{m,j+1}(t) \end{bmatrix} = \frac{d_{j-1}}{d_{j+1}} \cdot \left(R_{j+1} \cdot \begin{bmatrix} x_{m,j}(t) \\ 0 \\ z_{m,j}(t) \end{bmatrix} + \frac{x_{f,j+1}}{d_{j-1}} \right). \tag{4.11}$$

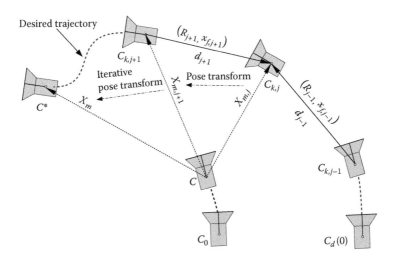

Figure 4.7: Pose transform from $\mathcal{C}_{k,j}$ to $\mathcal{C}_{k,j+1}$.

The orientation angle of $\mathcal{C}$ with respect to $\mathcal{C}_{k,j+1}$ is given by:

$$\theta_{j+1}(t) = \theta_j(t) - \theta_j^{j+1}, \tag{4.12}$$

where θ_j^{j+1} is the orientation angle of $\mathcal{C}_k(j+1)$ with respect to $\mathcal{C}_{k,j}$, which can be calculated using (4.6). After the iterative pose transform, the pose information of the current frame $\mathcal{C}$ is expressed with respect to the final frame $\mathcal{C}^*$, and the scale factor d is the distance between the last two key frames.

Remark 4.2   The key frames except the start frame $\mathcal{C}_0$ are selected from the desired trajectory; thus, they can be selected and processed offline. In the function POSE-TRANSFORM in Algorithm 2, $T(\mathcal{C}_{k,j-1}, \mathcal{C}_{k,j+1}, \mathcal{C}_{k,j})$ for these frames on the desired trajectory can be calculated offline to gain time efficiency.

Remark 4.3   Since there exist enough corresponding feature points in neighboring key frames, the calculations of $T(\mathcal{C}_{k,j-1}, \mathcal{C}_{k,j+1}, \mathcal{C}_{k,j})$ are sufficiently accurate, and the accumulated error caused by the iterative pose transform procedure is small. Besides, as the robot is closing the final frame, the accumulated error becomes smaller.

Remark 4.4   From (4.11) and (4.12), the pose information expressed in $\mathcal{C}_{k,j+1}$ is a continuous function of that expressed in $\mathcal{C}_{k,j}$. According to the iterative pose transformation, the pose information expressed in $\mathcal{C}^*$ is a continuous function of that expressed in $\mathcal{C}_{k,i}$. Besides, the scale factor d in the translation vector x_h is the distance between the last two key frames, and does not change in the process. As a result, the pose estimation system based on key frame strategy is continuous.

4.3.3   Dealing with Measuring Uncertainties

Compared to conventional positioning sensors, vision sensors are more noisy because the captured images are influenced by environmental conditions and camera characteristics. Besides, there also exist errors in image processing algorithms, affecting the dynamic performance and stability of visual control systems, causing steady state errors.

   To deal with the problems caused by the measuring uncertainties, previous approaches are divided into the following aspects: designing robust image feature extraction algorithms, using observers for the estimation of extracted feature points, designing robust controllers, and using redundant degrees of freedom to reduce the influence of measuring uncertainties for active vision systems.

4.3.3.1   Robust Image Feature Extraction

Image noise has great influence on the extraction of image features, especially for local image features based on image gradients. Error extraction and error matching of image features directly cause the measuring error of system states. Song [194] uses random sample consensus (RANSAC) algorithm to eliminate outliers in the feature matching, improving the measuring accuracy. In RANSAC-based methods, parameters are estimated iteratively from the noisy image measurements. Besides, commonly used methods include Hough transform, least-square method, and M-Estimator method [99].

4.3.3.2 Robust Observers

For the image feature vectors with noise, observers can be used to obtain the optimal state estimation, reducing the influence of noise. For systems with Gaussian noise, Kalman filters can be used for the estimation of states. For nonlinear systems, extended Kalman filters are used. Since the statistical characteristics of noises are required in Kalman filters, a self tuning Kalman filter is proposed in [204], the covariance matrix is estimated by linear regression on the measurements. An adaptive iterative Kalman filter is proposed in [103]. For the systems with non-Gaussian noise, particle filters [100] can be used for state estimation by randomly sampling to obtain the probabilistic distributions.

4.3.3.3 Robust Controllers

The above two types of methods deal with measuring uncertainties in the visual perception process. In [43], a robust statistical controller is proposed to deal with image noise. The reliability of each image feature points is computed based on the corresponding statistical characteristics, and is used to compute the weight of the image feature. This method reduces the influence of noisy image feature points and ensures the continuity of the control law with the existence of image measuring error. For the systems with redundant feature points, adequate weighting functions can be used to ensure the continuity of the control law with partial loss of feature points [78].

4.3.3.4 Redundant Degrees of Freedom

For active vision systems, the motion of cameras can be actuated by the pan-tilt mechanism, increasing redundant degrees of freedom for the control system. In [36], the upper and lower bounds of the steady state control error are estimated with the existence of image noise. Then, linear matrix inequality (LMI) based optimization is used to obtain the optimal configurations of image feature points and target objects. Besides, the representability of images captured at different poses is different for motion estimation. As a result, the representability can be considered in the control process to improve measuring accuracy. In [101], the minimum singular value of the image Jacobian matrix is used to measure the representability. In [195], "Active Structure from Motion" strategy is used to adjust the robot trajectory, obtaining more efficient information and more accurate 3D reconstruction.

Chapter 5

Road Scene 3D Reconstruction

5.1 Introduction

Using the vision system in Figure 4.2, the road scene can be reconstructed by the images $\mathbf{I}$ and $\mathbf{I}'$ captured by cameras C and C', respectively. Motivated by homography-based approaches, the drivable road region is reconstructed from a geometry point of view. Based on the proposed two-view geometry, the scene can be described by the parallax information with respect to the reference plane. A row-wise image registration algorithm is proposed to estimate the projective parallax information, based on which position and height information of points with respect to the reference plane can be reconstructed. Using the reconstructed height map, the scene can be segmented where the height changes abruptly and the road region is detected by a diffusion strategy. The proposed approach is general in the sense that it works for both planar and nonplanar road surfaces, and few priors are needed about the road appearance and structure. Experiments are made to show the effectiveness and robustness of the proposed approach. Compared to previous works, the main contributions are listed below:

- A two-view geometric model is exploited to describe general scenes with respect to a reference plane, which is suitable for road detection and reconstruction problems and is the first time to be used for road reconstruction.

■ The searching region for 3D reconstruction is significantly reduced owing to the two-view geometric model, improving the computational efficiency and the ability to deal with textureless or repetitively textured areas.

■ The computational efficiency of 3D reconstruction is mainly affected by parameter errors of the vision system, not the extrinsic parameters; thus, the measuring accuracy can be improved conveniently by increasing the baseline between the two cameras without increasing computational burdens significantly.

■ The geometric reconstruction approach is extendable to any calibrated perspective two-camera systems as long as the two cameras have corresponding views.

5.2 Algorithm Process

As shown in Figure 5.1, the road detection and reconstruction process is performed as follows: first, two images $\mathbf{I}, \mathbf{I}'$ are captured by the cameras $\mathcal{C}, \mathcal{C}'$; second, the projective parallax information $\mathbf{B}$ corresponding to $\mathbf{I}$ is reconstructed by the row-wise image registration based on the proposed two-view geometry; third, the height information $\mathbf{D}$ corresponding to $\mathbf{I}$ with respect to the grounding plane is calculated from $\mathbf{B}$; finally, the road region is segmented based on the height information, and a road map $\mathbf{R}$ is obtained.

As illustrated above, the key process of road reconstruction is the row-wise image registration that reconstructs parallax information from the two images $\mathbf{I}$

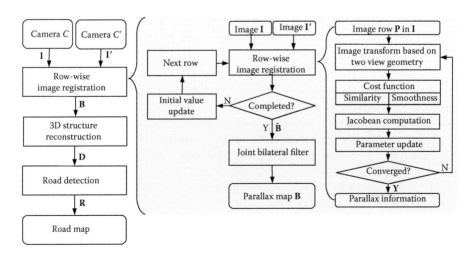

Figure 5.1: Working process of the road detection and reconstruction.

and $\mathbf{I}'$. For each image row $\mathbf{P}$ in $\mathbf{I}$, the optimal parallax information is searched iteratively using the Levenberg–Marquardt algorithm, which is actually based on the gradients of the objective function. The optimization ends when the convergence condition is satisfied, i.e., $\Delta \Phi < \Gamma_\Phi$. As mentioned above, the initial value of the current row is set as the optimal value of the previous row on the assumption of scene continuity. In fact, for the road reconstruction task, the images don't need to be processed in whole, and only the regions under the horizon need to be processed so that the computational efficiency can be further improved. The horizon in the image can be roughly determined by calibrated extrinsic parameters. In this chapter, the images are processed in whole to better show the effectiveness of 3D reconstruction, and the computational complexity is acceptable as will be stated in the experiments.

5.3 3D Reconstruction

Given two images $\mathbf{I}$ and $\mathbf{I}'$ captured by the cameras $\mathcal{C}$ and $\mathcal{C}'$, respectively, the projective parallax is estimated so that the warped image of $\mathbf{I}'$ matches with $\mathbf{I}$ best, which can be formulated as an image registration problem based on the proposed two-view geometry. Then, the height information can be directly calculated based on the vehicle geometry. In this section, the 3D reconstruction process is described in detail.

5.3.1 *Image Model*

Image points on one image row in $\mathbf{I}$ are denoted as $\mathbf{P} \triangleq \begin{bmatrix} p_1 \cdots p_N \end{bmatrix}$, where $p_i = \begin{bmatrix} u_i & v_i & 1 \end{bmatrix}^T$ is the homogeneous coordinate of the pixel on the i-th column, and N is the count of image columns. Similarly, the corresponding points in $\mathbf{I}'$ are denoted as $\mathbf{P}' \triangleq \begin{bmatrix} p_1' \cdots p_N' \end{bmatrix}$, where $p_i' = \begin{bmatrix} u_i' & v_i' & 1 \end{bmatrix}^T$ is the corresponding point of p_i. It can be obtained from (2.6), (4.1) that

$$
\begin{aligned}
p_i' &= \frac{z_i}{z_i'} \left(G p_i + \beta_i \cdot A' \frac{x_f}{d} \right) \\
&= \frac{z_i}{z_i'} \begin{bmatrix} g_{11} u_i + g_{12} v_i + g_{13} + \beta_i \dfrac{\alpha_u'(x_{fx} - x_{fy} \cot \theta') + u_0' x_{fz}}{d} \\[2mm] g_{21} u_i + g_{22} v_i + g_{23} + \beta_i \dfrac{\alpha_v' x_{fy} / \sin \theta' + v_0' x_{fz}}{d} \\[2mm] g_{31} u_i + g_{32} v_i + g_{33} + \beta_i \dfrac{x_{fz}}{d} \end{bmatrix},
\end{aligned}
\tag{5.1}
$$

where g_{kl} is the element on the k-th row and l-th column of G, and β_i is the projective parallax at point p_i. Since the intrinsic and extrinsic parameters of the two-camera system can be calibrated beforehand, i.e., A, A', R, x_f are known and constant, and then G can be directly calculated by (2.16).

Then the warping function $\begin{bmatrix} u'_i & v'_i \end{bmatrix}^T = w(\beta_i, p_i)$ from image $\mathbf{I}$ to image $\mathbf{I'}$ can be defined as:

$$\begin{bmatrix} u'_i & v'_i \end{bmatrix}^T = w(\beta_i, p_i)$$
$$= \begin{bmatrix} \dfrac{g_{11}u_i + g_{12}v_i + g_{13} + \beta_i(\alpha'_u(x_{fx} - x_{fy}\cot\theta') + u'_0 x_{fz})/d}{g_{31}u_i + g_{32}v_i + g_{33} + \beta_i x_{fz}/d} \\[2ex] \dfrac{g_{21}u_i + g_{22}v_i + g_{23} + \beta_i(\alpha'_v x_{fy}/\sin\theta' + v'_0 x_{fz})/d}{g_{31}u_i + g_{32}v_i + g_{33} + \beta_i x_{fz}/d} \end{bmatrix}. \quad (5.2)$$

The intensities of the image points in $\mathbf{P}$ and $\mathbf{P'}$ are denoted as $\mathbf{S} \triangleq \begin{bmatrix} s_1 \cdots s_N \end{bmatrix}^T$ and $\mathbf{S'} \triangleq \begin{bmatrix} s'_1 \cdots s'_N \end{bmatrix}^T$, respectively. Note that if point p'_i is out of the range of image $\mathbf{I'}$, s'_i is undefined and will be given an arbitrary value. Denote $\Delta \triangleq \begin{bmatrix} \delta_1 \cdots \delta_N \end{bmatrix}^T$ as a mask vector, where $\delta_i = 0$ if s'_i is undefined, and $\delta_i = 1$ otherwise. The similarity between these two sets of points can be measured by correlation coefficient, which is defined as:

$$c(\mathbf{S}, \mathbf{S'}) = \frac{\sum\limits_{i=1}^{N} \delta_i (s_i - \bar{s})(s'_i - \bar{s}')}{\sqrt{\sum\limits_{i=1}^{N} \delta_i (s_i - \bar{s})^2} \sqrt{\sum\limits_{i=1}^{N} \delta_i (s'_i - \bar{s}')^2}}, \quad (5.3)$$

where $\bar{s}$ and $\bar{s}'$ are the mean values of $\mathbf{S}$ and $\mathbf{S'}$, respectively, and can be calculated as:

$$\bar{s} = \sum_{i=1}^{N} \delta_i s_i \bigg/ \sum_{i=1}^{N} \delta_i \, , \bar{s}' = \sum_{i=1}^{N} \delta_i s'_i \bigg/ \sum_{i=1}^{N} \delta_i \, . \quad (5.4)$$

Correlation coefficient c varies between -1 and 1, and a larger c means the two sets of points match better. The case $c = 1$ means $\mathbf{S}$ and $\mathbf{S'}$ perfectly coincide, and the case $c = -1$ means $\mathbf{S}$ and the negative of $\mathbf{S'}$ perfectly coincide.

5.3.2 Parameterization of Projective Parallax

For each image row, the projective parallax is assumed to be continuous. In this chapter, one-dimensional cubic B-spline function is used to parameterize the parallax β on each row. As shown in Figure 5.2, the B-spline curve is defined by a control lattice $\{c_{j-2}\}_{j\in[1,M]}$ overlaid uniformly on the image row, where M is the number of control points. The overlaying rule is that the control point c_0 coincides with 0 and the control point c_{M-3} coincides with p_N, and then the interval κ between two neighbouring control points is

$$\kappa = \frac{N}{M-3}. \quad (5.5)$$

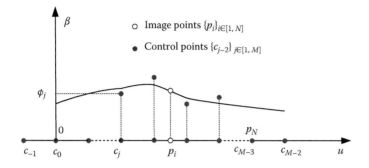

Figure 5.2: Spline parameterization.

Denoting $\Phi \triangleq \left[\phi_{-1} \cdots \phi_{M-2}\right]^T$ as the values of the control points $\{c_{j-2}\}_{j \in [1,M]}$, the approximate function is defined as:

$$\beta_i = \sum_{k=0}^{3} f_k(t_i) \phi_{k+l_i} \tag{5.6}$$

where

$$l_i = \left\lfloor \frac{i}{\kappa} \right\rfloor - 1, t_i = \frac{i}{\kappa} - \left\lfloor \frac{i}{\kappa} \right\rfloor \tag{5.7}$$

with $\lfloor \cdot \rfloor$ denoting the round down function, and $\{f_k(t)\}_{k=0}^{3}$ are the cubic B-spline basis functions defined as follows [123]:

$$\begin{aligned}
f_0(t) &= (1-t)^3/6, \\
f_1(t) &= (3t^3 - 6t^2 + 4)/6, \\
f_2(t) &= (-3t^3 + 3t^2 + 3t + 1)/6, \\
f_3(t) &= t^3/6.
\end{aligned} \tag{5.8}$$

Then, $\beta \triangleq \left[\beta_1 \cdots \beta_N\right]^T$ can be expressed in terms of Φ as:

$$\beta = F \cdot \Phi, \tag{5.9}$$

where F is a $N \times M$ matrix, with its i-th row F_i defined as:

$$F_i = \begin{bmatrix} \underbrace{0 \cdots 0}_{\lfloor \frac{i}{\kappa} \rfloor - 2} & f(t_i) & \underbrace{0 \cdots 0}_{M - \lfloor \frac{i}{\kappa} \rfloor - 2} \end{bmatrix}, \tag{5.10}$$

where $f(t_i) = \begin{bmatrix} f_0(t_i) & f_1(t_i) & f_2(t_i) & f_3(t_i) \end{bmatrix}$.

Take the derivative of β with respect to u:

$$\frac{\partial \beta}{\partial u} = F' \cdot \Phi, \tag{5.11}$$

where F' is a $N \times M$ matrix, with its i-th row defined as:

$$F_i' = \begin{bmatrix} \underbrace{0 \cdots 0}_{\lfloor \frac{i}{\kappa} \rfloor - 2} & f'(t_i)/\kappa & \underbrace{0 \cdots 0}_{M - \lfloor \frac{i}{\kappa} \rfloor - 2} \end{bmatrix}, \tag{5.12}$$

where $f'(t_i) = \frac{\partial f(t_i)}{\partial t_i}$.

5.3.3 Objective Function Definition and Linearization

The row-wise image registration aims to find β, so that points $\mathbf{P}$ match best with points $\mathbf{P}'$ by the warping function w defined in (5.2). Using the B-spline-based parameterization of β, the row-wise image registration can be formulated into the maximization of the following objective function:

$$E(\Phi) \triangleq c(\mathbf{S}, \mathbf{I}'(w(F \cdot \Phi, \mathbf{P}))) - \lambda_1 r_1(\Phi) - \lambda_2 r_2(\Phi), \tag{5.13}$$

where c represents the similarity, $r_1(\Phi)$ and $r_2(\Phi)$ are penalties of the parallax for the smoothness and the closeness to the previous row, respectively. $\lambda_1, \lambda_2 \in \mathbb{R}^+$ represent the trade-offs, a larger λ_1 enforces the spline curve to be smoother, and a larger λ_2 enforces the parallax on this row to be closer to that on the previous row. $r_1(\Phi), r_2(\Phi)$ are defined as:

$$r_1(\Phi) \triangleq \frac{1}{2} \sum_{i=1}^{N} \left(\frac{\partial \beta_i}{\partial u} \right)^2 e^{-\lambda_b \|\nabla \mathbf{I}_i\|_2}$$

$$r_2(\Phi) \triangleq \frac{1}{2} \sum_{i=1}^{N} (\beta_i - \beta_i^-)^2, \tag{5.14}$$

where $\beta^- \triangleq \begin{bmatrix} \beta_1^- \cdots \beta_N^- \end{bmatrix}^T$ is the parallax on the previous row, $\nabla \mathbf{I}_i \triangleq \begin{bmatrix} \partial \mathbf{I}/\partial u & \partial \mathbf{I}/\partial v \end{bmatrix}|_{(u,v)=(u_i,v_i)}$ denotes the image gradient at point p_i, and $\lambda_b \in \mathbb{R}^+$ is a scale factor. The term $e^{-\lambda_b \|\nabla \mathbf{I}_i\|_2}$ in $r_1(\Phi)$ acts as a weight factor in consideration of discontinuities in the image. For the registration of the first image row, the term $r_2(\Phi)$ is neglected.

The above optimization problem is generally solved by gradient-based algorithms. The gradient of the objective function $E(\Phi)$ with respect to Φ is given by:

$$J \triangleq \frac{\partial E(\Phi)}{\partial \Phi} = \frac{\partial c}{\partial \Phi} - \frac{\partial r_1}{\partial \Phi} - \frac{\partial r_2}{\partial \Phi}. \tag{5.15}$$

From (5.14) it is easy to obtain that:

$$\frac{\partial r_1}{\partial \Phi} = \sum_{i=1}^{N} F_i' \Phi F_i' e^{-\lambda_b \|\nabla \mathbf{I}_i\|_2},$$

$$\frac{\partial r_2}{\partial \Phi} = \sum_{i=1}^{N} (\beta_i - \beta_i^-) F_i. \tag{5.16}$$

The gradient of $c(\mathbf{S}, \mathbf{I}'(w(F \cdot \Phi, \mathbf{P})))$ with respect to Φ can be decomposed as:

$$\frac{\partial c}{\partial \Phi} = \frac{\partial c}{\partial \mathbf{S}'} \cdot \frac{\partial \mathbf{S}'}{\partial \beta} \cdot \frac{\partial \beta}{\partial \Phi}. \tag{5.17}$$

According to (5.3), it can be inferred that

$$\frac{\partial c}{\partial \mathbf{S}'} = \left[\frac{\partial c}{\partial s_1'} \cdots \frac{\partial c}{\partial s_N'} \right], \tag{5.18}$$

where

$$\frac{\partial c}{\partial s_i'} = \frac{\delta_i (s_i - \bar{s})}{\sqrt{\sum\limits_{j=1}^{N} \delta_j (s_j' - \bar{s}')^2 \sum\limits_{j=1}^{N} \delta_j (s_j - \bar{s})^2}}$$

$$- \frac{\delta_i (s_i' - \bar{s}') \sum\limits_{j=1}^{N} \delta_j (s_j - \bar{s})(s_j' - \bar{s}')}{\sum\limits_{j=1}^{N} \delta_j (s_j' - \bar{s}')^2 \sqrt{\sum\limits_{j=1}^{N} \delta_j (s_j - \bar{s})^2 \sum\limits_{j=1}^{N} \delta_j (s_j' - \bar{s}')^2}}. \tag{5.19}$$

Since $s_i' = \mathbf{I}'(w(\beta_i, p_i))$, and then it can be inferred that

$$\frac{\partial \mathbf{S}'}{\partial \beta} = diag \left(\left[\frac{\partial s_1'}{\partial \beta_1} \cdots \frac{\partial s_N'}{\partial \beta_N} \right] \right) \tag{5.20}$$

and $\frac{\partial s_i'}{\partial \beta_i}$ can be expressed as

$$\frac{\partial s_i'}{\partial \beta_i} = \frac{\partial s_i'}{\partial w_i} \cdot \frac{\partial w_i}{\partial \beta_i}, \tag{5.21}$$

where

$$\frac{\partial s_i'}{\partial w_i} = \left[\frac{\partial \mathbf{I}'}{\partial u'} \quad \frac{\partial \mathbf{I}'}{\partial v'} \right] \Big|_{[u',v']^T = w(\beta_i, p_i)} \tag{5.22}$$

and

$$\frac{\partial w_i}{\partial \beta_i} = \begin{bmatrix} \dfrac{(\alpha_u'(x_{fx} - x_{fy} \cot \theta') + u_0' x_{fz}) \bar{k}_i' - x_{fz} \bar{u}_i'}{d \left(\bar{k}_i' + \beta_i x_{fz}/d \right)^2} \\[2mm] \dfrac{(\alpha_v' x_{fy}/ \sin \theta' + v_0' x_{fz}) \bar{k}_i' - x_{fz} \bar{v}_i'}{d \left(\bar{k}_i' + \beta_i x_{fz}/d \right)^2} \end{bmatrix}, \tag{5.23}$$

in which $\begin{bmatrix} \bar{u}_i' & \bar{v}_i' & \bar{k}_i' \end{bmatrix}^T = Gp_i$.
Besides, from (5.9), it can be inferred that

$$\frac{\partial \beta}{\partial \Phi} = F. \tag{5.24}$$

5.3.4 Iterative Maximization

Levenberg–Marquardt algorithm is used to search for the local maxima of $E(\Phi)$, and the parameter update rule for the k-th iteration is given by:

$$\Phi_{k+1} = \Phi_k + \alpha\Delta\Phi, \tag{5.25}$$

where α is the step size, and $\Delta\Phi$ is the search direction defined as:

$$\Delta\Phi = (J^T J + \xi I)^{-1} J^T (E_{\max} - E(\Phi_k)), \tag{5.26}$$

where $E_{\max}$ is the maximum value of the function $E(\Phi)$. From (5.13), it can be inferred that $E_{\max} = 1$. I is an identity matrix, and ξ is a weighting factor that regulates the performance of the optimizer in terms of its convergence speed and stability. The convergence speed increases with a smaller ξ, while the stability improves with a larger ξ. Thus, the selection of ξ is a trade-off between convergence speed and stability. The optimization stops when $\Delta\Phi < \Gamma_\Phi$, where $\Gamma_\Phi > 0$ is a threshold.

Compared to global optimization methods having better convergence performance [202], the gradient-based optimization method is typical of low computational complexity, which is crucial for real applications. Due to the complexity of image information, the optimizer is easy to converge to local maxima. To deal with this problem, the following strategies are used to guarantee the initial value is near the optimum:

■ The row-wise image registration process is performed from the bottom row to the top row;

■ The optimal solution of the previous row is used as the initial value of the next row;

■ The initial value of the bottom row is set as zero.

Remark 5.1   Since the grounding plane of the vehicle is selected as the reference plane, then the parallax of the road surface on the bottom row is likely to be near zero. Hence, it is advisable to perform the image registration from the bottom row and set the initial value as zero.

Remark 5.2   The ideal condition is that the road surface near the vehicle coincides with the reference plane expressed by n and d. The robustness to d, n relies on the convergence region of the optimizer. Owing to the above strategy of initial value setting, as long as the image registration of the bottom row succeeds, other rows are barely influenced.

Remark 5.3   In classical stereo matching, if the baseline increases, one has to search over a larger space to find corresponding points and there exist more ambiguities. This problem doesn't exist in the proposed image registration method owing to the above strategies.

5.3.5 Post Processing

After the image registration process for every image row, the parallax map corresponding to $\mathbf{I}$ is obtained, which is denoted as $\hat{\mathbf{B}}$. Due to the existence of image noise, there are some mismatches in $\hat{\mathbf{B}}$. To obtain a smooth and accurate parallax map for road detection, $\hat{\mathbf{B}}$ is processed by a joint bilateral filter, which generally contains a spatial and a range filter kernel [221]. Denote p as a pixel in the image and q as a pixel in the neighbourhood $N(p)$ of p; denote the filtered map of $\hat{\mathbf{B}}$ as $\mathbf{B}$, which can be calculated as:

$$\mathbf{B}(p) = \frac{\sum_{q \in N(p)} \left(f_S(p,q) \cdot f_R(\mathbf{I}(p), \mathbf{I}(q)) \cdot \hat{\mathbf{B}}(p) \right)}{\sum_{q \in N(p)} \left(f_S(p,q) \cdot f_R(\mathbf{I}(p), \mathbf{I}(q)) \right)}, \tag{5.27}$$

where f_S and f_R are the spatial and range filter kernels, respectively. The range function is computed based on the input image $\mathbf{I}$, so that the texture of the parallax map $\mathbf{B}$ is enforced to be similar with image $\mathbf{I}$. There are several choices for the spatial and range filter kernels; in this chapter, a Gaussian kernel is used.

Once the parallax map $\mathbf{B}$ is obtained, the height map $\mathbf{D}$ can be directly calculated using the vehicle geometry in (4.3) and will be used in the following road detection step.

5.4 Road Detection

5.4.1 Road Region Segmentation

Given the estimated parallax map $\mathbf{B}$, the height map $\mathbf{D}$ corresponding to image $\mathbf{I}$ can be calculated using (4.3). $\mathbf{D}_u \triangleq \partial \mathbf{D}/\partial u$ and $\mathbf{D}_v \triangleq \partial \mathbf{D}/\partial v$ are denoted as the derivatives of $\mathbf{D}$ with respect to u and v, respectively. As shown in Figure 5.3, assuming that the road is continuous, the road region can be segmented where the road height changes abruptly, i.e., $|\partial \mathbf{D}_u/\partial u| > \Gamma_u$ in Figure 5.3a and $|\partial \mathbf{D}_v/\partial v| > \Gamma_v$ in Figure 5.3b, where Γ_u, Γ_v are thresholds.

Remark 5.4   Figure 5.3 only shows a standard scenario for road detection. In fact, the above condition works for other roads as long as there is an abrupt height change on boundaries, either planar or nonplanar roads.

5.4.2 Road Region Diffusion

In the classical configuration in Figure 4.2, the bottom middle point in the image is assumed to be drivable. As shown in Figure 5.4, the road region is obtained by a

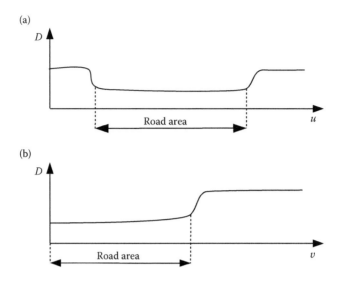

Figure 5.3: Road region segmentation. (a) Row-wise road segmentation and (b) column-wise road segmentation.

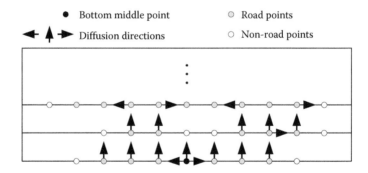

Figure 5.4: Road region diffusion.

diffusion process from the bottom middle point. The diffusion operations consist of left/right and up diffusion processes, which are indicated in Figure 5.4. A detailed implementation is given in Algorithm 3, the diffusion is processed from the bottom row to the top row. For each row, left/right edge points are picked out, from which left/right diffusion is performed until $|\partial \mathbf{D}_u / \partial u| > \Gamma_u$. Then, the road points on this row are diffused to the next row according to the condition that $|\partial \mathbf{D}_v / \partial v| < \Gamma_v$. The diffusion process stops when it reaches the top row or the number of road region set on the current row is less than a threshold Γ_c.

Denote **R** as the road map corresponding to the image **I**, and each image point is labeled with a value in $\{-1, 0, 1\}$, which mean undefined, non-road, and road, respectively. The road map is initialized with -1, and the bottom middle point is set as 1. In Algorithm 1, $size\_u$ and $size\_v$ denote the number of columns and rows in image **I**, respectively, and $\mathbf{R}(v)$ means the v-th row of the road map **R**.

Remark 5.5   As stated in Remark 5.1, for other configurations, the prior knowledge of drivable road point can be determined beforehand based on the extrinsic parameters.

Algorithm 3 Road region diffusion

Input: Derivative maps: $\mathbf{D}_u, \mathbf{D}_v$; Thresholds: $\Gamma_u, \Gamma_v, \Gamma_c$;
Output: Road region map: **R**;

> Initialization: $\mathbf{R} \leftarrow -1, \mathbf{D}_{uu} \leftarrow \frac{\partial \mathbf{D}_u}{\partial u}, \mathbf{D}_{vv} \leftarrow \frac{\partial \mathbf{D}_v}{\partial v}; v \leftarrow size\_v, \mathbf{R}_v \leftarrow \mathbf{R}(v), \mathbf{R}_v(\lfloor \frac{size\_u}{2} \rfloor) \leftarrow 1;$
> **repeat**
> > $\{u_l\}, \{u_r\} \leftarrow \text{GETEDGEPOINTS}(\mathbf{R}_v);$
> > $\mathbf{R}_v \leftarrow \text{LEFTRIGHTDIFFUSE}(\{u_l\}, \{u_r\}, \mathbf{R}_v);$
> > $\mathbf{R}_v(\mathbf{R}_v = -1) \leftarrow 0; \mathbf{R}(v) \leftarrow \mathbf{R}_v; v \leftarrow v - 1;$
> > $\mathbf{R}_v \leftarrow \text{UPDIFFUSE}(\mathbf{R}_v, \mathbf{D}_{vv});$
>
> **until** $v \leq 1$ or $count(\mathbf{R}_v = 1) < \Gamma_c$
> **function** $\text{GETEDGEPOINTS}(\mathbf{R}_v)$
> > Find $\{u_l\}$ satisfying $\mathbf{R}_v(u_l) = 1$ and $\mathbf{R}_v(u_l - 1) = -1;$
> > Find $\{u_r\}$ satisfying $\mathbf{R}_v(u_r) = 1$ and $\mathbf{R}_v(u_r + 1) = -1;$
> > **return** $\{u_l\}, \{u_r\}$
>
> **end function**
> **function** $\text{LEFTRIGHTDIFFUSE}(\{u_l\}, \{u_r\}, \mathbf{R}_v, \mathbf{D}_{uu})$
> > **for** each u_l in $\{u_l\}$ **do**
> > > Find the left nearest u_n satisfying $|\mathbf{D}_{uu}(v, u_n)| > \Gamma_u$ or $\mathbf{R}_v(u_n) \neq -1$ or $u_n \leq 1;$
> > > $\mathbf{R}_v((u_n, u_l)) \leftarrow 1;$
> >
> > **end for**
> > **for** each u_r in $\{u_r\}$ **do**
> > > Find the right nearest u_n satisfying $|\mathbf{D}_{uu}(v, u_n)| > \Gamma_u$ or $\mathbf{R}_v(u_n) \neq -1$ or $u_n \geq u_{max};$
> > > $\mathbf{R}_v((u_r, u_n)) \leftarrow 1;$
> >
> > **end for**
> > **return** $\mathbf{R}_v$
>
> **end function**
> **function** $\text{UPDIFFUSE}(\mathbf{R}_v, \mathbf{D}_{vv})$
> > $\mathbf{R}_{v-1} \leftarrow \mathbf{R}(v-1);$
> > **for** each u satisfying $\mathbf{R}_v(u) = 1$ **do**
> > > **if** $|\mathbf{D}_{vv}(v-1, u)| < \Gamma_v$ **then**
> > > > $\mathbf{R}_{v-1}(u) \leftarrow 1;$
> > >
> > > **else**
> > > > $\mathbf{R}_{v-1}(u) \leftarrow 0;$
> > >
> > > **end if**
> >
> > **end for**
> > **return** $\mathbf{R}_{v-1}$
>
> **end function**

5.5 Experimental Results

For experimental evaluation, the KITTI vision benchmark [79] is used with both intrinsic and extrinsic parameters of the two-camera system calibrated.

5.5.1 Row-Wise Image Registration

Based on previous development, image registration is row-wisely performed from the bottom row to the top row. Actually, the parameters used in the optimization can be tuned empirically according to the previously mentioned rules in Section 5.3.4, e.g., the tradeoff factors λ_1, λ_2 for regularization in (5.13) can be chosen around 1 for smoothness and accuracy; the scale factor λ_b in (5.14) should take into account the magnitude of image intensity for normalization; the threshold Γ_Φ is chosen by inspecting the convergence property of the optimizer and is generally around 10^{-5}; α and ξ in (5.25) and (5.26) are tuned to guarantee a fast convergence speed avoiding the divergence caused by image noise; the distance and normal of the reference plane d and n in the coordinate of the camera C can be obtained by calibration and their robustness will be analyzed later. The parameters are set as follows:

$$M = 50, \ \lambda_1 = 1, \ \lambda_2 = 0.6, \ \lambda_b = 0.05, \ \alpha = 1,$$
$$\Gamma_\Phi = 10^{-5}, \ \xi = 0.5, d = 1.65 \text{ m}, \ n = \begin{bmatrix} 0 & 1 & 0 \end{bmatrix}^T. \tag{5.28}$$

As shown in Figure 5.5, take one image row $\mathbf{S}$ in $\mathbf{I}$, for example, when the optimizer converges, the warped image of $\mathbf{S}'$ coincide with $\mathbf{S}$ mostly. While the two image signals won't coincide perfectly due to the different characteristics of cameras and the existence of image noise. After the row-wise image registration, the parallax map $\mathbf{B}$ corresponding to image $\mathbf{I}$ is obtained. For better visualization, the parallax map is normalized to a gray scale image according to the parallax value as $\bar{\beta} = \sqrt{(\beta - \beta_{min})/(\beta_{max} - \beta_{min})}$, with β_{min} and β_{max} are the minimum and maximum values in the parallax map $\mathbf{B}$. The normalized parallax map is shown in Figure 5.6, from which the road region can be clearly distinguished.

To show the robustness to the selection of reference plane, experiments are made for d varying from 1.3 to 2.0 m. For those $d \in [1.4, 1.9]$ m, the image registration is performed successfully. Two extreme conditions ($d = 1.4$ m and $d = 1.9$ m) are shown in Figure 5.7. For those results with different d, the lines of β are of the similar shape, with offsets with respect to each other. It can be inferred from Section 5.4.1 that the offset doesn't affect the road detection process. When the error is large, i.e., $d \leq 1.3$ m or $d \geq 2.0$ m, the optimizer converges to local maxima where some points are mismatched, and the image registration fails in some regions. In this case, the error threshold of d for successful image registration is about ± 0.25 m, which is enough in real applications. Actually, the robust region also depends on the intensity distribution of the image data, and the robustness can be further improved by other strategies of initial value setting.

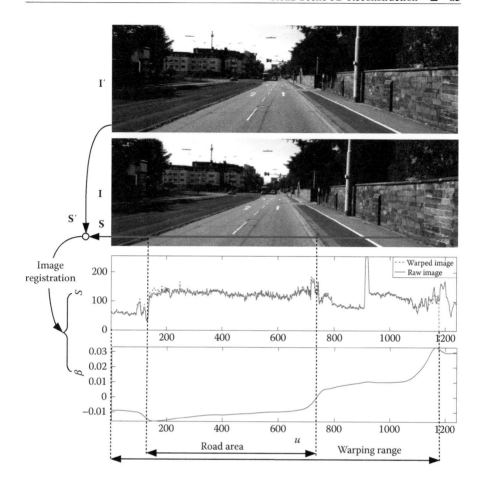

Figure 5.5: Image registration results.

Figure 5.6: Normalized parallax map corresponding to image I in Figure 5.5.

(a)

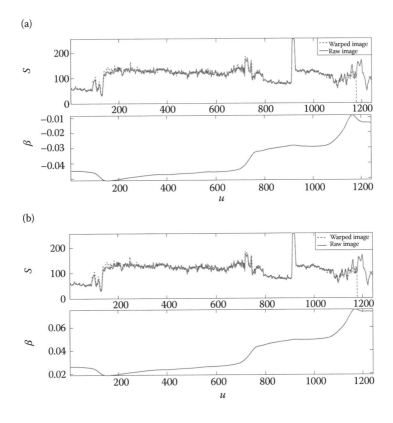

(b)

Figure 5.7: Robustness to *d* for the 3D reconstruction of images in Figure 5.5. (a) *d* = 1.4 m and (b) *d* = 1.9 m.

For comparison, the widely used semi-global matching method [93] is employed to obtain disparity maps. The resulting disparity map is shown in Figure 5.8. Note that the disparity map is only semi-dense, i.e., there exist many holes that are mismatched and eliminated by stereo validation. This phenomenon gets worse for textureless roads, as will be shown in the fifth scene in Figure 5.16. In real world, the road regions are likely to be textureless or repetitively textured, arising problems for classical stereo matching algorithms, as they are designed for general scenes and usually need to search over a large region to determine the correspondences and there exist many ambiguities. In comparison, the 3D reconstruction method, in this chapter, is designed specifically for road scenes. The regions of interest are believed to be continuous and near the reference plane; thus, the search region can be greatly reduced to deal with repetitive textures. Textureless regions can also be well dealt with by the spline-based approximation of the parallax information. Dense and smooth reconstruction result is generated

Figure 5.8: Disparity map by Semi-Global Matching [93] corresponding to image I in Figure 5.5.

from the two-view images based on which the road region is easy to be detected without the need of complex detection strategies.

5.5.2 Road Reconstruction

Based on the parallax map **B** obtained from image registration, the road is detected according to the road segmentation and diffusion rules in Section 5.4. The parameters are set as $\Gamma_u = 0.005, \Gamma_v = 0.01, \Gamma_c = 10$. The road detection result is shown in Figure 5.9, where road region is painted with red color. From the image, it can be observed intuitively that the detection result is accurate enough.

To show the advantages of the algorithm towards previous homography-based methods, comparisons are made using the same input images. As stated in Section 4.1, homography-based methods generally rely on the matching of feature points [133] or the warping error of two images [5]. To show the best performance of homography-based methods, the plane parameters are refined by try-and-error and set as $d = 1.74$ m, $n = \begin{bmatrix} 0.02 & 0.99 & 0 \end{bmatrix}^T$. To evaluate feature point based methods, the commonly used Speeded-Up Robust Features extractor and descriptor [11] is used. Figure 5.10 shows the matched feature points

Figure 5.9: Road detection result corresponding to image I in Figure 5.5.

Figure 5.10: Feature point correspondences between I and I' in Figure 5.5.

Figure 5.11: Homography-based warping error between I and I' in Figure 5.5.

between two images, from which it is clear that the feature points mainly lie in background regions and the points on the road are rather sparse, which are not enough to be used as prior knowledge for road segmentation. Figure 5.11 shows the warping error of two images based on homography transform. However, the responses in two images won't coincide perfectly due to the different characteristics of cameras and the existence of image noise, as also indicated in Figure 5.5. As shown in the enlarged region in Figure 5.11, this phenomenon is quite obvious around the lane markings with large intensity changes, causing mismatches when using the direct substraction of two images. Besides, these methods also suffer from textureless regions where the error measurement is not distinctive and the SNR (Signal to Noise Ratio) is low.

Using (4.3), the geometric information (x, z, D) for each image point in the coordinate of C can be calculated. Then, the 3D road map can be reconstructed by transforming every image point on the road to the Cartesian space of C, as shown

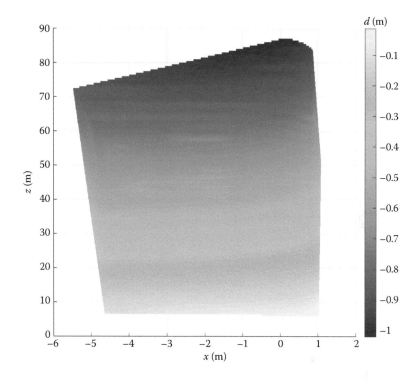

Figure 5.12: 3D road map corresponding to image I in Figure 5.5.

in Figure 5.12, with different colors indicating different height with respect to the reference plane. To evaluate the precision of road reconstruction, range data from a Velodyne laser scanner is used [79] and projected to the image plane corresponding to **I** based on the calibrated extrinsic parameters. The reconstruction error is evaluated row-wisely as shown in Figure 5.13, where $\bar{v}$ is defined as in Figure 5.3. Obviously, in the near regions, i.e., $\bar{v} < 110$, the accuracy is high and the measuring error is less than 0.1m. While in the far regions, the measuring error gradually increases to about 0.45 m due to the limitation of baseline and image resolution. The measuring accuracy can be further improved by increasing the baseline or image resolution. However, for most perspective cameras, the field of view is limited (about $50° \sim 60°$); thus, the length of baseline is also limited to guarantee enough corresponding views between two cameras.

5.5.3 Computational Complexity

The algorithm consists of 3D reconstruction and road detection processes, which are described in Figure 5.1. The size of images used in the experiment is

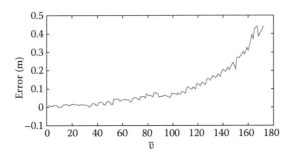

Figure 5.13: Road reconstruction error per image row with respect to the range data from the Velodyne laser scanner.

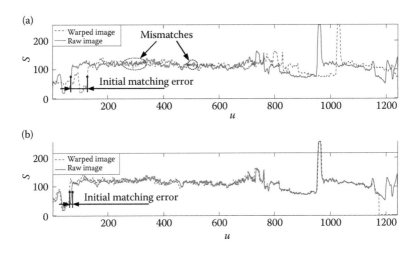

Figure 5.14: Initial states for image registration of the bottom row. (a) Using conventional epipolar geometry and (b) using the proposed two-view geometry.

1242×375. The algorithm is implemented in MATLAB$^®$ using a PC with Intel$^®$ Core^{TM} i5-4590 CPU and 4 GB RAM.

3D reconstruction generally relies on the matching of the two-view images. Figure 5.14 shows the initial states for row-wise image registration of the bottom row. Conventional stereo matching methods use the epipolar geometry to constrain the searching space, the initial state for image registration is shown in Figure 5.14a. The proposed two-view geometry is used, and the initial state for image registration is shown in Figure 5.14b. Obviously, using the two-view geometry, the initial matching error is much smaller and the searching region is reduced, so that the optimization process is more efficient. Figure 5.15a collects

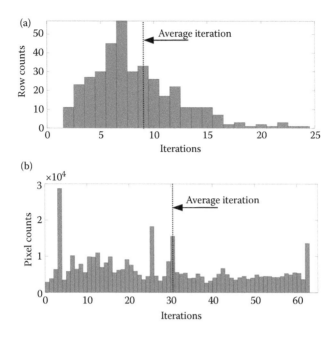

Figure 5.15: Histogram of iteration numbers for images in Figure 5.5. (a) Histogram of iteration numbers vs. row counts for row-wise image registration based on the proposed two-view geometry. (b) Histogram of iteration numbers vs. pixel counts for each pixel for pixel-wise stereo matching based on epipolar geometry.

the histogram of iteration numbers per row, and the iterations needed to finish the image registration per row is about nine times averagely. Comparably, stereo matching methods are performed pixel-wise, and the iteration numbers used to search the local optimum are collected in Figure 5.15b. Owing to the strategy used in iterative maximization in Section 5.4.1, the initial value is near the optimum and the iteration cost for row-wise image registration is much smaller than that used for conventional stereo matching. Another benefit is that for textureless or repetitively textured regions, the solution is easy to result in mismatches as shown in Figure 5.14a. Since the searching region is greatly reduced with the proposed two-view geometry, this problem can be well dealt with as shown in Figure 5.14b and the registration results in Figure 5.5.

Based on the two-view geometry in (4.1), it can be inferred from Figure 5.14 that the initial matching error is mainly reduced by the homography induced by the reference plane, the small matching errors induced by the parallax information is compensated by image registration. For points not on the reference plane, the matching error between two images is almost proportional to the length of baseline x_f, so that more iterations will be needed for the registration of the bottom image row with a larger baseline. As mentioned in Section 5.4.1, since the optimal value of the previous row is used as the initial value of the next row, as long as the registration of the bottom row succeeds, the following rows are not influenced significantly, so that the computational efficiency of the registration for the whole image is not influenced significantly.

Considering the time cost, the overall running time for the registration of the whole images is about 1.5 s on the experimental platform. For the bilateral filter, the MATLAB implementation provided by [32] is used without GPU acceleration, and the running time is about 0.3 s. The road region diffusion process can be formulated as a flood filling problem [193], which has been efficiently implemented on various platforms. The mask image Λ for flood filling is composed of

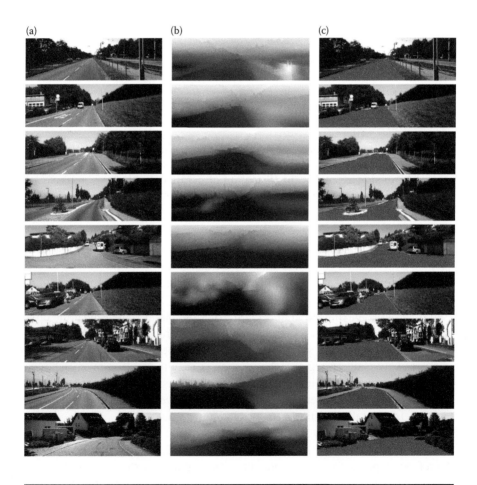

Figure 5.16: Road detection results. (a) Input images, (b) normalized parallax maps, and (c) detected road region.

two binary images Λ_u and Λ_v, which are obtained by $|\mathbf{D}_u| > \Gamma_u$ and $|\mathbf{D}_v| > \Gamma_v$, respectively. Then, the mask image is obtained by performing the "OR" operation of the two mask images. The overall running time for road detection process is about 0.1 s. Since this chapter mainly focuses on the theoretical development of the road reconstruction system, the algorithm is implemented in MATLAB for convenience. The computational efficiency can be further improved by a C++ implementation with GPU acceleration for real time applications.

5.5.4 Evaluation for More Scenarios

For evaluation, experiments are also conducted on other representative scenarios, as shown in Figure 5.16. As stated before, this chapter only considers drivable road region; lane markings are not considered. From the results, it is shown that the road regions are properly segmented where the height changes abruptly, such as road boundaries, obstacles, etc. Since the road region is detected using the diffusion strategy that begins from the bottom middle point, multi-lanes can also be detected as long as they are reachable, see the fourth scene in Figure 5.16. Besides, from the results, it can also be inferred that the proposed approach works for roads with random bend and slope, which cannot be handled by other homography-based methods as described by Section 4.1, see the fourth, fifth, and ninth scenes in Figure 5.16. Since this chapter concentrates on the detection of drivable road region that mainly rely on geometric information, road color and lane markings are not considered. While the ground truth data considers the current driving lane, in which road color and lane markings should be taken into account [176]; thus, the precision of detection is difficult to determine quantitatively. Nevertheless, the performance of road detection can be observed intuitively from the images.

Chapter 6

Recursive Road Detection with Shadows

6.1 Introduction

In this chapter, a pixel-wise illuminant invariant image transform is proposed based on the imaging model of cameras to deal with shadows, so that the transformed image is only related to the characteristics of cameras and captured objects ideally. For the application of road detection, most road features are available in the transformed images, which are fused with the raw images to obtain more details in both dark and bright regions. Then, the road is reconstructed from a geometry point of view, which is robust to the recovering accuracy of colors. With the two-camera system, the scene is reconstructed row-wisely based on the two-view geometry with respect to a reference plane, which is selected to be the grounding plane of the vehicle. This is quite suitable for road scenes in which the regions of interest are road regions that are approximately planar. Then, points in the scene can be described by the projective parallax with respect to the reference plane, and relative height information of each point can be calculated by the vehicle geometry. Unlike stereo vision based methods that first extract disparity maps of the whole images and then classify the road region from them, the road reconstruction is performed row-by-row recursively in this chapter, which consists of geometric reconstruction and road detection steps. For each row, the height information with respect to the grounding plane is estimated by the geometric reconstruction based on the two-view geometry model, and the road region is segmented using the distributions of height information and image intensity. This process is performed recursively from the bottom row until the road region on the current row is small enough, and the entire road region is obtained. The geometric reconstruction and road detection are processed inter-actively, i.e., the road region is segmented based on the reconstructed geomet-ric information, and will direct the geometric reconstruction of the next row, improving the performance of matching accuracy and computational efficiency. Besides, owing to the two-view geometric model with respect to the reference plane, the searching region for geometric reconstruction is significantly reduced to deal with the ambiguities caused by weak or repetitive textures.

6.2 Algorithm Process

In this chapter, a recursive framework is proposed for road reconstruction, as shown in Figure 6.1. To deal with the effects caused by shadows, illuminant invariant images are extracted and fused with raw images to recover details in both dark and bright regions. Then, a recursive framework is proposed for road reconstruction. The road region is reconstructed row-wisely from the bottom row to the row on which the road region disappears. For each row, the geometric information is first reconstructed, based on which the road region is segmented. In the geometric reconstruction step, the projective parallax with respect to the grounding plane is estimated using row-wise image registration based on the two-view geometry, and then the height information of each pixel can be directly calculated. In the road detection step, the road region is classified using the

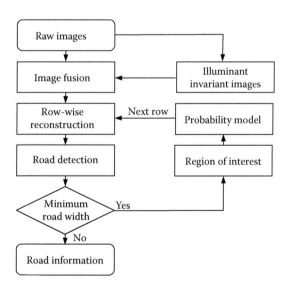

Figure 6.1: Algorithm process of the recursive road detection with shadows.

distributions of height and image intensity. Based on the road detection result of the current row, the region of interest (ROI) is extracted so that the next row doesn't need to be processed in whole. Besides, the probability distribution of road boundaries is constructed, which will be used in the image registration of the next row. This process is performed recursively until the road region of the current row is small enough and the entire road region can be obtained by combining the row-wise detection results.

Owing to the recursive framework, the images are not needed to be processed in whole to improve computational efficiency. For each image row, the geometric reconstruction is performed based on the two-view geometry with respect to a reference plane, and reconstruction results of previous rows are used as prior knowledge to determine the initial value, ROI and probability model of road boundary of the next row, reducing the search space and computational complexity. Besides, owing to the image fusion with illuminant invariant images, the algorithm works well with shadows. In the following, relevant models are developed first, based on which the road reconstruction algorithm is proposed.

6.3   Illuminant Invariant Color Space

In outdoor scenes, the lighting condition is complex and the effect on appearance caused by shadows influences the reliability of the road detection system. In the following, an illuminant invariant color space is developed to eliminate the effects of different lighting conditions.

6.3.1   Imaging Process

The physical process of camera imaging can be represented by the following expression [153]:

$$L = gl \int_{\Lambda} Q(\lambda) S(\lambda) W(\lambda) d\lambda, \qquad (6.1)$$

where L denotes the camera response, λ denotes the wavelength of light, Λ denotes the range of wavelength, g denotes the geometric factor of the scene, l denotes the intensity of illuminant, $Q(\lambda)$ denotes the reflectance property at the point, $S(\lambda)$ denotes the spectral power distribution of the light source, and $W(\lambda)$ denotes the spectral radiant at the point, which can be generally approximated as [153]:

$$W(\lambda) = w_1 \lambda^{-5} e^{-\frac{w_2}{T\lambda}}, \qquad (6.2)$$

where w_1, w_2 are constants and T is the color temperature. To generate the output image, cameras generally use gamma functions to emphasize details in the dark and/or bright regions, then the image intensity I_w is given by:

$$I_w = L^{\gamma}. \qquad (6.3)$$

where γ is a positive constant. For each image channel $n \in \{r, g, b\}$, assuming the spectral sensitivity of the corresponding image sensor is narrow enough (i.e., $S(\lambda)$ is Dirac delta function), then the image intensity I_{wn} is given by:

$$I_{wn} = \left(w_1 g l Q_n \lambda_n^{-5}\right)^{\gamma_n} e^{-\frac{w_2 \gamma_n}{T \lambda_n}},\tag{6.4}$$

where $Q_n = Q(\lambda_n)$, λ_n and γ_n are the central wavelength and gamma factor of channel n, respectively, which can be determined by calibration in practice [83].

6.3.2 Illuminant Invariant Color Space

From (6.4), the image intensity is affected by scene geometry, light condition, and object reflectance property. Then, the aim is to design a color space that is invariant to the light source and only relates to the characteristics of objects. For the raw image intensities $\{I_{wn}\}$, they are transformed to the illuminant invariant color space $\{H_n\}$ as follows:

$$H_r = \frac{I_{wr}}{I_{wg}^{\gamma_{gr}} I_{wb}^{\gamma_{br}}}, H_g = \frac{I_{wg}}{I_{wr}^{\gamma_{rg}} I_{wb}^{\gamma_{bg}}}, H_b = \frac{I_{wb}}{I_{wr}^{\gamma_{rb}} I_{wg}^{\gamma_{gb}}},\tag{6.5}$$

where $\gamma_{gr}, \gamma_{br}, \gamma_{rg}, \gamma_{bg}, \gamma_{rb}, \gamma_{gb} \in [0, 1]$ are constant factors that are chosen according to the camera's characteristics. Take H_r for example, (6.5) can be rewritten as:

$$H_r = (w_1 g l)^{e_1} e^{-\frac{w_2}{T} e_2} \frac{(\lambda_r^{-5} Q_r)^{\gamma_r}}{(\lambda_g^{-5} Q_g)^{\gamma_g \gamma_{gr}} (\lambda_b^{-5} Q_b)^{\gamma_b \gamma_{br}}},\tag{6.6}$$

where $e_1 = \gamma_r - \gamma_g \gamma_{gr} - \gamma_b \gamma_{br}, e_2 = \frac{\gamma_r}{\lambda_r} - \frac{\gamma_g}{\lambda_g} \gamma_{gr} - \frac{\gamma_b}{\lambda_b} \gamma_{br}$. Obviously, if $e_1, e_2 \to 0$, H_r is only affected by the characteristics of the camera and object. Then, γ_{gr}, γ_{br} can be chosen to minimize $\left\| \begin{bmatrix} e_1 & e_2 \end{bmatrix} \right\|$. Without loss of generality, the other channels can be designed similarly.

6.3.3 Practical Issues

There are several practical issues to implement the above algorithm. Without loss of generality, assuming the images are sampled to 8-bits, i.e., the range of pixel intensity is $[0, 255]$. If a pixel in the scene is saturated, i.e., it has one or more channels that are equal to 255, then the true color of the corresponding world point is not recoverable. Fortunately, this issue can be partially relieved by using the mapping based on gamma function defined as in (6.3), which has been considered in the algorithm. Due to the discrete quantization process, there are some pixels that have one or more channels that are equal to 0, and singularity exists in (6.5). To avoid singularity, the images are expressed in floating point in the algorithm and is added with a small value.

6.4 Road Reconstruction Process

In this section, the algorithm process shown in Figure 6.1 will be introduced in detail, mainly including image fusion, geometric reconstruction, road detection, and model update for recursive processing.

6.4.1 Image Fusion

In outdoor scenes, the captured images $\{\mathbf{I}_{wn}\}, \{\mathbf{I}'_{wn}\}$ may suffer from shadows, which is crucial for road detection because textures in shadow regions are very weak and the intensities are very low. To this end, illuminant invariant images $\{\mathbf{H}_n\}, \{\mathbf{H}'_n\}$ corresponding to $\{\mathbf{I}_{wn}\}, \{\mathbf{I}'_{wn}\}$ are obtained from (6.5), and then they are fused to obtain more details in both dark and bright regions:

$$\mathbf{I}_n = \rho \mathbf{I}_{wn} + (1-\rho)\mathbf{H}_n, \mathbf{I}'_n = \rho \mathbf{I}'_{wn} + (1-\rho)\mathbf{H}'_n, \tag{6.7}$$

where $\rho \in (0,1)$ is the weighting factor. In the following, the fused images $\{\mathbf{I}_n\}, \{\mathbf{I}'_n\}$ are converted to gray scale images, which are denoted as $\mathbf{I}$ and $\mathbf{I}'$, respectively.

6.4.2 Geometric Reconstruction

For each image row $\mathbf{P}$ in $\mathbf{I}$, the geometric reconstruction can be formulated into a row-wise image registration problem solved by iterative optimization.

6.4.2.1 Geometric Modeling

Image warping function

Using the general two-view geometry, image points $\mathbf{P} \triangleq [p_1 \cdots p_N]$ on one image row in image $\mathbf{I}$ can be related to $\mathbf{P}' \triangleq [p'_1 \cdots p'_N]$ in image $\mathbf{I}'$, with N being the count of image columns. Their relationship can be expressed by the the warping function $[u'_i \quad v'_i]^T = w(\beta_i, p_i)$, which can be obtained as follows:

$$\begin{bmatrix} u'_i \\ v'_i \end{bmatrix} = \begin{bmatrix} \dfrac{g_{11}u_i + g_{12}v_i + g_{13} + \beta_i(\alpha'_u(x_{fx} - x_{fy}\cot\theta') + u'_0 x_{fz})/d}{g_{31}u_i + g_{32}v_i + g_{33} + \beta_i x_{fz}/d} \\[3mm] \dfrac{g_{21}u_i + g_{22}v_i + g_{23} + \beta_i(\alpha'_v x_{fy}/\sin\theta' + v'_0 x_{fz})/d}{g_{31}u_i + g_{32}v_i + g_{33} + \beta_i x_{fz}/d} \end{bmatrix}, \tag{6.8}$$

where β_i is the projective parallax at point p_i and g_{kl} is the element at the (k,l) of the projective homography G, which can be calculated by the calibrated camera parameters.

Image similarity

Denote the intensities of the image points in $\mathbf{P}$ and $\mathbf{P}'$ as $\mathbf{S} \triangleq \left[s_1 \cdots s_N \right]^T$ and $\mathbf{S}' \triangleq \left[s_1' \cdots s_N' \right]^T$, respectively. From the warping function (6.8), the corresponding points in $\mathbf{P}'$ is possible to be out of the range of $\mathbf{I}'$. Hence, a mask vector $\Delta \triangleq \left[\delta_1 \cdots \delta_N \right]^T$ is defined with $\delta_i = 0$ if p_i' is out of the range of image $\mathbf{I}'$ and $\delta_i = 1$ otherwise. The similarity between these two sets of points can be measured by the correlation coefficient, which is defined as:

$$c(\mathbf{S}, \mathbf{S}') = \frac{\sum_{i=1}^{N} \delta_i (s_i - \bar{s})(s_i' - \bar{s}')}{\sqrt{\sum_{i=1}^{N} \delta_i (s_i - \bar{s})^2} \sqrt{\sum_{i=1}^{N} \delta_i (s_i' - \bar{s}')^2}}, \tag{6.9}$$

where $\bar{s}$ and $\bar{s}'$ are the mean values of valid points in $\mathbf{S}$ and $\mathbf{S}'$, respectively. The correlation coefficient c varies from -1 to 1, and a larger c means $\mathbf{S}$ and $\mathbf{S}'$ match better.

Parameterization of projective parallax

For each image row, the projective parallax is assumed to be continuous; thus, one-dimensional cubic B-spline function is used for parameterization, as shown in Figure 6.2. The B-spline curve is defined by a control lattice $\{c_{j-2}\}_{j \in [1,M]}$ overlaid uniformly on the image row, with M being the number of control points, and the interval between two neighboring control points is $\kappa = \frac{N}{M-3}$. Denoting $\Phi \triangleq \left[\phi_{-1} \cdots \phi_{M-2} \right]^T$ as the values of control points, and $\beta \triangleq \left[\beta_1 \cdots \beta_N \right]^T$ as the parallax values of image points. Since the spline curve is defined locally, (i.e., each image point is affected by four neighboring control points); then, it

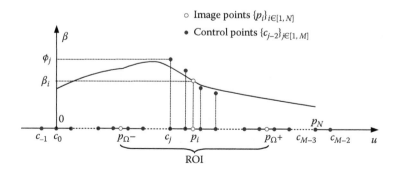

Figure 6.2: Spline function based parameterization.

is convenient to segment for the ROI. Define the ROI of current row as $\Omega = [\Omega^-, \Omega^+]$; then, the image coordinates in the ROI is defined as:

$$P_\Omega \triangleq \begin{bmatrix} p_{\Omega^-} \cdots p_{\Omega^+} \end{bmatrix}, \tag{6.10}$$

where the number of image points is $N_\Omega = \Omega^+ - \Omega^- + 1$. The parallax values of the image points in the ROI are defined as:

$$\beta_\Omega \triangleq \begin{bmatrix} \beta_{\Omega^-} \cdots \beta_{\Omega^+} \end{bmatrix}^T, \tag{6.11}$$

where the count of control points is M_Ω. Then the approximating function is defined as:

$$\beta_\Omega = F_\Omega \cdot \Phi_\Omega, \tag{6.12}$$

where F_Ω is a $N_\Omega \times M_\Omega$ matrix, with its i-th row $F_{\Omega i}$ defined as:

$$F_i \triangleq \begin{bmatrix} \underbrace{0 \cdots 0}_{\lfloor \frac{i}{\kappa} \rfloor - 2} & f(t_i) & \underbrace{0 \cdots 0}_{M_\Omega - \lfloor \frac{i}{\kappa} \rfloor - 2} \end{bmatrix}, \tag{6.13}$$

in which $f(t_i) = \begin{bmatrix} f_0(t_i) & f_1(t_i) & f_2(t_i) & f_3(t_i) \end{bmatrix}$ and $t_i = \frac{i}{\kappa} - \lfloor \frac{i}{\kappa} \rfloor$, with $\lfloor \cdot \rfloor$ denoting the round down function and $\{f_k(t)\}_{k=0}^3$ being the cubic B-spline basis functions [123].

Objective function

Based on the proposed models, the row-wise image registration can be formulated into the maximization of the following objective function:

$$E(\Phi_\Omega) \triangleq c(S_\Omega, I'(w(F_\Omega \cdot \Phi_\Omega, P_\Omega))) - \lambda_1 r_1(\Phi_\Omega) - \lambda_2 r_2(\Phi_\Omega). \tag{6.14}$$

In the objective function, c is the correlation coefficient assessing the quality of image matching. Due to the existence of image noise and different characteristics of cameras, the intensities of pixels in two cameras corresponding to the same points may not be the same. Then, a larger c does not necessarily mean a better image registration quality. Assuming the environment is continuous, $r_1(\Phi_\Omega)$ and $r_2(\Phi_\Omega)$ are introduced as regularization terms to better deal with image noise and improve the numerical stability; they act as the penalties of β for the smoothness and the closeness to the previous row, respectively. $\lambda_1, \lambda_2 \in \mathbb{R}^+$ are weighting factors, acting as trade-offs between image matching accuracy and smoothness. $r_1(\Phi_\Omega)$ and $r_2(\Phi_\Omega)$ are defined as:

$$r_1(\Phi_\Omega) \triangleq \frac{1}{2} \sum_{i=1}^N \left(\frac{\partial \beta_i}{\partial u} \right)^2 e^{-\lambda_3 \eta_i \|\nabla I_i\|_2},$$

$$\tag{6.15}$$

$$r_2(\Phi_\Omega) \triangleq \frac{1}{2} \sum_{i=1}^N (\beta_i - \beta_i^-)^2.$$

In (6.15), the term $e^{-\lambda_3 \eta_i \|\nabla \mathbf{I}_i\|_2}$ in $r_1(\Phi)$ represents as a weighting factor in consideration of discontinuities at boundaries, where $\lambda_3 \in \mathbb{R}^+$ is a scale factor, $\nabla \mathbf{I}_i$ denotes the image gradient at point p_i, and η_i denotes the pseudo probability of being road boundary that is computed by the road detection result of the previous row. As a result, the smoothness penalty gets smaller at points that are more likely to be road boundary. In (6.15), the term $\beta^- \triangleq \left[\beta_1^- \cdots \beta_N^- \right]^T$ in $r_2(\Phi)$ denotes the parallax of the previous row. For the registration of the first image row, the term $r_2(\Phi)$ is neglected and the term η_i is initialized as 1 everywhere.

Iterative optimization

Levenberg–Marquardt algorithm is used to search for the optimum iteratively, the parameter update rule for the k-th iteration is given by:

$$\Phi_{\Omega,k+1} = \Phi_{\Omega,k} + \alpha \Delta \Phi_\Omega \tag{6.16}$$

where α is the step size and $\Delta \Phi_\Omega$ is the search direction defined as:

$$\Delta \Phi_\Omega = (J_\Omega^T J_\Omega + \xi I)^{-1} J_\Omega^T (E_{max} - E(\Phi_{\Omega,k})), \tag{6.17}$$

in which $J_\Omega = \frac{\partial E_\Omega}{\partial \Phi_\Omega}$ and E_{max} is the maximum value of $E(\Phi)$. From (6.14) it can be inferred that $E_{max} = 1$. I is an identity matrix and ξ is a weighting factor that regulates the performance of the optimizer in terms of its convergence speed and stability. The convergence speed increases with a smaller ξ, while the stability improves with a larger ξ.

Optimization process

By using the gradient-based optimization method, the convergence region is limited; the following strategies are used so that the initial values of the optimizer are near the optimal solution:

- ■ The image registration is performed recursively from the bottom row, and the initial values of β on the bottom row are set as zero.

- ■ Assuming the road surface is continuous, the optimal solution of the previous row is used as the initial value of the next row.

Since the reference plane is selected as the grounding plane of the vehicle, the parallax of the road region on the bottom row is likely to be near zero. As a result, owing to the above strategies, the image registration for continuous road regions can be well performed. From the perspective of optimization, the performance of the optimizer is mainly affected by the parameter errors, not the parameters such as the length of baseline, which is one of the main advantages to conventional stereo vision based methods.

Remark 6.1 In practice, the two-camera system is generally an integrated system so that its intrinsic and extrinsic parameters are almost constant in the process and can be calibrated beforehand. The main concern is that the road surface near the vehicle doesn't coincide with the reference plane defined by d, n due to the roughness of the road, and this will be studied specifically in the experiments.

Remark 6.2   It is obvious that if the registration of the bottom row succeeds, the registration of the following rows will be easy. To improve the robustness to parameter error, feature point matches can be used for the initialization of the optimizer, which is not in the scope of this chapter.

6.4.3   Road Detection

Road segmentation

After the row-wise geometric reconstruction, the height information with respect to the grounding plane can be calculated using the vehicle-road geometric model. As shown in Figure 6.3, since road boundaries generally occur at points that both image intensity and height change significantly, the following criterion is proposed for the possibility of being road boundaries:

$$C_u \triangleq \nabla_u D \cdot \nabla_u \mathbf{I}$$
$$C_v \triangleq \nabla_v D \cdot \nabla_v \mathbf{I}, \tag{6.18}$$

where ∇_u and ∇_v mean the partial derivatives along u and v directions, respectively.

Road region growing

For the configuration in Figure 4.2, the bottom middle point in $\mathbf{I}$ is assumed to be road point; then, the road region can be obtained by the following region growing strategy with appropriate thresholds τ_u, τ_v:

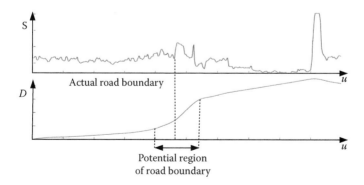

Figure 6.3: Road boundary features.

- For each point on the current row, label it as road point if its corresponding point on the previous row is road point and $C_v < \tau_v$ (neglected for the bottom row);

- For each edge point, label its neighboring points as road points iteratively on the condition that $C_u < \tau_u$;

Here, edge points are road points that have neighboring non-road points.

Remark 6.3   In practice, the initialization of road points can be determined beforehand by the configuration of the vision system, and the above region growing method works for any other configurations. In this chapter, only the most common scenario is considered.

6.4.4   Recursive Processing

Probability model

In the definition of objective function (6.15), the probability model of road boundaries is embodied as the weighting factor η_i. Assuming the road boundaries are continuous, the probability distribution can be estimated by the road detection result of the previous row, and is described by the pseudo probability density function as:

$$\eta_i = \varepsilon + (1 - \varepsilon) \sum_{\psi \in \Psi} e^{-\frac{(i - \psi)^2}{2\sigma^2}}, \tag{6.19}$$

where Ψ is the set of road boundaries of the previous row and σ is the standard deviation. Note that under some circumstances, the road boundaries are not continuous, ε is a factor to take into account these circumstances.

Region of interest (ROI)

In real world, it can be assumed that most road boundaries are continuous, thus it is advisable to impose the ROI in the geometric reconstruction based on the road detection result of the previous row, so that the image row doesn't need to be processed in whole. From the set of road boundaries Ψ of the previous row, the ROI Ω is defined as:

$$\Omega = [\max(\min(\Psi) - \mu, 1), \ \min(\max(\Psi) + \mu, N)], \tag{6.20}$$

where $\mu \in \mathbb{R}^+$ is a slack constant and the ROI of the bottom row is set as the whole image row. However, there're exceptions, a typical circumstance is that the road region is larger than the ROI, i.e., the left most or the right most detected road boundary coincides with the boundary of ROI. In that case, the ROI is enlarged to the whole image row and the road reconstruction of the row is processed again.

Adaptive threshold

In this chapter, the distributions of image intensities and height information are exploited to determine the road boundary. In real scenes, the best threshold for left and right road boundaries may be different. Based on the road segmentation strategy mentioned above, the thresholds for the next image row are calculated by the values $\bar{C}_{u,l}, \bar{C}_{u,r}$ at the segmenting point of the current image row:

$$\tau_{u,l} = \Gamma \cdot C_{u,l}, \tau_{u,r} = \Gamma \cdot C_{u,r} \tag{6.21}$$

Besides, to avoid unreasonable threshold setting, the thresholds are limited by upper and lower bounds.

6.5 Experiments

To evaluate the proposed approach, experiments are done on the KITTI benchmark [80]. The input images are captured by two on-vehicle cameras and both intrinsic and extrinsic parameters are calibrated.

6.5.1 Illuminant Invariant Transform

As shown in Figure 6.4, the proposed algorithm is evaluated using the scene with shadows. In the raw image, the left side is covered with heavy shadows, causing great challenge for road detection. Since the camera parameters λ_n, γ_n are not provided in the benchmark, the parameters for illuminant invariant transform are selected by trail and error, which are given by:

$$\gamma_{gr} = 0.6, \gamma_{br} = 0.6, \gamma_{rg} = 0.48,$$
$$\gamma_{bg} = 0.57, \gamma_{rb} = 0.4, \gamma_{gb} = 0.4. \tag{6.22}$$

As shown in Figure 6.4, the resulting invariant image $\{\mathbf{H}\}$ consists of three color channels $\mathbf{H}_r, \mathbf{H}_g, \mathbf{H}_b$, in which effects caused by shadows are removed mostly, and the fused images are used for road reconstruction with $\rho = 0.5$, as defined in (6.7). In Figure 6.5, several image transform methods [2,45,91,149] widely used in road detection or visual localization are evaluated for comparison. The results of the methods in [2,45] are similar, as shown in Figure 6.5a and 6.5c, a one-channel invariant image is generated, in which shadows are removed mostly but the resulting image contrast is low and the color information is lost; the result of the method in [149] has a higher image contrast while is more noisy. In Figure 6.5b, a three-channel log-chromaticity image is generated, in which shadows are not removed properly. Compared to previous works on illuminant invariant transform, the proposed method performs better in color recovery and has the computational complexity $O(N)$. The scene structures can be clearly classified from the resulting illuminant invariant image. Besides, this algorithm can be used for other road detection systems that rely on appearance information.

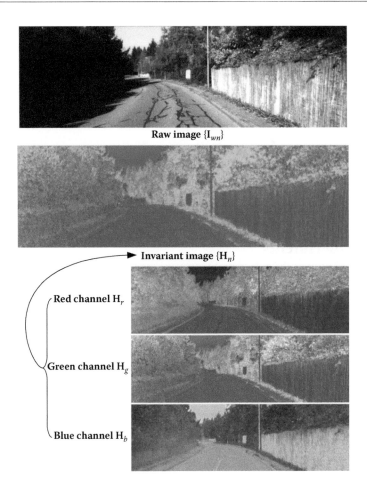

Figure 6.4: Illuminant invariant transform.

6.5.2 Road Reconstruction

Geometric reconstruction

Based on the calibrated extrinsic parameters, the parameters of the reference plane are set as $d = 1.65$ m, $n = \begin{bmatrix} 0 & 1 & 0 \end{bmatrix}^T$, and parameters for geometric reconstruction are set as follows:

$$M = 70, \ \lambda_1 = 0.5, \ \lambda_2 = 0.1, \ \lambda_3 = 0.05,$$
$$\alpha = 1, \ \xi = 1.5, \ \varepsilon = 0.2, \ \sigma = 10, \ \mu = 50. \tag{6.23}$$

For example, take one image row **S** in **I**, the result of image registration is illustrated in Figure 6.6. The probability distribution η of road boundary and ROI Ω are constructed from the road detection result of the previous row, and only points

Figure 6.5: State of the art methods for image transform in road detection or visual localization. (a) Illuminant invariant image in [2], (b) log-chromaticity image in [91], (c) illuminant invariant image in [45], and (d) illuminant invariant image in [149].

in the region Ω are required to be processed. After the registration, $\mathbf{S}$ coincide with the warped points of $\mathbf{S}'$ mostly, while they cannot be matched accurately due to the existence of image noise. Based on the distributions of image intensity and reconstructed height information, the criterion C_u can be directly calculated from (6.18), based on which the road boundaries can be classified.

To show the robustness to the selection of the reference plane, multiple planes with d varying from 1.2 to 2.0 m are used as the reference plane for row-wise registration. As shown in Figure 6.7, the geometric reconstruction is well performed for $d \in [1.3, 1.9]$ m and the resulting β curves have similar shapes, with offsets among each other. Since only the gradient of the height information is used for road segmentation, the offsets do not affect the road detection task. In addition, the robustness to the normal vector n is not easy to be determined quantitatively since it has three degrees of freedom. While from the reconstruction result, it can be seen that β is inclined, which means that the reference plane does not match the road surface perfectly and the road detection is not affected.

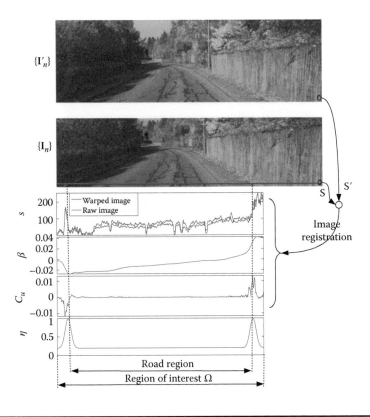

Figure 6.6: Row-wise geometric reconstruction.

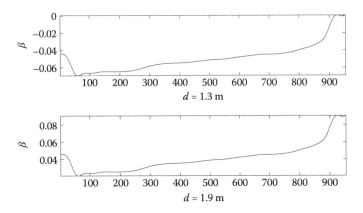

Figure 6.7: Row-wise geometric reconstruction with different *d*.

Road detection

After the geometric reconstruction, the road region is obtained row-wisely based on the proposed region growing strategy with $\tau_u = 0.02, \tau_v = 0.05$, which is shown as the red region in Figure 6.8a. Using the vehicle geometry, the road surface is reconstructed geometrically, as shown in Figure 6.8b, which is not planar actually. Due to the limitation of baseline and the existence of shadows, the road region too far from the vehicle cannot be distinguished, and the pavement on the left side (a little higher than the road) can only be recognized near the vehicle. This condition can be improved by increasing the length of baseline, which doesn't affect the efficiency of geometric reconstruction as mentioned above.

To show the necessity of image fusion with illuminant invariant image, road detection is performed using the same parameters without image fusion, and the result is shown in Figure 6.9. From the result, the existence of strong shadow effect is troublesome to the road detection. First, in the shadowed region, the image texture is not descriptive enough for accurate image registration, since smoothness terms play bigger roles in the objective function (6.14); second, the low image contrast also causes a low image gradient at road boundaries, causing wrong boundary segmentation according to (6.18).

Optimization process

The proposed method reconstructs the scene and detects the road region simultaneously. Figure 6.10 shows the iterations cost to finish the geometric

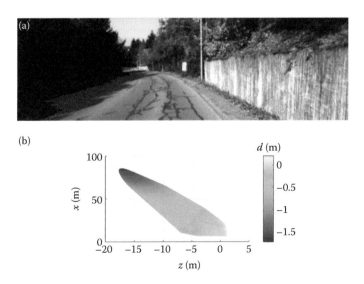

Figure 6.8: Road detection results. (a) Detected road region and (b) reconstructed road surface.

Figure 6.9: Road detection result without image fusion.

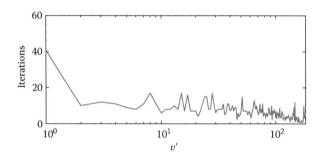

Figure 6.10: Iterations to finish row-wise geometric reconstruction.

reconstruction, in which v' denotes the row number counting from the bottom. It is shown that in the beginning the optimizer needs about 40 iterations to converge. For the following rows, owing to the strategy that the initial value is set as the optimal solution of the previous row, the optimizer only needs about 10 iterations or less to converge, so that the computational efficiency is greatly improved. Besides, the ROI is updated according to the road boundaries of previous rows. Due to the perspective effect, the size of ROI generally decreases from the bottom to the top row, as shown in Figure 6.11, reducing the required computational complexity.

6.5.3 Comparisons with Previous Methods

To show the effectiveness of the proposed method, intermediate results of previous methods are provided for comparisons intuitively, mainly including homography-based methods and stereo matching methods.

As discussed in Section 4.1, previous homography-based works usually rely on the image warping error [5], the feature point correspondences [44], or both of them [170]. To show the best performance of these methods, instead of automatically estimating the homography matrix online as done in [5,44,170], the

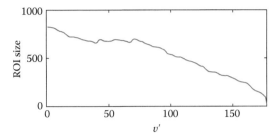

Figure 6.11: ROI sizes of each row for geometric reconstruction.

homography is directly calculated using the extrinsic parameters of the two-camera system and the plane parameters adjusted from the reconstruction result in Figure 6.8b. Then, the image $\mathbf{I}$ is warped by the homography transform and the error between the image $\mathbf{I}'$ and the warped image of $\mathbf{I}$ is shown in Figure 6.12a. Then, the road region is classified from the image warping error, as shown

Figure 6.12: Homography-based road detection by image warping error. (a) Warping error of two images and (b) detected road region.

in Figure 6.12b. From the result, it can be seen that even though the parameters in the homography are adjusted carefully, the two images cannot coincide perfectly due to the different imaging characteristics of the cameras, and the detected road region is noisy. Besides, this method also suffers from low image contrast, and the shadowed region is detected as road region mistakenly. In Figure 6.13, speeded-up robust feature feature points [11] are extracted and matched from the images; then, the correspondences are validated according to the homography transform. The resulting road points are illustrated in Figure 6.13. From the result, it can be seen that the feature points are rather sparse and mainly lie in textured areas, while the textureless regions (e.g., the shadowed region) cannot be classified. The above two conditions have shown the disadvantages of homography-based methods using ideal parameter setup. Besides, the main disadvantage of homography-based methods is that it requires planar scenes, while most outdoor roads are not strictly planar. This is one of the main improvements of this approach by applying a general two-view geometry that works for both planar and nonplanar scenes.

Stereo matching methods

Previous works based on stereo vision rely on the disparity maps that are generated by stereo matching of the binocular images [121,163,215]. To evaluate the performance of stereo matching, the widely used Semi-Global Stereo Matching method [93] is employed. As shown in Figure 6.14, there're some mismatched

Figure 6.13: Classified road points (yellow) based on homography transform.

Figure 6.14: Disparity map extracted by Semi-Global Matching [93].

and unmatched points in the resulting disparity map, raising difficulties for reliable road detection. The unmatched points in the disparity maps can be possibly approximated by a bilateral filter [32]. In [186], undirected graph is constructed using the sparse disparity map, and then the road area is segmented from the graph. To be convenient, the results provided in [186] are used for comparison, and the results are shown in Figure 6.15. It can be seen that if the count of mismatched and unmatched points are large, the method would be problematic, besides the last two scenes in Figure 6.17c. In those circumstances, the road detection would be complex due to the poor quality of stereo matching. In fact, classical stereo matching methods are designed for general scenes, and thus need to search a large region for the correspondences and there exist more ambiguities caused by weak or repetitive textures. Comparably, the searching region of the proposed algorithm is greatly reduced by the two-view geometric model (this can be seen from Figure 6.10 that only a few iterations are needed for reconstruction), and the parallax information is regulated by a spline curve, so that dense and smooth reconstruction results can be obtained.

6.5.4 Other Results

Since this chapter considers the drivable road region that can be reached by the vehicle, the algorithm mainly works with geometric information and appearance

Figure 6.15: Comparison of road detection results by the proposed method (left) vs. stereo matching based method [186] (right).

Figure 6.16: Road regions with and without appearance priors. (a) Ground truth for road region provided in [80] and (b) detected road region.

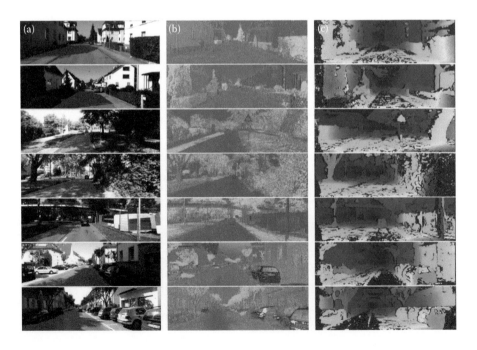

Figure 6.17: Road detection results. (a) Detected road regions, (b) illuminant invariant images, and (c) disparity maps extracted by stereo matching [93].

priors are not needed, as stated in Section 2.1. However, in the KITTI benchmark, the definition of road region takes into consideration of appearance priors, such as lane markings, color difference with background scenes, etc. Figure 6.16 shows a typical scene for this ambiguity. From Figure 6.16b, it can be seen that the height difference between the pavement and the road surface is so small that the pavement is also recognized to be drivable. For quantitative evaluation, 100 images are selected from the training set of the KITTI benchmark that do not have the ambiguity of road region; then, the performance is evaluated in the perspective image space and achieves the following indexes [75]:

$$K_p = 0.904, K_r = 0.959, K_f = 0.927, K_a = 0.971, \qquad (6.24)$$

where K_p, K_r, K_f, and K_a denote the indexes for precision, recall, F-measure, and accuracy, respectively. The detection results of several representative scenes are provided in Figure 6.17. In Figure 6.17b, illuminant invariant images are extracted, in which effects caused by shadows are removed mostly. In Figure 6.17a, road regions are detected and painted with red color. It can be seen that most road regions are segmented properly where the height information changes abruptly, despite of the existence of shadows. In the future works, the drivable road region will be further segmented into multiple driving areas taking into appearance priors such as lane markings and road colors.

Chapter 7

Range Identification of Moving Objects

7.1 Introduction

In this chapter, we present a unique nonlinear estimation strategy to simultaneously estimate the velocity and structure of a moving object using a single camera. Roughly speaking, satisfaction of a persistent excitation condition (similar to [35] and others) allows the determination of the inertial coordinates for feature points on the object. A homography-based approach is utilized to develop the object kinematics in terms of reconstructed Euclidean information and image-space information for the fixed camera system. The development of object kinematics relies on the work presented in [29] and [152], and requires *a priori* knowledge of a single vector of one feature point of the object. A novel nonlinear integral feedback estimation method developed in our previous efforts [39] is then employed to identify the linear and angular velocity of the moving object. An adaptive nonlinear estimator is proposed in this chapter with the design of measurable and unmeasurable signal filters to reconstruct the range and the 3D Euclidean coordinates of feature points. Identifying the velocities of the object facilitates the development of a measurable error system that can be used to formulate a nonlinear least squares adaptive update law. A Lyapunov-based analysis is then presented that indicates if a persistent excitation condition is satisfied then the range and the 3D Euclidean coordinates of each feature point can be determined.

7.2 Geometric Modeling

7.2.1 Geometric Model of Vision Systems

We define an orthogonal coordinate frame, denoted by $\mathcal{F}$, attached to the object and an inertial coordinate frame, denoted by $\mathcal{I}$, whose origin coincides with the optical center of the fixed camera (see Figure 7.1). Let the 3D coordinates of the ith feature point on the object be denoted as the constant $s_i \in \mathbb{R}^3$ relative to the object reference frame $\mathcal{F}$, and $\bar{m}_i(t) \triangleq \begin{bmatrix} x_i & y_i & z_i \end{bmatrix}^T \in \mathbb{R}^3$ relative to the inertial coordinate system $\mathcal{I}$.

It is assumed that the object is always in the field of view of the camera, and hence, the distances from the origin of $\mathcal{I}$ to all the feature points remain positive (i.e., $z_i(t) > f_0$, where f_0 is the focal length of the camera). To relate the coordinate systems, let $R(t) \in SO(3)$ and $x_f(t) \in \mathbb{R}^3$ denote the rotation and translation, respectively, between $\mathcal{F}$ and $\mathcal{I}$. Also, let three of the non-collinear feature points on the object, denoted by $O_i, \forall i = 1, 2, 3$, define the plane π shown in Figure 7.1. Now, consider the object to be at some fixed reference position and orientation, denoted by $\mathcal{F}^*$, as defined by a reference image of the object. We can similarly define the constant terms $\bar{m}_i^*$, R^* and x_f^*, and the plane π^* for the object at the reference position. From Figure 7.1, the following relationships can be developed

$$\bar{m}_i = x_f + R s_i \qquad \bar{m}_i^* = x_f^* + R^* s_i. \tag{7.1}$$

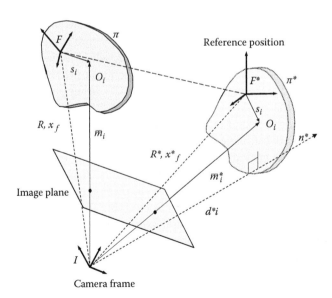

Figure 7.1: Geometric relationships between a fixed camera and a moving object.

After eliminating s_i in (7.1), we have

$$\bar{m}_i = \bar{x}_f + \bar{R}\bar{m}_i^*, \tag{7.2}$$

where $\bar{R}(t) \in SO(3)$ and $\bar{x}_f(t) \in \mathbb{R}^3$ are rotational and translational variables, respectively, defined as follows:

$$\bar{R} = R(R^*)^T \qquad \bar{x}_f = x_f - \bar{R}x_f^*, \tag{7.3}$$

As also illustrated in Figure 7.1, $n^* \in \mathbb{R}^3$ denotes the constant normal to the plane π^* expressed in the coordinates of $\mathcal{I}$, and the constant projections of $\bar{m}_i^*$ along the unit normal n^*, denoted by $d_i^* \in \mathbb{R}$ are given by:

$$d_i^* = n^{*T}\bar{m}_i^*. \tag{7.4}$$

We define normalized Euclidean coordinates, denoted by $m_i(t), m_i^* \in \mathbb{R}^3$, for the feature points as follows:

$$m_i \triangleq \frac{\bar{m}_i}{z_i} \qquad m_i^* \triangleq \frac{\bar{m}_i^*}{z_i^*}. \tag{7.5}$$

As seen by the camera, each of these feature points have projected pixel coordinates, denoted by $p_i(t), p_i^* \in \mathbb{R}^3$, expressed relative to $\mathcal{I}$ as follows:

$$p_i = \begin{bmatrix} u_i & v_i & 1 \end{bmatrix}^T \qquad p_i^* = \begin{bmatrix} u_i^* & v_i^* & 1 \end{bmatrix}^T. \tag{7.6}$$

The projected pixel coordinates of the feature points are related to the normalized Euclidean coordinates by the pinhole model of [66] such that

$$p_i = Am_i \qquad p_i^* = Am_i^*, \tag{7.7}$$

where $A \in \mathbb{R}^{3\times3}$ is a known, constant, upper triangular, and invertible intrinsic camera calibration matrix [150]. From (7.2), (7.4), (7.5) and (7.7), the relationship between image coordinates of the corresponding feature points in $\mathcal{F}$ and $\mathcal{F}^*$ can be expressed as follows:

$$p_i = \underbrace{\frac{z_i^*}{z_i}}_{\alpha_i} \underbrace{A\left(\bar{R} + \bar{x}_{hi}(n^*)^T\right)A^{-1}}_{G} p_i^*, \tag{7.8}$$

where $\alpha_i \in \mathbb{R}$ denotes the depth/range ratio, and $\bar{x}_{hi}(t) = \dfrac{\bar{x}_f(t)}{d_i^*} \in \mathbb{R}^3$ denotes the scaled translation vector. The matrix $G(t) \in \mathbb{R}^{3\times3}$ defined in (7.8) is a full rank homogeneous collineation matrix defined upto a scale factor [150]. If the structure of the moving object is planar, all feature points lie on the same plane, and hence, the distances d_i^* defined in (7.4) is the same for all feature points, henceforth denoted as d^*. It is assumed that the constant rotation matrix R^* is known[1]. Given four corresponding feature points that are coplanar but non-collinear (if the structure of the object is not planar, the virtual parallax method described in [150] could be utilized), $R(t)$, $\bar{R}(t)$, and $\alpha_i(t)$ can be calculated based on (7.8) (see [29] and [31] for details).

7.2.2 Object Kinematics

To quantify the translation of $\mathcal{F}$ relative to the fixed coordinate system $\mathcal{F}^*$, we define $e_v(t) \in \mathbb{R}^3$ in terms of the image coordinates and the depth/range ratio of the feature point O_1 as follows:

$$e_v \triangleq \begin{bmatrix} u_1 - u_1^* & v_1 - v_1^* & -\ln(\alpha_1) \end{bmatrix}^T, \tag{7.9}$$

where $\ln(\cdot) \in \mathbb{R}$ denotes the natural logarithm. In (7.9) and in the subsequent development, any point O_i on π could have been utilized; however, to reduce the notational complexity, we have elected to select the feature point O_1. Following the development in [29], the translational kinematics can be obtained as follows:

$$\dot{e}_v = \frac{\alpha_1}{z_1^*} A_{e1} R \left[v_e - [s_1]_\times \omega_e \right], \tag{7.10}$$

where the notation $[s_1]_\times$ denotes the 3×3 skew symmetric form of s_1, $v_e(t)$, $\omega_e(t) \in \mathbb{R}^3$ denote (unknown coordinates), the unknown linear and angular velocity of the object expressed in the local coordinate frame $\mathcal{F}$, respectively, and $A_{ei}(t) \in \mathbb{R}^{3 \times 3}$ is a known, invertible matrix [30]. Similarly, to quantify the rotation of $\mathcal{F}$ relative to $\mathcal{F}^*$, we define $e_\omega(t) \in \mathbb{R}^3$ using the axis–angle representation [197] as Follows:

$$e_\omega \triangleq u\phi, \tag{7.11}$$

where $u(t) \in \mathbb{R}^3$ represents a unit rotation axis, and $\phi(t) \in \mathbb{R}$ denotes the rotation angle about $u(t)$ that is assumed to be confined to the region $-\pi < \phi(t) < \pi$. After taking the time derivative of (7.11), the following expression can be obtained (see [29] for further details):

$$\dot{e}_\omega = L_\omega R e_e, \tag{7.12}$$

In (7.12), the Jacobian-like matrix $L_\omega(t) \in \mathbb{R}^{3 \times 3}$ is measurable and only singular for multiples of 2π (i.e., out of the assumed workspace) [29]. From (7.10) and (7.12), the kinematics of the object can be expressed as:

$$\dot{e} = Jv, \tag{7.13}$$

---

[1]The constant rotation matrix R^* can be obtained *apriori* using various methods (e.g., a second camera, Euclidean measurement).

where $e(t) \triangleq \begin{bmatrix} e_v^T & e_\omega^T \end{bmatrix}^T \in \mathbb{R}^6$, $v(t) \triangleq \begin{bmatrix} v_e^T & \omega_e^T \end{bmatrix}^T \in \mathbb{R}^6$, and $J(t) \in \mathbb{R}^{6 \times 6}$ is a Jacobian-like matrix defined as:

$$J = \begin{bmatrix} \dfrac{\alpha_1}{z_1^*} A_{e1} R & -\dfrac{\alpha_1}{z_1^*} A_{e1} R [s_1]_\times \\ 0_3 & L_\omega R \end{bmatrix}, \tag{7.14}$$

where $0_3 \in \mathbb{R}^{3 \times 3}$ denotes a zero matrix.

7.2.3 Range Kinematic Model

To facilitate the development of the range estimator, we first define the extended image coordinates, denoted by $p_{ei}(t) \triangleq \begin{bmatrix} u_i & v_i & -\ln(\alpha_i) \end{bmatrix}^T \in \mathbb{R}^3$, for any feature point O_i. Following the development of (7.10) in [39], it can be shown that the time derivative of p_{ei} is given by

$$\dot{p}_{ei} = \frac{\alpha_i}{z_i^*} A_{ei} R \begin{bmatrix} v_e + [\omega_e]_\times s_i \end{bmatrix} = W_i V_{vw} \theta_i, \tag{7.15}$$

where $W_i(.) \in \mathbb{R}^{3 \times 3}, V_{vw}(t) \in \mathbb{R}^{3 \times 4}$ and $\theta_i \in \mathbb{R}^4$ are defined as follows

$$W_i \triangleq \alpha_i A_{ei} R \qquad V_{vw} \triangleq \begin{bmatrix} v_e & [\omega_e]_\times \end{bmatrix}, \tag{7.16}$$

$$\theta_i \triangleq \begin{bmatrix} \dfrac{1}{z_i^*} & \dfrac{s_i^T}{z_i^*} \end{bmatrix}^T. \tag{7.17}$$

The elements of $W_i(.)$ are known and bounded, and an estimate of $V_{vw}(t)$, denoted by $\hat{V}_{vw}(t)$, is available by appropriately re-ordering the vector $\hat{v}(t)$ given in (7.22).

Our objective is to identify the unknown constant θ_i defined in (7.17). To facilitate this objective, we define a parameter estimation error signal, denoted by $\tilde{\theta}_i(t) \in \mathbb{R}^4$, as follows:

$$\tilde{\theta}_i(t) \triangleq \theta_i - \hat{\theta}_i(t), \tag{7.18}$$

where $\hat{\theta}_i(t) \in \mathbb{R}^4$ is the parameter estimator.

Based on (7.15), the multiplication term before $\theta_i(t)$ including measurable variable $W_i(t)$ and unmeasurable variable $V_{vw}(t)$. The subsequent least-squares update law design for $\hat{\theta}_i(t)$ usually requires the multiplication term before θ_i be measurable. To deal with this issue, we design a measurable filter $W_{fi}(t) \in \mathbb{R}^{3 \times 4}$ to get a filtered signal of $W_i \hat{V}_{vw}$ and a non-measurable filter signal $\eta_i(t) \in \mathbb{R}^3$ to get a filtered signal of $W_i \tilde{V}_{vw} \theta_i$ as follows:

$$\dot{W}_{fi} = -\beta_i W_{fi} + W_i \hat{V}_{vw} \tag{7.19}$$

$$\dot{\eta}_i = -\beta_i \eta_i + W_i \tilde{V}_{vw} \theta_i, \tag{7.20}$$

where $\beta_i \in \mathbb{R}$ is a scalar positive gain, and $\tilde{V}_{vw}(t) \triangleq V_{vw}(t) - \hat{V}_{vw}(t) \in \mathbb{R}^{3 \times 4}$ is an estimation error signal. $W_{fi}(t)$ and $\eta_i(t)$ can be interpreted as intermediate mathematical variables utilized to facilitate subsequent Lyapunov based estimator design.

Remark 7.1 In the analysis provided in [39], it was shown that a filter signal $r(t) \in \mathbb{R}^6$ defined as $r(t) = \tilde{e}(t) + \dot{\tilde{e}}(t) \in \mathcal{L}_\infty \cap \mathcal{L}_2$. From this result, it is easy to show that the signals $\tilde{e}(t), \dot{\tilde{e}}(t) \in \mathcal{L}_2$ [53]. Since $J(t) \in \mathcal{L}_\infty$ and invertible, it follows that $J^{-1}(t) \dot{\tilde{e}}(t) \in \mathcal{L}_2$. Hence $\tilde{v}(t) \triangleq v(t) - \hat{v}(t) \in \mathcal{L}_2$, and it is easy to show that $\left\| \tilde{V}_{vw}(t) \right\|_\infty^2 \in \mathcal{L}_1$, where the notation $\|.\|_\infty$ denotes the induced ∞-norm of a matrix [115].

7.3   Motion Estimation

7.3.1   *Velocity Identification of the Object*

Remark 7.2 In the subsequent analysis, it is assumed that a single vector $s_1 \in \mathbb{R}^3$ between two feature points is known. With this assumption, each element of $J(t)$ is known with the possible exception of the constant $z_1^* \in \mathbb{R}$. The reader is referred to [39] where it is shown that z_1^* can also be computed given s_1.

In [39], an estimator was developed for online asymptotic identification of the signal $\dot{e}(t)$. Designating $\hat{e}(t)$ as the estimate for $e(t)$, the estimator was designed as follows:

$$\hat{e} \triangleq \int_{t_0}^t (K + I_6)\tilde{e}(\tau)d\tau + \int_{t_0}^t \rho \, \text{sgn}(\tilde{e}(\tau)) \, d\tau + (K + I_6)\tilde{e}(t), \qquad (7.21)$$

where $\tilde{e}(t) \triangleq e(t) - \hat{e}(t) \in \mathbb{R}^6$ is the estimation error for the signal $e(t)$, and $K, \rho \in \mathbb{R}^{6 \times 6}$, are positive definite constant diagonal gain matrices, $I_6 \in \mathbb{R}^{6 \times 6}$ is the 6×6 identity matrix, t_0 is the initial time, and $\text{sgn}(\tilde{e}(t))$ denotes the standard signum function applied to each element of the vector $\tilde{e}(t)$. The reader is referred to [39], and the references therein for analysis pertaining to the development of the above estimator. In essence, it was shown in [39] that $\hat{e}(t) \rightarrow \dot{e}(t)$ as $t \rightarrow \infty$. Since $J(t)$ is known and invertible, the six degree of freedom velocity of the moving object can be identified as follows:

$$\hat{v}(t) = J^{-1}(t)\hat{e}(t), \text{ and hence } \hat{v}(t) \rightarrow v(t) \text{ as } t \rightarrow \infty. \qquad (7.22)$$

7.3.2   *Range Identification of Feature Points*

Motivated by the Lyapunov based convergence and stability analysis, we design the following estimator:

$$\dot{\hat{p}}_{ei} = \beta_i \tilde{p}_{ei} + W_{fi}\dot{\hat{\theta}}_i + W_i\hat{V}_{vw}\hat{\theta}_i, \tag{7.23}$$

where $\hat{p}_{ei}(t) \in \mathbb{R}^3$ and $\tilde{p}_{ei}(t) \triangleq p_{ei}(t) - \hat{p}_{ei}(t) \in \mathbb{R}^3$ denotes the measurable estimation error signal for the extended image coordinates of the feature points. From (7.20) and (7.23), it can be shown that

$$\tilde{p}_{ei} = W_{fi}\tilde{\theta}_i + \eta_i + \mu e^{-\beta_i t}, \tag{7.24}$$

where $\mu \triangleq \tilde{p}_{ei}(0) - \eta_i(0) - W_{fi}(0)\tilde{\theta}_i(0) \in \mathbb{R}$ is an unknown constant based on the initial conditions. Motivated by the Lyapunov based convergence and stability analysis, we select the following least-squares update law for $\hat{\theta}_i(t)$:

$$\dot{\hat{\theta}}_i = L_i W_{fi}^T \tilde{p}_{ei}, \tag{7.25}$$

where $L_i(t) \in \mathbb{R}^{4 \times 4}$ is an estimation gain that is recursively computed as follows:

$$\frac{d}{dt}(L_i^{-1}) = W_{fi}^T W_{fi}, \tag{7.26}$$

In the subsequent analysis, it is required that $L_i^{-1}(0)$ in (7.26) be positive definite. This requirement can be satisfied by selecting the appropriate nonzero initial values.

Theorem 7.1
The update law defined in (7.25) ensures that $\tilde{\theta}_i(t) \to 0$ as $t \to \infty$ provided that the following persistent excitation condition [192] holds (see Remark 5 in [26] for more details on the condition):

$$\gamma_1 I_4 \leq \int_{t_0}^{t_0+T} W_{fi}^T(\tau)W_{fi}(\tau)d\tau \leq \gamma_2 I_4 \tag{7.27}$$

and provided that the gains β_i in (7.19) and (7.20) satisfy the following inequality:

$$\beta_i > k_{1i} + k_{2i}\|W_i\|_\infty^2 \qquad k_{1i} > 4, \tag{7.28}$$

where $t_0, \gamma_1, \gamma_2, T, k_{1i}, k_{2i} \in \mathbb{R}$ are positive constants.

Proof: See Appendix A.1 for the proof of Theorem 7.1.

The range estimation $\hat{z}_i(\tau,t)$ for $z_i(\tau)$ can be calculated based on $\alpha_i(\tau)$ in (7.8) and $\hat{\theta}_i(t)$ in (7.25) as follows

$$\hat{z}_i(\tau,t) = \frac{1}{\alpha_i(\tau)\left[\hat{\theta}_i(t)\right]_1}, \tag{7.29}$$

where the bracket term in the denominator denotes the first element of the vector $\hat{\theta}_i(t)$ and τ denotes the time when the object arrives at the position and orientation where the range needs to be identified. Note that $\alpha_i(\tau)$ is calculated through the homography decomposition at the time τ. It is clear that $\hat{z}_i(\tau,t)$ can be utilized to estimate the range $z_i(\tau)$ and $\hat{z}_i(t,t)$ tracks the time-varying range signal $z_i(t)$ when $\hat{\theta}_i(t)$ converges to θ_i as $t \to \infty$.

After utilizing (7.5), (7.7), (7.17) and the definition of $\alpha_i(t)$ in (7.8), the estimates for the 3D Euclidean coordinates of the feature points on the object relative to the camera frame, denoted by $\hat{m}_i(\tau,t) \in \mathbb{R}^3$, can be calculated as follows:

$$\hat{m}_i(\tau,t) = \frac{1}{\alpha_i(\tau)\left[\hat{\theta}_i(t)\right]_1} A^{-1} p_i(\tau). \tag{7.30}$$

7.4 Simulation Results

For simulations, the image acquisition hardware and the image processing step were both replaced with a software component that generated feature trajectories utilizing object kinematics. We selected a planar object with four feature points initially 2 m away along the optical axis of the camera as the body undergoing

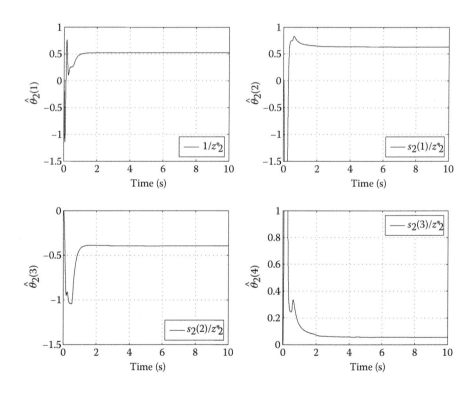

Figure 7.2: The estimated parameters for θ_2.

motion. The velocity of the object along each of the six degrees of freedom were set to $0.2\sin(t)$. The coordinates of the object feature points in the object's coordinate frame $\mathcal{F}$ were arbitrarily chosen to be the following:

$$s_1 = \begin{bmatrix} 1.0 & 0.5 & 0.1 \end{bmatrix}^T \qquad s_2 = \begin{bmatrix} 1.2 & -0.75 & 0.1 \end{bmatrix}^T$$

$$s_3 = \begin{bmatrix} 0.0 & -1.0 & 0.1 \end{bmatrix}^T \qquad s_4 = \begin{bmatrix} -1.0 & 0.5 & 0.1 \end{bmatrix}^T.$$

The object's reference orientation R^* relative to the camera was selected as $diag(1,-1,-1)$, where $diag(.)$ denotes a diagonal matrix with arguments as the diagonal entries. The estimator gain β_i was set to 20 for all i feature points. As an example, Figure 7.2 depicts the convergence of $\hat{\theta}_2(t)$ from which the Euclidean coordinates of the second feature point s_2 could be computed as shown in (7.17). Similar graphs were obtained for convergence of the estimates for the remaining feature points.

7.5 Conclusions

This chapter presents an adaptive nonlinear estimator to identify the range and the 3D Euclidean coordinates of feature points on an object under motion using a single camera. The estimation error is proved to converge to zero provided that a persistent excitation condition holds. Lyapunov-based system analysis methods and homography-based vision techniques are used in the development and analysis of the identification algorithm. Simulation results demonstrate the performance of the estimator.

Chapter 8

Motion Estimation of Moving Objects

8.1 Introduction

This chapter proposes an approach to asymptotically estimate the velocity and range of feature points on a moving object using a static-moving camera system. The normalized Euclidean coordinates of feature points and the homography-based techniques are utilized to construct two auxiliary state vectors, which are expressed in terms of the static camera and the moving camera, respectively. The velocity and range are estimated in a sequential way based on these two measurable auxiliary state vectors. First, a nonlinear observer is designed to estimate the up-to-a-scale velocity of feature points. Then, the range of feature points can be identified by an adaptive estimator. Thanks to the introduction of the static-moving camera system, motion constraint and *a priori* geometric knowledge are not required for the moving object. The asymptotic convergence of estimation errors is proved with Lyapunov-based methods.

8.2 System Modeling

8.2.1 Geometric Model

As illustrated in Figure 8.1, $\mathcal{I}$ denotes the inertial coordinate frame whose origin coincides with the optical center of the static camera; $\mathcal{F}$ denotes the moving camera coordinate frame; the orthogonal coordinate frame attached to the moving object is denoted by $\mathcal{F}_l$. To incorporate the homography-based techniques, a

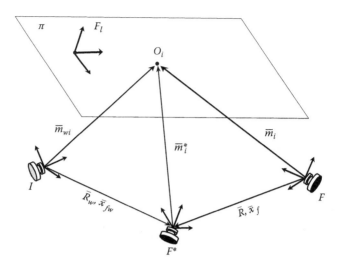

Figure 8.1: Coordinate frames of the static-moving camera system.

reference image captured by the moving camera is introduced to express a fixed pose $\mathcal{F}^*$ relative to the object. It should be noted that an arbitrary moving camera image containing the object can be selected as the reference image, which is easy to implement. To relate the coordinate frames, let $\bar{R}_w(t) \in SO(3)$ and $\bar{x}_{fw}(t) \in \mathbb{R}^3$ denote the rotation and translation, respectively, between $\mathcal{F}^*$ and $\mathcal{I}$. $\bar{R}(t) \in SO(3)$ and $\bar{x}_f(t) \in \mathbb{R}^3$ denote the rotation and translation between $\mathcal{F}^*$ and $\mathcal{F}$, respectively.

The 3-D coordinates of the ith feature point are denoted as $\bar{m}_{wi}(t) \triangleq \left[x_{wi}(t) \;\; y_{wi}(t) \;\; z_{wi}(t) \right]^T$, $\bar{m}_i(t) \triangleq \left[x_i(t) \;\; y_i(t) \;\; z_i(t) \right]^T$, $\bar{m}_i^* \triangleq \left[x_i^* \;\; y_i^* \;\; z_i^* \right]^T \in \mathbb{R}^3$ expressed in terms of $\mathcal{I}$, $\mathcal{F}$, and $\mathcal{F}^*$, respectively. Meanwhile, the normalized Euclidean coordinates of the feature points can be expressed in $\mathcal{I}$, $\mathcal{F}$, and $\mathcal{F}^*$, respectively, as follows:

$$m_{wi} \triangleq \frac{\bar{m}_{wi}}{z_{wi}} = \begin{bmatrix} \frac{x_{wi}}{z_{wi}} & \frac{y_{wi}}{z_{wi}} & 1 \end{bmatrix}^T$$

$$m_i \triangleq \frac{\bar{m}_i}{z_i} = \begin{bmatrix} \frac{x_i}{z_i} & \frac{y_i}{z_i} & 1 \end{bmatrix}^T \tag{8.1}$$

$$m_i^* \triangleq \frac{\bar{m}_i^*}{z_i^*} = \begin{bmatrix} \frac{x_i^*}{z_i^*} & \frac{y_i^*}{z_i^*} & 1 \end{bmatrix}^T.$$

Each feature point will have a image coordinate expressed in terms of $\mathcal{I}$, $\mathcal{F}$, and $\mathcal{F}^*$, denoted by $p_{wi}(t)$, $p_i(t)$, $p_i^* \in \mathbb{R}^3$, respectively. The image coordinates of the feature points are related to $m_{wi}(t)$, $m_i(t)$, and m_i^* by the following pinhole lens models [66]

$$p_{wi} = A_w m_{wi} \qquad p_i = A m_i \qquad p_i^* = A m_i^*, \qquad (8.2)$$

where $A_w, A \in \mathbb{R}^{3 \times 3}$ are two known, constant, and invertible intrinsic camera calibration matrices. Based on homography techniques, the relationships between image coordinates of the corresponding feature points in $\mathcal{I}$ and $\mathcal{F}^*$, as well as $\mathcal{F}$ and $\mathcal{F}^*$, can be developed [86,150].

$$p_{wi} = \alpha_{wi} A_w H_w A^{-1} p_i^* \qquad p_i = \alpha_i A H A^{-1} p_i^*, \qquad (8.3)$$

where $\alpha_{wi}(t) \triangleq \frac{z_i^*}{z_{wi}}$, $\alpha_i(t) \triangleq \frac{z_i^*}{z_i} \in \mathbb{R}$ denote invertible depth ratios, $H_w(t), H(t) \in \mathbb{R}^{3 \times 3}$ denote homography matrices. If the structure of the moving object is planar, four corresponding feature points that are coplanar but non-collinear can be extracted to calculate $\alpha_{wi}(t)$, $\alpha_i(t)$, $H_w(t)$, and $H(t)$ based on (8.3). Then, $\bar{R}_w(t)$ and $\bar{R}(t)$ can be decomposed from $H_w(t)$ and $H(t)$, respectively [86]. For the non-coplanar case, the virtual parallax method described in [150] can be utilized.

Assumption 8.1 *The feature points remain in the field of view of the static camera and the moving camera. In addition, $\bar{m}_{wi}(t)$, $\bar{m}_i(t)$, and $\bar{m}_i^*$ are bounded, i.e., $\bar{m}_{wi}(t)$, $\bar{m}_i(t)$, $\bar{m}_i^* \in \mathcal{L}_\infty$ and $z_{wi}(t)$, $z_i(t)$, $z_i^* > 0$.*

Remark 8.1 Since the feature points are in the field of view of both cameras and the image coordinates are constrained, $p_{wi}(t)$, $p_i(t)$ are bounded, i.e., $p_{wi}(t)$, $p_i(t) \in \mathcal{L}_\infty$. Based on (8.1), (8.2) and Assumption 8.1 that $z_{wi}(t)$, $z_i(t)$ are bounded, it can be concluded that $\bar{m}_{wi}(t)$, $\bar{m}_i(t) \in \mathcal{L}_\infty$. Since the reference image is captured by the moving camera and $z_{wi}(t)$, $z_i(t) > 0$, it can be inferred that $\bar{m}_i^*$ is bounded and $\alpha_{wi}(t)$, $\alpha_i(t) > 0$. Based on (8.2) and the fact that A_w, A are invertible, $m_{wi}(t)$, $m_i(t) \in \mathcal{L}_\infty$ can be obtained. Furthermore, from (8.1) and Assumption 8.1 that $z_{wi}(t)$, $z_i(t)$ are bounded, it can be concluded that $\bar{m}_{wi}(t)$, $\bar{m}_i(t) \in \mathcal{L}_\infty$. Since the reference image is captured by the moving camera and $z_{wi}(t)$, $z_i(t) > 0$, it can be inferred that $\bar{m}_i^*$ is bounded and $\alpha_{wi}(t)$, $\alpha_i(t) > 0$. It can be concluded from (8.1) and Assumption 8.1 that $\bar{m}_{wi}(t)$, $\bar{m}_i(t)$ are bounded and the feature points remain in the field of view of the static and moving cameras (see Remark 1 in [48] for further details). Since the reference image is captured by the moving camera, it can be inferred that $\bar{m}_i^*$ is bounded and $\alpha_{wi}(t)$, $\alpha_i(t) > 0$.

8.2.2 Camera Motion and State Dynamics

To facilitate the subsequent development, two measurable auxiliary state vectors $\theta_{wi}(t)$, $\theta_i(t) \in \mathbb{R}^3$ are constructed as follows:

$$\begin{aligned}
\theta_{wi} &\triangleq \begin{bmatrix} \theta_{wi,1} & \theta_{wi,2} & \theta_{wi,3} \end{bmatrix}^T = \begin{bmatrix} \frac{x_{wi}}{z_{wi}} & \frac{y_{wi}}{z_{wi}} & -\ln(\alpha_{wi}) \end{bmatrix}^T \\
\theta_i &\triangleq \begin{bmatrix} \theta_{i,1} & \theta_{i,2} & \theta_{i,3} \end{bmatrix}^T = \begin{bmatrix} \frac{x_i}{z_i} & \frac{y_i}{z_i} & -\ln(\alpha_i) \end{bmatrix}^T,
\end{aligned} \qquad (8.4)$$

where $\ln(\cdot) \in \mathbb{R}$ denotes the natural logarithm. It is clear that $\theta_{wi}(t)$ and $\theta_i(t)$ can be obtained based on (8.2) and the homography techniques. After taking the time derivative of $\bar{m}_{wi}(t)$ and $\bar{m}_i(t)$, the motion dynamics of the moving object with respect to these two cameras can be expressed as: [24,38]

$$\dot{\bar{m}}_{wi} = v_{wi} \qquad \dot{\bar{m}}_i = (v_i - v) - [\omega]_\times \bar{m}_i, \tag{8.5}$$

where $v_{wi}(t) \in \mathbb{R}^3$ denotes the unknown linear velocity of the ith feature point expressed in terms of $\mathcal{I}$. $v(t)$, $\omega(t) \in \mathbb{R}^3$ denote the linear and angular velocities of the moving camera, respectively. $v_i(t) \in \mathbb{R}^3$ is the unknown linear velocity of the ith feature point. Specifically, $v_i(t)$, $v(t)$, and $\omega(t)$ are all expressed in terms of $\mathcal{F}$. For any vector $x \in \mathbb{R}^3$, $[x]_\times$ denotes the 3×3 skew-symmetric expansion of x [86]. Based on the geometry relationships described by Figure 8.1, the following expression can be obtained:

$$v_i = \bar{R}\bar{R}_w^T v_{wi}, \tag{8.6}$$

Taking the time derivative of (8.4) and substituting from (8.5), the dynamics of the auxiliary state vectors can be determined by

$$\dot{\theta}_{wi} = \alpha_{wi} J_{wi} \zeta_{wi}$$
$$\dot{\theta}_i = \alpha_i J_i \zeta_i - \alpha_i J_i v \eta_i + J_i(-[\omega]_\times m_i), \tag{8.7}$$

where $J_{wi}(t)$, $J_i(t) \in \mathbb{R}^{3 \times 3}$ are defined as

$$J_{wi} = \begin{bmatrix} 1 & 0 & -\theta_{wi,1} \\ 0 & 1 & -\theta_{wi,2} \\ 0 & 0 & 1 \end{bmatrix} \qquad J_i = \begin{bmatrix} 1 & 0 & -\theta_{i,1} \\ 0 & 1 & -\theta_{i,2} \\ 0 & 0 & 1 \end{bmatrix}. \tag{8.8}$$

From (8.8), it can be inferred that $J_{wi}(t)$ and $J_i(t)$ are measurable and invertible. In (8.7), the up-to-a-scale velocities $\zeta_{wi}(t)$, $\zeta_i(t) \in \mathbb{R}^3$, and the inverse range $\eta_i \in \mathbb{R}$ are defined as follows:

$$\zeta_{wi} \triangleq \frac{v_{wi}}{z_i^*} \qquad \zeta_i \triangleq \frac{v_i}{z_i^*} \qquad \eta_i \triangleq \frac{1}{z_i^*}. \tag{8.9}$$

Now, our objective is to identify the unknown scaled velocities $\zeta_{wi}(t)$, $\zeta_i(t)$, and the unknown inverse range η_i by using the measurable signals $\alpha_{wi}(t)$, $\alpha_i(t)$, $\bar{R}_w(t)$, $\bar{R}(t)$, $\theta_{wi}(t)$, and $\theta_i(t)$.

Assumption 8.2 *The unknown linear velocity of the feature points $v_{wi}(t)$ is continuous differentiable and bounded up to its second-order time derivative.*

Assumption 8.3 *The velocities $v(t)$, $\omega(t)$ of the moving camera are assumed as available. Practically, these velocity signals can be measured by sensors such as an inertial measurement unit equipped with the moving camera.*

Remark 8.2   According to (8.4), (8.7), Assumptions 8.1, and 8.2, it is clear that $\theta_{wi}(t)$, $\dot{\theta}_{wi}(t) \in \mathcal{L}_\infty$. Differentiating the first equation of (8.7), $\ddot{\theta}_{wi}(t) \in \mathcal{L}_\infty$ can be obtained based on $\theta_{wi}(t)$, $\dot{\theta}_{wi}(t)$, $v_{wi}(t)$, $\dot{v}_{wi}(t) \in \mathcal{L}_\infty$. Furthermore, $\dddot{\theta}_{wi}(t) \in \mathcal{L}_\infty$ can be obtained by differentiating $\ddot{\theta}_{wi}(t)$ and utilizing $\theta_{wi}(t)$, $\dot{\theta}_{wi}(t)$, $\ddot{\theta}_{wi}(t)$, $v_{wi}(t)$, $\dot{v}_{wi}(t)$, $\ddot{v}_{wi}(t) \in \mathcal{L}_\infty$.

Remark 8.3   To aid the subsequent design and analysis for the observers, the vector functions $\mathrm{Sgn}(\cdot)$, $\mathrm{Tanh}(\cdot) \in \mathbb{R}^n$ and the matrix function $\mathrm{Cosh}(\cdot) \in \mathbb{R}^{n \times n}$ are defined as:

$$\mathrm{Sgn}(x) \triangleq \begin{bmatrix} \mathrm{sgn}(x_1) & \mathrm{sgn}(x_2) \cdots \mathrm{sgn}(x_n) \end{bmatrix}^T$$
$$\mathrm{Tanh}(x) \triangleq \begin{bmatrix} \tanh(x_1) & \tanh(x_2) \cdots \tanh(x_n) \end{bmatrix}^T$$
$$\mathrm{Cosh}(x) \triangleq \mathrm{diag}\begin{bmatrix} \cosh(x_1) & \cosh(x_2) \cdots \cosh(x_n) \end{bmatrix},$$

where $x = \begin{bmatrix} x_1 & x_2 \cdots x_n \end{bmatrix}^T \in \mathbb{R}^n$, and $\mathrm{sgn}(\cdot), \tanh(\cdot), \cosh(\cdot)$ are the standard signum, hyperbolic tangent, and hyperbolic cosine functions, respectively. Throughout this chapter, $\| \cdot \|$ denotes the Euclidean norm for vectors and the spectral norm for matrices [95].

8.3   Motion Estimation

8.3.1   *Scaled Velocity Identification*

In this section, a nonlinear observer is developed to identify the auxiliary state vector $\theta_{wi}(t)$ and its dynamics $\dot{\theta}_{wi}(t)$. Based on the estimated value of $\dot{\theta}_{wi}(t)$, the scaled velocity $\zeta_{wi}(t)$ can be determined by the first equation of (8.7). Then, another scaled velocity $\zeta_i(t)$ can be obtained by using (8.6) and (8.9) with the rotation matrices decomposed from the homography.

To quantify the observation error of $\theta_{wi}(t)$, $\tilde{\theta}_{wi}(t) \in \mathbb{R}^3$ is defined as follows:

$$\tilde{\theta}_{wi} \triangleq \theta_{wi} - \hat{\theta}_{wi}, \tag{8.10}$$

where $\hat{\theta}_{wi}(t) \in \mathbb{R}^3$ denotes the estimate for $\theta_{wi}(t)$. Motivated by the subsequent Lyapunov based convergence and stability analysis, a nonlinear observer for $\theta_{wi}(t)$ is designed as:

$$\dot{\hat{\theta}}_{wi} = \lambda_{w1} \mathrm{Tanh}\,(u_{wi}) + \lambda_{w2} \tilde{\theta}_{wi}, \tag{8.11}$$

where $u_{wi}(t) \in \mathbb{R}^3$ is a Filippov solution [69] to the following differential equation:

$$\dot{u}_{wi} = \mathrm{Cosh}^2\,(u_{wi}) \left(\beta_w \mathrm{Sgn}(\tilde{\theta}_{wi}) + \frac{\lambda_{w2}+1}{\lambda_{w1}} \mathrm{Tanh}(\tilde{\theta}_{wi}) + \frac{\lambda_{w2}+1}{\lambda_{w1}} \tilde{\theta}_{wi} \right). \tag{8.12}$$

In (8.11) and (8.12), λ_{w1}, λ_{w2}, $\beta_w \in \mathbb{R}$ are positive constant gains. Inspired by [39,57,217], the robust integral of the signum of the error term (i.e., $\text{Sgn}(\tilde{\theta}_{wi})$ in (8.12)) is introduced into the observer to facilitate the following analysis. Denote $\hat{\zeta}_{wi}(t)$, $\hat{\zeta}_i(t) \in \mathbb{R}^3$ as the estimates of the scaled velocities $\zeta_{wi}(t)$, $\zeta_i(t)$, respectively. Based on the first equation of (8.7) and (8.10), it is clear that if $\tilde{\theta}_{wi}(t)$, $\dot{\tilde{\theta}}_{wi}(t) \to 0$ as $t \to \infty$, then $\zeta_{wi}(t)$ can be identified by the following expression:

$$\hat{\zeta}_{wi} = \frac{1}{\alpha_{wi}} J_{wi}^{-1} \dot{\hat{\theta}}_{wi}. \tag{8.13}$$

In addition, from (8.6) and (8.9), it can be concluded that $\zeta_i = \bar{R}\bar{R}_w^T \zeta_{wi}$ indicating $\zeta_i(t)$ can be estimated by:

$$\hat{\zeta}_i = \bar{R}\bar{R}_w^T \hat{\zeta}_{wi} = \bar{R}\bar{R}_w^T \left(\frac{1}{\alpha_{wi}} J_{wi}^{-1} \dot{\hat{\theta}}_{wi} \right). \tag{8.14}$$

To facilitate the subsequent development, a filtered observation error, denoted by $r_{wi}(t) \in \mathbb{R}^3$, is introduced as follows:

$$r_{wi} = \dot{\tilde{\theta}}_{wi} + \text{Tanh}(\tilde{\theta}_{wi}) + \tilde{\theta}_{wi}. \tag{8.15}$$

Taking the time derivative of (8.11) and utilizing (8.12), (8.15) to yield

$$\ddot{\tilde{\theta}}_{wi} = \lambda_{w2} r_{wi} + \lambda_{w1} \beta_w \text{Sgn}(\tilde{\theta}_{wi}) + \text{Tanh}(\tilde{\theta}_{wi}) + \tilde{\theta}_{wi}. \tag{8.16}$$

Based on (8.10) and (8.16), the time derivative of (8.15) can be obtained as follows:

$$\dot{r}_{wi} = \left(-\lambda_{w2} r_{wi} - \text{Tanh}(\tilde{\theta}_{wi}) - \tilde{\theta}_{wi} \right) + \tilde{N}_w + \left(N_w - \lambda_{w1} \beta_w \text{Sgn}(\tilde{\theta}_{wi}) \right), \tag{8.17}$$

where $\tilde{N}_w(t)$, $N_w(t) \in \mathbb{R}^3$ are two auxiliary variables defined as:

$$\tilde{N}_w \triangleq \left(\text{Cosh}^{-2}(\tilde{\theta}_{wi}) + I_3 \right) \dot{\tilde{\theta}}_{wi} \qquad N_w \triangleq \ddot{\tilde{\theta}}_{wi}, \tag{8.18}$$

where $I_3 \in \mathbb{R}^{3\times3}$ denotes the 3×3 identity matrix. From Remark 8.1 and (8.18), it can be concluded that $N_w(t)$, $\dot{N}_w(t) \in \mathcal{L}_\infty$. Based on (8.15), (8.18) and the fact that $\|\text{Cosh}^{-2}(\tilde{\theta}_{wi})\| \leq 1$, $\tilde{N}_w(t)$ can be upper bounded as:

$$\|\tilde{N}_w\| \leq 2 \|r_{wi} - \text{Tanh}(\tilde{\theta}_{wi}) - \tilde{\theta}_{wi}\| \leq 6 \|Z_{wi}\|, \tag{8.19}$$

where $Z_{wi}(t) \in \mathbb{R}^9$ is a composited error vector defined as:

$$Z_{wi} \triangleq \begin{bmatrix} \tilde{\theta}_{wi}^T & \text{Tanh}^T(\tilde{\theta}_{wi}) & r_{wi}^T \end{bmatrix}^T. \tag{8.20}$$

The structure of $r_{wi}(t)$, $\dot{r}_{wi}(t)$ given in (8.15), (8.17) is motivated by the need to inject and cancel terms in the following convergence analysis and will become apparent later.

Theorem 8.1
The observer defined in (8.11) and (8.12) can ensure that the observation error $\tilde{\theta}_{wi}(t)$ and its corresponding dynamics $\dot{\tilde{\theta}}_{wi}(t)$ asymptotically converge to zero in the sense that

$$\tilde{\theta}_{wi}(t), \dot{\tilde{\theta}}_{wi}(t) \to 0 \quad as \quad t \to \infty \tag{8.21}$$

provided that the following conditions hold:

$$\lambda_{w1}\beta_w \geq \|N_w\|_{\mathcal{L}_\infty} + \|\dot{N}_w\|_{\mathcal{L}_\infty} \tag{8.22}$$

$$\lambda_{w2} > 10, \tag{8.23}$$

where $\|\cdot\|_{\mathcal{L}_\infty}$ is the $\mathcal{L}_\infty$ norm [115]. Furthermore, given the result in (8.21), the scaled velocities $\zeta_{wi}(t)$ and $\zeta_i(t)$ can be determined by (8.13) and (8.14).

Proof 8.1   To prove Theorem 8.1, a non-negative function $V_1(t) \in \mathbb{R}$ is defined as follows:

$$V_1 \triangleq \frac{1}{2}\tilde{\theta}_{wi}^T\tilde{\theta}_{wi} + \sum_{j=1}^{3} \ln\left(\cosh(\tilde{\theta}_{wi,j})\right) + \frac{1}{2}r_{wi}^T r_{wi} + \Delta_w, \tag{8.24}$$

where $\Delta_w(t) \in \mathbb{R}$ is an auxiliary function given by:

$$\Delta_w \triangleq \xi_w - \int_0^t \Lambda_w(\tau)d\tau. \tag{8.25}$$

In (8.25), $\xi_w, \Lambda_w(t) \in \mathbb{R}$ are auxiliary terms defined as follows:

$$\begin{aligned}
\xi_w &\triangleq -\tilde{\theta}_{wi}^T(0)N_w(0) + \lambda_{w1}\beta_w\tilde{\theta}_{wi}^T(0)\text{Sgn}\left(\tilde{\theta}_{wi}(0)\right) \\
\Lambda_w &\triangleq r_{wi}^T\left(N_w - \lambda_{w1}\beta_w\text{Sgn}(\tilde{\theta}_{wi})\right).
\end{aligned} \tag{8.26}$$

After substituting (8.15) into the second equation of (8.26) and integrating in time, the following expression can be obtained:

$$\begin{aligned}
\int_0^t \Lambda_w(\tau)d\tau &= \int_0^t \left(\dot{\tilde{\theta}}_{wi} + \text{Tanh}(\tilde{\theta}_{wi}) + \tilde{\theta}_{wi}\right)^T \\
&\quad \times \left(N_w - \lambda_{w1}\beta_w\text{Sgn}(\tilde{\theta}_{wi})\right)d\tau \\
&= \tilde{\theta}_{wi}^T N_w\big|_0^t - \lambda_{w1}\beta_w\tilde{\theta}_{wi}^T\text{Sgn}(\tilde{\theta}_{wi})\big|_0^t \\
&\quad + \int_0^t \text{Tanh}^T(\tilde{\theta}_{wi})\left(N_w - \lambda_{w1}\beta_w\text{Sgn}(\tilde{\theta}_{wi})\right)d\tau \\
&\quad + \int_0^t \tilde{\theta}_{wi}^T\left(N_w - \dot{N}_w - \lambda_{w1}\beta_w\text{Sgn}(\tilde{\theta}_{wi})\right)d\tau.
\end{aligned} \tag{8.27}$$

Since $\text{Tanh}^T(\tilde{\theta}_{wi})\text{Sgn}(\tilde{\theta}_{wi}) \geq \|\text{Tanh}(\tilde{\theta}_{wi})\|$, $\tilde{\theta}_{wi}^T\text{Sgn}(\tilde{\theta}_{wi}) \geq \|\tilde{\theta}_{wi}\|$, (8.27) can be upper bounded as follows:

$$\int_0^t \Lambda_w(\tau)d\tau \leq \|\theta_{wi}(t)\|\left(\|N_w(t)\| - \lambda_{w1}\beta_w\right) + \xi_w$$

$$+ \int_0^t \|\text{Tanh}(\tilde{\theta}_{wi})\|\left(\|N_w\| - \lambda_{w1}\beta_w\right)d\tau \tag{8.28}$$

$$+ \int_0^t \|\tilde{\theta}_{wi}\|\left(\|N_w\| + \|\dot{N}_w\| - \lambda_{w1}\beta_w\right)d\tau,$$

where ξ_w is defined in (8.26). Based on (8.25) and (8.28), it can be inferred that $\Lambda_w(t) \geq 0$ on condition that (8.22) holds.

Taking the time derivative of (8.24) and substituting from (8.15), (8.17), (8.25), and (8.26), the following expression can be obtained:

$$\dot{V}_1 = -\tilde{\theta}_{wi}^T\tilde{\theta}_{wi} - 2\text{Tanh}^T(\tilde{\theta}_{wi})\tilde{\theta}_{wi} - \text{Tanh}^T(\tilde{\theta}_{wi})\text{Tanh}(\tilde{\theta}_{wi})$$
$$- \lambda_{w2}r_{wi}^T r_{wi} + r_{wi}^T\tilde{N}_w. \tag{8.29}$$

Based on (8.19) and the fact that $0 \leq \text{Tanh}^T(\tilde{\theta}_{wi})\text{Tanh}(\tilde{\theta}_{wi}) \leq \text{Tanh}^T(\tilde{\theta}_{wi})\tilde{\theta}_{wi}$, (15.19) can be upper bounded as:

$$\dot{V}_1 \leq -\left(1 - \frac{9}{\lambda_{w2} - 1}\right)\|Z_{wi}\|^2, \tag{8.30}$$

where $Z_{wi}(t)$ is defined in (8.20). It is not difficult to conclude that if λ_{w2} is selected to satisfy the sufficient condition given in (8.23), then $1 - \frac{9}{\lambda_{w2}-1} > 0$ is guaranteed.

Using signal chasing, it can be determined that $\tilde{\theta}_{wi}(t)$, $r_{wi}(t)$, $\dot{\tilde{\theta}}_{wi}(t) \in \mathcal{L}_\infty \cap \mathcal{L}_2$ and $\dot{r}_{wi}(t) \in \mathcal{L}_\infty$. Furthermore, Barbalat's lemma [192] can be employed to conclude that $\tilde{\theta}_{wi}(t)$, $r_{wi}(t) \to 0$ as $t \to \infty$ on condition that λ_{w1}, β_w, and λ_{w2} are chosen appropriately to ensure (8.22) and (8.23) hold. Given (8.15), linear analysis techniques [53] can be used to determine that $\dot{\tilde{\theta}}_{wi}(t) \to 0$ as $t \to \infty$.

8.3.2 Range Identification

In this section, the auxiliary state vector $\theta_i(t)$ constructed in (8.4) and the scaled velocity estimate $\hat{\zeta}_i(t)$ are utilized to identify the range information of the object. Before presenting the estimator for η_i, it is first shown that $\theta_i(t)$ and $\dot{\theta}_i(t)$ can be identified, using same method as shown in Section 8.3.1, by the following expression:

$$\dot{\hat{\theta}}_i = \lambda_1\text{Tanh}(u_i) + \lambda_2\tilde{\theta}_i, \tag{8.31}$$

where $\lambda_1, \lambda_2 \in \mathbb{R}$ are positive constant gains, $\hat{\theta}_i(t) \in \mathbb{R}^3$ is the estimate for $\theta_i(t)$, $\tilde{\theta}_i(t) \triangleq \theta_i - \hat{\theta}_i \in \mathbb{R}^3$ is the observation error, and the dynamics of $u_i(t) \in \mathbb{R}^3$ is given by:

$$\dot{u}_i = \mathrm{Cosh}^2(u_i)\left(\beta\,\mathrm{Sgn}(\tilde{\theta}_i) + \frac{\lambda_2+1}{\lambda_1}\mathrm{Tanh}(\tilde{\theta}_i) + \frac{\lambda_2+1}{\lambda_1}\dot{\tilde{\theta}}_i\right), \tag{8.32}$$

where $\beta \in \mathbb{R}$ is a positive constant gain. Denote $N(t) \triangleq \ddot{\theta}_i \in \mathbb{R}^3$ as an auxiliary variable. Similar to Theorem 1, it can be concluded that if λ_1, β, and λ_2 are chosen to satisfy $\lambda_1\beta \geq \|N\|_{\mathcal{L}_\infty} + \|\dot{N}\|_{\mathcal{L}_\infty}$ and $\lambda_2 > 10$, then the estimator defined in (8.31) and (8.32) can ensure that $\tilde{\theta}_i(t), \dot{\tilde{\theta}}_i(t) \to 0$ as $t \to \infty$.

Based on the second equation of (8.7) and the estimation signals $\hat{\theta}_i(t), \hat{\zeta}_i(t)$, the adaptive estimator for η_i, denoted by $\hat{\eta}_i(t) \in \mathbb{R}$, is designed as follows:

$$\dot{\hat{\eta}}_i = \kappa\frac{v^T}{\alpha_i}J_i^{-1}\left(\alpha_i J_i\hat{\zeta}_i - \alpha_i J_i v\hat{\eta}_i + J_i(-[\omega]_\times m_i) - \dot{\hat{\theta}}_i\right), \tag{8.33}$$

where $\kappa \in \mathbb{R}$ is a positive constant gain. To facilitate the subsequent analysis, two auxiliary observation errors $\tilde{\zeta}_i(t) \in \mathbb{R}^3$, $\tilde{\eta}_i(t) \in \mathbb{R}$ are introduced:

$$\tilde{\zeta}_i \triangleq \zeta_i - \hat{\zeta}_i \qquad \tilde{\eta}_i \triangleq \eta_i - \hat{\eta}_i. \tag{8.34}$$

Then, (8.33) can be rewritten as:

$$\dot{\hat{\eta}}_i = \kappa v^T\left(v\tilde{\eta}_i - \tilde{\zeta}_i + \frac{1}{\alpha_i}J_i^{-1}\dot{\tilde{\theta}}_i\right). \tag{8.35}$$

Remark 8.4   In the analysis provided in Theorem 8.1, it is shown that $\dot{\tilde{\theta}}_{wi}(t) \in \mathcal{L}_2$. Based on (8.7), (8.13), (8.14), and (8.34), it can be obtained that $\tilde{\zeta}_i(t) = \bar{R}\bar{R}_w^T\left(\frac{1}{\alpha_{wi}}J_{wi}^{-1}\dot{\tilde{\theta}}_{wi}\right)$. Since $\bar{R}(t), \bar{R}_w(t), \alpha_{wi}(t), J_{wi}(t) \in \mathcal{L}_\infty$ and $J_{wi}(t)$ is invertible, it follows that $\bar{R}\bar{R}_w^T\left(\frac{1}{\alpha_{wi}}J_{wi}^{-1}\dot{\tilde{\theta}}_{wi}\right) \in \mathcal{L}_2$. Hence, $\tilde{\eta}_i(t) \in \mathcal{L}_2$. In addition, by using the same analysis method as shown in Theorem 8.1, it is not difficult to prove that $\dot{\tilde{\theta}}_i(t) \in \mathcal{L}_2$.

Theorem 8.2
The estimator designed in (8.33) ensures that $\tilde{\eta}_i(t) \to 0$ as $t \to \infty$ provided that the following persistent excitation condition holds:

$$\sigma_1 \leq \int_{t_0}^{t_0+\delta} v(\tau)^T v(\tau)d\tau \leq \sigma_2, \tag{8.36}$$

where $t_0, \sigma_1, \sigma_2, \delta \in \mathbb{R}$ are positive constants.

Proof 8.2 Let $V_2(t) \in \mathbb{R}$ denote a non-negative function defined as follows:

$$V_2 = \frac{1}{2\kappa} \tilde{\eta}_i^2. \tag{8.37}$$

Taking the time derivative of (8.37) and substituting from (8.35), the following expression can be obtained:

$$
\begin{aligned}
\dot{V}_2 &= -(v\tilde{\eta}_i)^T \left(v\tilde{\eta}_i - \tilde{\zeta}_i + \frac{1}{\alpha_i} J_i^{-1} \dot{\tilde{\theta}}_i \right) \\
&\le -\|v\tilde{\eta}_i\|^2 + \|v\tilde{\eta}_i\| \cdot \frac{\|J_i^{-1}\|}{\alpha_i} \cdot \|\dot{\tilde{\theta}}_i\| + \|v\tilde{\eta}_i\| \cdot \|\tilde{\zeta}_i\| \\
&\le -(1 - \varepsilon_1 - \varepsilon_2)\|v\tilde{\eta}_i\|^2 + \frac{\varepsilon_3^2 \|\dot{\tilde{\theta}}_i\|^2}{4\varepsilon_1} + \frac{\|\tilde{\zeta}_i\|^2}{4\varepsilon_2}
\end{aligned}
\tag{8.38}
$$

where ε_1, ε_2, $\varepsilon_3 \in \mathbb{R}$ are positive constants selected to ensure that

$$1 - \varepsilon_1 - \varepsilon_2 \ge \mu > 0 \qquad \varepsilon_3 \ge \frac{\|J_i^{-1}\|}{\alpha_i}, \tag{8.39}$$

where $\mu \in \mathbb{R}$ is a positive constant. It should be noted that since $J_i(t)$ and $\alpha_i(t)$ are bounded, there exists ε_3 satisfying the inequality of (8.39). Based on $\dot{\tilde{\theta}}_i(t)$, $\tilde{\zeta}_i(t) \in \mathcal{L}_2$ (see Remark 8.4), it can be obtained that

$$\int_0^\infty \left(\frac{\varepsilon_3^2 \|\dot{\tilde{\theta}}_i\|^2}{4\varepsilon_1} + \frac{\|\tilde{\zeta}_i\|^2}{4\varepsilon_2} \right) d\tau \le \varepsilon, \tag{8.40}$$

where $\varepsilon \in \mathbb{R}$ is a positive constant. From (8.38) (8.40), the following inequalities can be determined:

$$
\begin{aligned}
V_2(t) &\le V_2(0) + \varepsilon \\
\int_0^\infty \mu \|v\tilde{\eta}_i\|^2 d\tau &\le V_2(0) - V_2(\infty) + \varepsilon.
\end{aligned}
\tag{8.41}
$$

It can be concluded from (8.41) that $v(t)\tilde{\eta}_i(t) \in \mathcal{L}_2$, $V_2(t) \in \mathcal{L}_\infty$. Similarly, signal chasing can be utilized to prove that $\frac{d(v\tilde{\eta}_i)}{dt} \in \mathcal{L}_\infty$. Based on Barbalat's lemma [192], it can be obtained that $v(t)\tilde{\eta}_i(t) \to 0$ as $t \to \infty$. If the linear velocity of the moving camera $v(t)$ satisfies the persistent excitation condition given in (8.36), then it can be concluded that [180] $\tilde{\eta}_i(t) \to 0$ as $t \to \infty$.

Since $\zeta_{wi}(t)$, $\zeta_i(t)$, and η_i can be estimated, the velocities of the feature points $v_{wi}(t)$ and $v_i(t)$ can be calculated based on (8.9). Additionally, utilizing (8.1), (8.2), and $\alpha_{wi}(t)$, $\alpha_i(t)$ given in (8.3), the 3-D Euclidean coordinates of the feature points $\bar{m}_{wi}(t)$ and $\bar{m}_i(t)$ relative to the static camera and the moving camera can be recovered.

8.4   Simulation Results

Simulation studies are performed to illustrate the performance of the proposed estimators. For the simulation, the intrinsic camera calibration matrices are given as follows:

$$A_w = A = \begin{bmatrix} 500 & 0 & 640 \\ 0 & 400 & 480 \\ 0 & 0 & 1 \end{bmatrix}.$$

To incorporate homography-based techniques, four coplanar but non-collinear feature points are selected to calculate $\alpha_{wi}(t)$, $\alpha_i(t)$, $H_w(t)$, and $H(t)$. Utilizing classical decomposition algorithm [86], the signals $\bar{R}_w(t)$ and $\bar{R}(t)$ can be obtained. Without lost of generality, the initial values of $\bar{R}_w(t)$, $\bar{R}(t)$, $\bar{x}_{fw}(t)$, and $\bar{x}_f(t)$ are chosen as $\bar{R}_w(0) = \bar{R}(0) = I_3$, $\bar{x}_{fw}(0) = \bar{x}_f(0) = \begin{bmatrix} 0 & 0 & 4 \end{bmatrix}^T$. The Euclidean coordinate of ith feature point, expressed in terms of $\mathcal{F}^*$, is set to be $\bar{m}_i^* = \begin{bmatrix} 0.56 & -0.525 & 1 \end{bmatrix}^T$ [m]. The velocities of the object and the moving camera are selected, respectively, as

$$v_{wi} = \begin{bmatrix} 0.4\cos(2t) & 1 - 0.1t & 0.4 + 0.3\sin(0.55\pi t) \end{bmatrix}^T [m/s]$$
$$v = \begin{bmatrix} 0 & 0 & 0.5 + 0.1\sin(0.2t) \end{bmatrix}^T [m/s]$$
$$\omega = \begin{bmatrix} 0 & 0 & 0.1 + 0.1\sin(0.2t) \end{bmatrix}^T [rad/s].$$

In addition, the initial values of $\hat{\theta}_{wi}(t)$, $\hat{\theta}_i(t)$, $u_{wi}(t)$, $u_i(t)$, and $\hat{\eta}_i(t)$ are chosen as $\hat{\theta}_{wi}(0) = \theta_{wi}(0)$, $\hat{\theta}_i(0) = \theta_i(0)$, $u_{wi}(0) = u_i(0) = \begin{bmatrix} 0 & 0 & 0 \end{bmatrix}^T$, and $\hat{\eta}_i(0) = 1.6$.

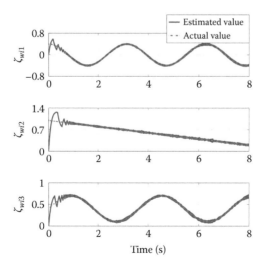

Figure 8.2: Scaled velocity identification $\zeta_{wi}(t)$.

The gains of the observers given in (8.11), (8.12), (8.31)–(8.33) are designed as $\lambda_{w1} = 10$, $\lambda_{w2} = 11$, $\lambda_1 = 4$, $\lambda_2 = 11$, $\beta_w = 0.05$, $\beta = 0.1$, $\kappa = 80$. To test the robustness of the observers, a random noise, obeying Gaussian distribution with a mean of 0 pixel and a standard deviation of 3 pixels, is injected into the coordinates of feature points.

The scaled velocity identification results are depicted in Figures 8.2, 8.3, and the range identification result is shown in Figure 8.4. It is clear that the estimated values all converge closely to the actual ones in the presence of image noise.

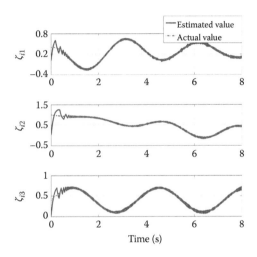

Figure 8.3: Scaled velocity identification $\zeta_i(t)$.

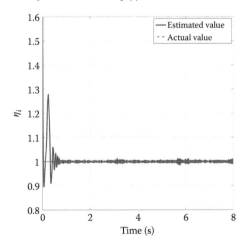

Figure 8.4: Range identification $\eta_i(t)$.

8.5   Conclusion

This chapter presents an approach for a static-moving camera system to asymptotically identify the velocity and range of feature points on a moving object. Based on the static camera, a nonlinear observer is designed first to identify the up-to-a-scale velocity of feature points. After that, utilizing the moving camera and the estimated scaled velocity, an adaptive estimator is given to identify the range of feature points. Lyapunov-based techniques are exploited to prove the asymptotic convergence for the velocity observer and the range estimator. Besides, motion constraint and *a priori* geometric knowledge are not required for the moving object. Simulation results are provided to demonstrate the performance of the proposed strategy.

VISUAL CONTROL III
OF ROBOTICS

Chapter 9

Introduction to Visual Control

9.1 Previous Works

9.1.1 Classical Visual Control Approaches

Visual control and navigation is one of the most important functionalities of robots, since vision sensors are cheap and provide rich environment information. Generally, robot tasks can be divided into pose regulation, path following, and trajectory tracking problems, while pose regulation and path following tasks can be further formulated into trajectory tracking problems by imposing time constraints. In this book, the visual trajectory tracking task is considered, which is performed by the classic idea of teach-by-showing with a monocular camera installed at an arbitrary position of the mobile robot. First, there is a teaching process during which a set of images are recorded to express the desired trajectory with time constraints. Then, the robot is regulated to track the desired trajectory using visual feedback. In visual control systems, system feedback is constructed based on visual information, which can be divided into position-based, image-based, and multiple-view geometry based methods [107].

9.1.1.1 Position Based Methods

Position-based methods use the vision system for positioning [159], so that the design of the vision system and controller is decoupled, and conventional robot control strategies can be used. However, the positioning process often requires prior knowledge about the environment, and the positioning precision is limited by the configuration of the vision system, such as image resolution, parameter error, etc. Besides, when the positions of the robot and landmarks need to be estimated simultaneously during the exploration of an unknown environment, the vision system would be quite complex, and this process can be generally formulated into the Simultaneous Localization and Mapping [62] problem. In general, multiple sensory information (e.g., odometry and inertial sensors) can be fused for better performance [128].

9.1.1.2 Image Based Methods

Image-based visual control methods offer good performance with lower computational complexity and better robustness to parameter errors. Classical image-based methods directly use the coordinates of feature points to construct error systems, while noise and temporary loss of feature points would have a direct influence on system states, affecting the stability and accuracy of the control system. In [78], a weighting matrix is used to reduce the effects of noise or losing of feature points. Besides, there are some methods that directly rely on the image information instead of extracted feature points, such as kernel based [111], mutual information based [47], and photometric [42] methods, which are more robust to partial occlusion and image noise. Generally, online optimization algorithms are used for controller development, but the convergence region cannot be theoretically guaranteed due to the complexity of image information, and the existence of local minima.

9.1.1.3 Multiple-View Geometry Based Method

To achieve better robustness to image noise and partial occlusion, a good choice is to design the visual servo system based on multiple-view geometry [86], such as epipolar geometry, homography, and trifocal tensor. They describe the constraints among multiple views, and are related to the relative pose information among the views. The estimation of multiple-view geometry are generally based on the matched features among the considered views. When the estimation is computed robustly (e.g., using Random Sample Consensus [72]), wrong feature matches can be filtered out, which are inevitably present when using automatic feature extractors and matchers. Besides, multiple-view geometry abstracts geometric information from noisy and redundant image information, making it convenient to develop control systems.

Epipolar-based methods generally use epipoles between the current view and the target one to construct error systems [10]. These methods are suitable for non-planar scenes, but become ill-conditioned for planar scenes, which commonly exist in man-made environments. Besides, the estimation of epipolar geometry degenerates with short baseline, and pose ambiguity exists when the epipoles tend to zero. A natural way to solve this problem is to use a two-step strategy, i.e., first, the robot is regulated to align with the target based on the epipoles between the current and target views, and then the longitudinal error is corrected by switching to another controller, such as image-based control [154], homography-based control [141], and epipolar-based control [15] that relies on the epipoles between the current and start views. However, the resulting control systems are not smooth and have the risk of singularity.

Homography-based methods offer good performance without the drawbacks of epipolar-based methods. Most of these methods exploit a combination of reconstructed Euclidean information and image space information, and hence they are also called 2.5-D visual servo control [152]. Rotational and translational error systems are decoupled and local minima problems are avoided. For the application of mobile robots, Fang [64] proposed a visual servo regulation controller for asymptotic pose regulation, Chen [31] proposed a visual servo tracking controller achieving asymptotic convergence to the desired trajectory. Generally, in the decomposition process, four solutions are obtained and an initial guess of the normal vector of the plane on which the feature points are located is needed to determine the unique one. If there is no dominant plane in the scene, the homography can be estimated by the virtual plane based method [150], but it is sensitive to image noise and feature point mismatches, and it may be difficult to guess the normal vector of the virtual plane beforehand. In [188,189], both homography and epipolar geometry are used to construct the error system for nonplanar scenes. However, due to the model complexity, only local stability is guaranteed.

Trifocal tensor describes the intrinsic geometry among three views independently of the observed scene. The use of three views provides robustness and enough information to determine the robot pose. Nevertheless, it has been rarely exploited in visual control until a series of papers [13,14,142] recently, and its advantages with respect to epipolar based methods are evaluated through experiments in [18]. For the application of visual servoing, trifocal tensor can be divided into 1-D and 2-D ones, which are defined based on the bearing angle and the image coordinates of the feature points, respectively. Pose regulation tasks of mobile robots are considered in [13,14,17,142]. In [142], an over-constrained controller is proposed with trifocal tensor elements directly used as the feedback, and may suffer from local minima problems. In [14], a two-step strategy is used to obtain a square error system in each step. In [13], super-twisting control is used to deal with parameter uncertainties and to generate smooth control inputs, and an adequate reference trajectory is designed to correct the lateral error which is not guaranteed to be convergent to zero in the controller. In [17], an observer for actual pose information is designed with an exciting controller, which also requires an adequate reference trajectory to guarantee the convergence of three degrees of freedom (DOF). Nevertheless, the works in [13,17] perform well for pose regulation tasks but cannot be directly used for the tracking of arbitrary trajectories. Path following tasks of mobile robots are considered in [12,16,177], with rotational controllers designed based on the heading error with respect to current target position [12,16] or the relative distance/angle to the desired path [177]. However, the above methods cannot be directly used for trajectory tracking tasks.

9.1.2 Main Existing Problems

9.1.2.1 Model Uncertainties

In visual servo systems, due to the unknown depth information in monocular cameras and the uncertain model parameters in robotics, there exist model uncertainties that influence the stability and performance of the systems. Previous works focus on adaptive control, robust control, and intelligent control methods to deal with model uncertainties.

Adaptive control methods use adaptive laws for online adjustment to guarantee the system robustness, and can be divided into parameter-adaptive and Jacobian-adaptive methods.

- Adaptive parameter methods: Since the motion characteristics of feature points rely on the depth and camera parameters, the parameters can be estimated by minimizing the reprojection error between the real and estimated position. A useful adaptive method is to use iterative searching methods [134,210] to online estimate unknown parameters. Different methods for depth parameter adaptation have been proposed in image based methods [76] and multiple-view geometry based methods [29]. If the camera parameters are unknown, depth independent Jacobian matrix based method is proposed in [136]. The control law is designed based on the depth independent part of the image Jacobian matrix, so that the camera parameters are linear parameterized in the resulting close-loop system. For the unknown depth parameter, adaptive law is further designed in [131]. The depth independent Jacobian based method can be generalized to visual control systems that rely on other image features [137]. For the task of trajectory tracking, various methods [138,209,212] are proposed that do not rely on the derivatives of image feature points.

- Adaptive Jacobian methods: This type of methods directly estimate the image Jacobian matrix online, such as the Broyden algorithm [102], dynamic Broyden algorithm [102], recursive least square algorithm [169], Kalman filter algorithm [147], etc. In [166], experiments are conducted to compare adaptive Jacobian methods with adaptive parameter methods, achieving comparable performance in terms of control precision and robustness. Adaptive Jacobian methods do not rely on the complex model development based on *a priori* knowledge, but are only effective in trained regions.

To guarantee system stability with parameter disturbances, another type of methods design robust controllers based on optimal parameter estimation and guarantee system stability within certain regions of parameter variation.

■ Lyapunov based methods: Conventional methods use Lyapunov method for controller design, dealing with model uncertainties caused by unknown depth information [96], camera parameter error [222], robot load variation [116], and quantization error in robot motion control [203].

■ Optimization based methods: There also exist optimization based methods that rely on the online optimization (e.g., H_2/H_∞ index [206], stability region [199], etc.) to obtain the optimal control inputs with parameter uncertainties.

■ Sliding mode control methods: Besides, sliding mode control is also widely used for visual control systems [15]. The sliding mode surface, which is independent on the uncertain parametrs and disturbances, is constructed, and the control laws are designed to enforce the convergence with respect to the surface. To deal with the non-smoothness of conventional sliding mode controllers, the super-twisting control method [13] can be used to generate smooth control inputs.

Even though robust control methods are not sensitive to parameter uncertainties and disturbances, relatively large control gains are needed to guarantee system stability, improving the risk of system vibration.

Intelligent control methods do not rely on accurate mathematical models, and are suitable for visual control with model uncertainties. Conventional intelligent control methods consist of computational intelligence based methods [211,219], fuzzy control methods [22,190], iterative learning control methods [108,109], etc.

■ Computational intelligence based methods: Artificial neural network [219] and genetic algorithm [211] are generally used for model fitting of visual control systems, and learned models are used for control development. This type of methods do not require complex system modeling but need sufficient offline training in workspace.

■ Fuzzy control methods: Fuzzy logic rules are used to describe the relationships among various variables, which are constructed using *a priori* knowledge or offline learning. In applications, fuzzy logics are used to design control laws [22] or parameter update laws [190]. However, for visual control systems with multiple DOFs, it is difficult to design fuzzy logics with the complex coupling among control variables.

■ Iterative learning methods: For repetitive tasks, control process in previous iterations can be used to compute suitable control inputs, achieving accurate trajectory tracking control. Direct iterative learning controllers compute feedforward terms using iterative learning law [109] and feedback assistant terms can be used to improve converge rate [106]. Indirect iterative learning controllers update model parameters using iterative learning law [108], obtaining accurate system models for trajectory tracking.

9.1.2.2 Field of View Constraints

In classical visual servo systems, it is required that there exist enough correspondences in the relevant views (i.e., the current and final views for most methods, and additionally the start view for trifocal tensor based methods), which is commonly described as the field of view constraints. Besides, existing feature point extraction algorithms suffer weak repeatability and precision with large displacements and view point changes [81], aggravating this limitation. Previous works dealing with the field of view constraints mainly focus on pose regulation and path following tasks, which are summarized as follows:

■ For pose regulation tasks, the field of view constraints in visual servo control have been studied using multiple methods. In [210], the size of control inputs is modified dynamically using weight matrices, which are obtained by optimization considering the constraints. Another kind of effective approaches combines path/trajectory planning and tracking [30,140]. In [30], a navigation function based method is proposed for trajectory planning for six DOF tasks, and an adaptive controller is exploited to track the desired trajectory. In [19,178,179], shortest paths for nonholonomic mobile robots with field-of-view constraints are proposed, which are generally composed of a set of extremal curves. Switched controllers [140] can be used for path following, nevertheless, the resulting system is not continuous and the path in the workspace is generally tortuous.

■ For path following tasks, a common strategy is to select a set of key images as subgoals and complete the tasks by a set of pose regulation tasks to these images [18,47]. This strategy only guarantees that the robot goes through these points where key images are selected, but the precise following of the whole path cannot be achieved. Besides, the switching of key images yields a hybrid system with complicated dynamics, and the stability of subsystems cannot guarantee the stability of the whole system [52]. This problem can be handled practically by the key image switching strategy in which the key image switches when the regulation controller with respect to the current target converges sufficiently. In [177], the trifocal tensor is constructed by the current image, the previous image and the key image on the desired path, and based on this the shortest distance towards the desired path is estimated.

It should be noted that the above methods dealing with the field of view constraints work for pose regulation and path following tasks but are not applicable for trajectory tracking tasks. For pose regulation tasks, the start and final images still need to have enough correspondences and the workspace is limited. For path following tasks, the subgoal strategy is acceptable to decompose the task into a series of pose regulation tasks at the cost of the potential risk of instability for the resulting hybrid system and poor dynamic performance (sufficient convergence toward subgoals is generally required to guarantee the stability of the hybrid system practically). In this chapter, the trajectory tracking task is considered, in which the tracking of both pose and velocities are required with time constraints. Thus, the subgoal strategy in path following tasks is not acceptable. Besides, the workspace of trajectory tracking tasks is larger than pose regulation tasks, i.e., the start and final images don't need to have enough correspondences.

9.1.2.3 Nonholonomic Constraints

For the trajectory tracking problem, the objective is to regulate the wheeled mobile robot to track a desired trajectory using visual feedback. Specifically, a set of images are captured beforehand during the teaching process to describe the desired trajectory. Then by comparing the current image information with the desired trajectory (i.e., the prerecorded images), the error signals can be constructed to facilitate the controller design. In [31], a homography-based method is proposed to achieve the trajectory tracking task with the requirement that the desired trajectory keeps moving all the time. To relax the field-of-view constraint, a trajectory tacking approach combining the trifocal tensor techniques with the key frame strategy is developed in [104]. Both in [31] and [104], the orientation and the scaled position information is extracted from the multiple images, and the unknown depth information is compensated by the feed-forward terms. In addition, based on an online adaptive estimated algorithm, the trajectory tracking problem is addressed in [213] without direct position measurement.

Differing from the trajectory tracking, for the regulation problem, the objective is to drive the wheeled mobile robot toward a desired stationary pose. A prerecorded image is utilized to express the desired stationary pose. According to Brockett's theorem [21], it is impossible to solve the regulation problem by using the continuous time-invariant controller in the presence of the nonholonomic constraint. To overcome this difficulty, different approaches have been developed, including continuous time-varying controllers [64,223], discontinuous time-invariant controllers [15], and switching controllers [143,145]. Considering the field-of-view constraint, a hybrid control strategy is proposed in [140], in which the elements of the homography matrix are directly used to construct the system errors. Moreover, with a pan platform, an active visual regulation scheme is developed in [65] to keep the feature points in the view of camera.

However, all the above works study the visual trajectory tracking and regulation tasks separately, and hence, none of them can address these two problems simultaneously. In fact, to avoid the switching between control strategies, it is preferable to accomplish the trajectory tracking task and the regulation task simultaneously with a single controller [214]. Additionally, to solve the visual servoing problem of mobile robots, it should be noted that not only the nonholonomic constraint but also the system uncertainties caused by the vision sensor should be carefully considered. So the existing unified controllers [59,98,124,214], which require that the mobile robot's full states are available, cannot be directly exploited to address this visual servoing problem. So far, there are a small amount of results about the unified tracking and regulation visual servoing of wheeled mobile robots. In [148], the orientation and position information of the mobile robot is calculated based on *a priori* geometric knowledge about the target object and the homography techniques. After that, an unified controller is designed to achieve uniformly ultimately bounded tracking and regulation. Recently, a model-free 2-1/2-D visual servoing strategy is developed in [127]. Specifically, the orientation angle and image signals are utilized to construct the composite system errors, and the continuous time-varying controller can achieve asymptotic tracking and regulation. However, the controller designed in [127] is complicated, and some control gains should be adjusted carefully to satisfy certain requirement.

9.2 Task Descriptions

In vision-based robotic systems, robot tasks are described in terms of visual information. Classical approaches exploit teach-by-showing strategy for task description, i.e., a series of images are captured beforehand to show the desired pose or trajectory and the robot is regulated using visual feedback. This strategy is suitable for repetitive tasks in industrial applications. However, robots are faced with complex tasks in real world. In the past few years, researchers have proposed several strategies for task description and execution. In real applications, vision-based robotic systems can be divided into two types: autonomous systems and semiautonomous systems.

9.2.1 Autonomous Robot Systems

Autonomous robot systems work using visual feedback without the human interaction and have been widely used in applications, such as the visual navigation of mobile robots and the end effector control of robot manipulators. The visual navigation task of mobile robots can be formulated into a visual trajectory tracking problem [31] or a series of visual pose regulation problems [54]. Off-line training is generally needed to obtain the desired image sequences. For robot manipulators, conventional approaches use teaching-playing strategy for task execution. Taking the assembling task for example, the engineers should first program the trajectory for teaching and the manipulator is actuated by the recorded control signals to finish the task. This process is simplified by using vision-based strategies, i.e., the engineer only needs to perform the task for demonstration under the perception of the camera, and image sequences are recorded to express the desired trajectory, then the manipulator is controlled to accomplish the task based on visual feedback [106].

Conventional visual servo systems generally use teach-by-showing strategy for task description. The set point or desired trajectory of the visual servo controller is given by the image(s) captured at the desired pose(s). This strategy is suitable for robotic manipulators and mobile robots that works repetitively in industrial scenes. However, this strategy is high-cost for the visual navigation of mobile robots in large-scale scenes. Considering the practicability, the following strategies are used in previous approaches:

- Set the desired state or trajectory by the images captured by other cameras, such as the teach-by-zooming strategy in [156].

- Set the desired state or trajectory by the images in other modalities (satellite images, etc.), such as mutual information based methods in [47].

- Express the visual control task based on geometric information [155].

In real life, when telling someone to go to some place, we can provide the picture or map of the desired place, or describe the geometric property of the scene. Actually, the above three strategies correspond to the behavioral habits of human. Besides, most visual control systems require the visibility of target in the images (i.e., the field-of-view constriant), which eliminates the reachable workspace of the robots. In [105], a visual navigation strategy using sparse landmarks is proposed. The field-of-view constraint is loosened using key frame strategy, optimizing the performed trajectory by nonholonomic robots in workspace. In [129], a global task space controller is proposed. Global continuous controllers are constructed using several feedback information effective in specific regions, making it possible to go through the dead zone of visual perception and the singular region in the control process.

9.2.2 Semiautonomous Systems

Currently most robotic tasks need the precise task description beforehand, but some complex tasks in real life cannot be described quantitatively. As a result, semiautonomous systems are constructed for complex tasks. High-level decision making is made by human in the control process, low-level motion control is performed by the visual servo controller. Then, the task is accomplished by the semiautonomous system consisting of the cooperation of human and machine.

In previous approaches, the following strategies are used for human-machine cooperation in visual control systems:

■ The system is controlled in a cascade framework, human is responsible for the high-level decision control, and the visual servo controller is responsible for the low-level motion control, such as the remote controller for under water robots [113] and the coordinating controller for semiautonomous driving wheel chairs [167].

■ The visual servo controller imposes motion constraints onto the manipulated object, reducing the DOF that need to be controlled by the human. The auxiliary control system makes it easier to manipulate for human and improves the control precision.

■ A switched control system consists of the visual servo controller and the human. The task is decomposed into the regions managed by the human and the robot, accomplishing the task cooperatively [130].

Besides, visual servo techniques have also been widely used in clinical applications, helping the manipulation of doctors. Readers can refer to the relevant reviews in [7,8].

9.3 Typical Visual Control Systems

For the visual control of robotics, the tasks can be divided into three types: pose regulation, trajectory tracking, and path following. In a pose regulation task, the control objective is defined by the image captured at the desired pose; in a path following or trajectory tracking, the control objective is defined by a series of images captured along the desired path, while the trajectory tracking requires the time constraint. In this section, typical visual servo systems are introduced.

9.3.1 Visual Control for General Robots

For the visual tracking of general robots, eye-to-hand and eye-in-hand configurations are illustrated in Figures 9.1 and 9.2b, respectively. The robot is considered as a fully-actuated rigid body (e.g., a manipulator) which has six DOF.

For the eye-to-hand case in Figure 9.1, the control objective is to ensure that the trajectory of $\mathcal{F}$ tracks $\mathcal{F}_d$ (i.e., $\bar{m}_i(t)$ tracks $\bar{m}_{di}(t)$), where the trajectory of $\mathcal{F}_d$ is constructed relative to the reference camera position/orientation given by $\mathcal{F}^*$. To develop a relationship between the planes, an inertial coordinate system, denoted by $\mathcal{I}$, is defined where the origin coincides with the center of a fixed camera. Orthogonal coordinate systems $\mathcal{F}$, $\mathcal{F}_d$, and $\mathcal{F}^*$ are attached to the planes π, π_d, and π^*, respectively (see Figure 9.1). To relate the coordinate systems,

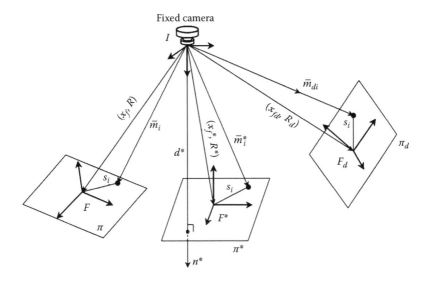

Figure 9.1: Coordinate frame relationships between a fixed camera and the plane defined by the current, desired, and reference feature points (i.e., π, π_d, and π^*).

let $R(t)$, $R_d(t)$, $R^* \in SO(3)$ denote the rotation between $\mathcal{F}$ and $\mathcal{I}$, $\mathcal{F}_d$ and $\mathcal{I}$, and $\mathcal{F}^*$ and $\mathcal{I}$, respectively, and let $x_f(t)$, $x_{fd}(t)$, $x_f^* \in \mathbb{R}^3$ denote the respective translation vectors expressed in the coordinates of $\mathcal{I}$. Then, the tracking control objective can be stated as follows: $\bar{R}(t) \to \bar{R}_d(t)$, $m_1(t) \to m_{d1}(t)$, and $z_1(t) \to z_{d1}(t)$ (and hence, $\bar{x}_h(t) \to \bar{x}_{hd}(t)$).

For the eye-in-hand case, the camera is held by a robot end-effector (not shown) and the observed target object could be static or dynamic, as shown in Figure 9.2a and b, respectively.

For the static case, the coordinate frames $\mathcal{F}$, $\mathcal{F}_d$, and $\mathcal{F}^*$ depicted in Figure 9.2a are attached to the camera and denote the actual, desired, and reference locations for the camera, respectively. $R(t)$, $R_d(t) \in SO(3)$ denote the rotation between $\mathcal{F}$ and $\mathcal{F}^*$ and between $\mathcal{F}_d$ and $\mathcal{F}^*$, respectively, and $x_f(t)$, $x_{fd}(t) \in \mathbb{R}^3$ denote translation vectors from $\mathcal{F}$ to $\mathcal{F}^*$ and $\mathcal{F}_d$ to $\mathcal{F}^*$ expressed in the coordinates of $\mathcal{F}$ and $\mathcal{F}_d$, respectively.

For the dynamic case as illustrated in Figure 9.2b, $\mathcal{I}$ denotes the inertial coordinate frame; $\mathcal{F}$ denotes the camera coordinate frame mounted on the following rigid body; the xy-axis of $\mathcal{F}_l$ defines the plane π which contains the coplanar feature points of the leading object; $\mathcal{F}^*$ denotes the desired fixed relative position and orientation of the moving camera with respect to $\mathcal{F}_l$. In Figure 9.2b, $\mathcal{F}$, $\mathcal{F}_l$, and $\mathcal{F}^*$ are all moving with respect to $\mathcal{I}$ while the relative position and orientation between $\mathcal{F}_l$ and $\mathcal{F}^*$ are fixed, and the angular and linear velocities of $\mathcal{F}_l$ are unknown. The known reference image is taken from the view of $\mathcal{F}^*$. To

(a)   (b)

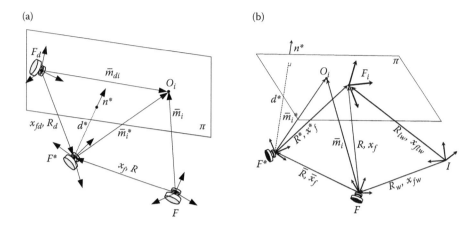

Figure 9.2: Task description in eye-in-hand configuration. (a) Static target object and (b) dynamic target object.

relate the coordinate frames, let $R(t) \in SO(3)$ and $x_f(t) \in \mathbb{R}^3$ denote the rotation and translation between $\mathcal{F}_l$ and $\mathcal{F}$ expressed in terms of $\mathcal{F}$, respectively. Similarly, $R^* \in SO(3)$ and $x_f^* \in \mathbb{R}^3$ denote the rotation and translation between $\mathcal{F}_l$ and $\mathcal{F}^*$ expressed in terms of $\mathcal{F}^*$, respectively. The objective is to develop a visual servo controller that ensures the following rigid body (i.e., camera), denoted by $\mathcal{F}$, tracks the leading object, denoted by $\mathcal{F}_l$, with a fixed relative rotation R^* and a fixed relative translation x_f^* in the sense that

$$R(t) \to R^*, x_f(t) \to x_f^* \qquad \text{as} \qquad t \to \infty.$$

9.3.2 Visual Control for Mobile Robots

For the visual control of mobile robots, eye-to-hand and eye-in-hand configurations are illustrated in Figure 9.3a and b, respectively. The general wheeled mobile robot has two differential driving wheels and moves on the motion plane with three DOF.

For the eye-to-hand case in Figure 9.3a, the planar motion of a WMR is assumed to be visible by a monocular camera mounted with a fixed position and orientation[1]. An inertial reference frame, denoted by $\mathcal{I}$, is attached to the camera. The origin of an orthogonal coordinate system $\mathcal{F}$ is coincident with the center of the WMR wheel axis. As illustrated in Figure 9.3a, the xy-axis of $\mathcal{F}$ defines the plane of motion where the x-axis of $\mathcal{F}$ is perpendicular to the wheel axis, and the y-axis is parallel to the wheel axis. The z-axis of $\mathcal{F}$ is perpendicular

---

[1]No assumptions are made with regard to the alignment of the WMR plane of motion and the focal axis of the camera.

(a)

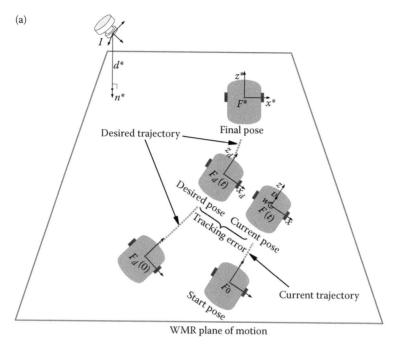

WMR plane of motion

(b)

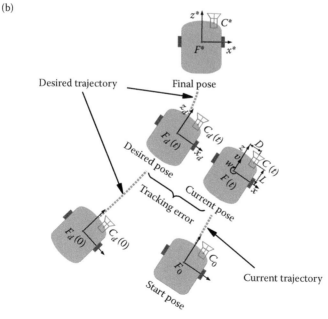

Figure 9.3: Task configuration of vision guided mobile robots. (a) Eye-to-hand configuration and (b) eye-in-hand configuration.

to the plane of motion and is located at the center of the wheel axis. The linear velocity of the WMR along the x-axis is denoted by $v_c(t) \in \mathbb{R}$, and the angular velocity $\omega_c(t) \in \mathbb{R}$ is about the z-axis of $\mathcal{F}$ (see Figure 9.3a). A desired trajectory is defined by a prerecorded time-varying trajectory of $\mathcal{F}_d$ that is assumed to be second-order differentiable where $v_{cd}(t)$, $\omega_{cd}(t) \in \mathbb{R}$ denote the desired linear and angular velocity of $\mathcal{F}_d$, respectively. A fixed orthogonal coordinate system, denoted by $\mathcal{F}^*$, represents a fixed (i.e., a single snapshot) reference position and orientation of the WMR relative to the fixed camera.

For the eye-in-hand case in Figure 9.3b, a monocular camera is installed rigidly at an arbitrary position of the robot, with its heading direction parallel to that of the robot. Denote the current coordinate systems of the robot and the camera as $\mathcal{F}(t)$ and $\mathcal{C}(t)$, respectively. The origin of $\mathcal{F}(t)$ is located at the center of the robot wheel axis. The xz-axis of $\mathcal{F}(t)$ defines the plane of motion, with the x-axis parallel to the wheel axis and the z-axis perpendicular to the wheel axis. The y-axis of $\mathcal{F}(t)$ is perpendicular to the plane of motion. The origin of $\mathcal{C}(t)$ is located at an arbitrary position ($x = D$, $z = L$) with respect to $\mathcal{F}(t)$ and the axes of $\mathcal{C}(t)$ are parallel to those of $\mathcal{F}(t)$. The wheeled mobile robot has two control inputs, i.e., the linear velocity $v(t) \in \mathbb{R}$ and the angular velocity $\omega(t) \in \mathbb{R}$, with their positive directions indicated in Figure 9.3.

The visual trajectory tracking task is performed by the classical teach-by-showing strategy. In the teaching process, the robot moves from $\mathcal{F}_d(0)$ along the desired trajectory $\mathcal{F}_d(t)$ to $\mathcal{F}^*$, during which a set of images are captured to express the desired trajectory. In the tracking process, the robot is placed at an arbitrary pose $\mathcal{F}_0$ and is regulated to track the desired trajectory $\mathcal{F}_d(t)$ using visual feedback. Since $\mathcal{F}(t)$ coincides with $\mathcal{F}_d(t)$ if and only if $\mathcal{C}(t)$ coincides with $\mathcal{C}_d(t)$, the tracking process can be described in the camera's coordinate.

Assumption 9.1 *In this task, the start view $\mathcal{C}_0$ is required to share visual information with the start view of the desired trajectory $\mathcal{C}_d(0)$.*

Remark 9.1 The above assumption is natural for such a trajectory tracking task, which has a much larger workspace than previous works that require the correspondence between the start view and target view.

Remark 9.2 A special case is the pose regulation task, in which views in the desired trajectory $\mathcal{C}_d(t)$ coincide with the final view $\mathcal{C}^*$.

Remark 9.3 In the teaching process, a set of images are recorded with specific time intervals to express the desired trajectory. To generate continuous and sufficiently smooth trajectory, sufficiently smooth spline function can be utilized to fit the desired trajectory (position and orientation) as functions of time.

Chapter 10

Visual Tracking Control of General Robotic Systems

10.1 Introduction

In contrast to the approaches in [46] and [158] in which a path is planned as a means to reach a desired setpoint, hybrid tracking controllers are developed in this chapter where the robot end-effector is required to track a prerecorded time-varying reference trajectory. To develop the hybrid controllers, a homography-based visual servoing approach is utilized. The motivation for using this approach is that the visual servo control problem can be incorporated with a Lyapunov-based control design strategy to overcome many practical and theoretical obstacles associated with more traditional, purely image-based approaches. Specifically, one of the challenges of this problem is that the translation error system is corrupted by an unknown depth-related parameter. By formulating a Lyapunov-based argument, an adaptive update law is developed to actively compensate for the unknown depth parameter. In addition, the proposed approach facilitates: (i) translational/rotational control in the full six DOFs task-space without the requirement of an object model, (ii) partial servoing on pixel data that yields improved robustness and increases the likelihood that the centroid of the object remains in the camera field-of-view [152], and (iii) the use of an image Jacobian that is only singular for multiples of 2π, in contrast to the state-dependent singularities present in the image Jacobians associated with many of the purely image-based controllers. The homography-based controllers in this chapter target both the fixed camera and the camera-in-hand configurations. The control development for the fixed camera problem is presented in detail, and the camera-in-hand problem is included as an extension.

10.2 Visual Tracking with Eye-to-Hand Configuration

10.2.1 Geometric Modeling

10.2.1.1 Vision System Model

To make the subsequent development more tractable, four target points located on an object (i.e., the end-effector of a robot manipulator) denoted by $O_i \ \forall i = 1$, 2, 3, 4 are considered to be coplanar[1] and not colinear. Based on this assumption, consider a fixed plane, denoted by π^*, that is defined by a reference image of the object. In addition, consider the actual and desired motion of the plane containing the end-effector target points, denoted by π and π_d, respectively (see Figure 9.1). The Euclidean coordinates of the target points on π, π_d, and π^* can be expressed in terms of $\mathcal{I}$, respectively, as follows:

$$\bar{m}_i(t) \triangleq \begin{bmatrix} x_i(t) & y_i(t) & z_i(t) \end{bmatrix}^T$$

$$\bar{m}_{di}(t) \triangleq \begin{bmatrix} x_{di}(t) & y_{di}(t) & z_{di}(t) \end{bmatrix}^T \tag{10.1}$$

$$\bar{m}_i^* \triangleq \begin{bmatrix} x_i^* & y_i^* & z_i^* \end{bmatrix}^T$$

under the standard assumption that the distances from the origin of $\mathcal{I}$ to the target points remains positive (i.e., $z_i(t)$, $z_{di}(t)$, $z_i^* > \varepsilon$, where ε denotes an arbitrarily small positive constant). As also illustrated in Figure 9.1, $n^* \in \mathbb{R}^3$ denotes the constant unit normal to the plane π^* expressed in the coordinates of $\mathcal{I}$, and $s_i \in \mathbb{R}^3$ denotes the constant coordinates of the i-th target point. The constant distance from the origin of $\mathcal{I}$ to π^* along the unit normal is denoted by $d^* \in \mathbb{R}$ and is defined as follows:

$$d^* \triangleq n^{*T} \bar{m}_i^*. \tag{10.2}$$

From the geometry between the coordinate frames depicted in Figure 9.1, the following relationships can be developed:

$$\bar{m}_i = \left(\bar{R} + \frac{\bar{x}_f}{d^*} n^{*T} \right) \bar{m}_i^* \qquad \bar{m}_{di} = \left(\bar{R}_d + \frac{\bar{x}_{fd}}{d^*} n^{*T} \right) \bar{m}_i^*, \tag{10.3}$$

where $\bar{R}(t)$, $\bar{R}_d(t) \in SO(3)$ and $\bar{x}_f(t)$, $\bar{x}_{fd}(t) \in \mathbb{R}^3$ are new rotational and translational variables, respectively, defined as follows (see Appendix B.1 for further

---

[1]It should be noted that if four coplanar target points are not available then the subsequent development can exploit the classic eight-points algorithm [150] with no four of the eight target points being coplanar.

insight into the geometrical significance of the new rotational and translational variables):

$$\begin{aligned} \bar{R} &= R\left(R^*\right)^T & \bar{R}_d &= R_d\left(R^*\right)^T \\ \bar{x}_f &= x_f - \bar{R}x_f^* & \bar{x}_{fd} &= x_{fd} - \bar{R}_d x_f^* \end{aligned} \quad . \tag{10.4}$$

Remark 10.1 The subsequent development requires that the constant rotation matrix R^* be known. The constant rotation matrix R^* can be obtained *a priori* using various methods (e.g., a second camera, Euclidean measurements). The subsequent development is also based on the assumption that the target points do not become occluded.

10.2.1.2 Euclidean Reconstruction

The relationship given by (10.3) provides a means to quantify a translation and rotation error between $\mathcal{F}$ and $\mathcal{F}^*$ and between $\mathcal{F}_d$ and $\mathcal{F}^*$. Since the Euclidean position of $\mathcal{F}$, $\mathcal{F}_d$, and $\mathcal{F}^*$ cannot be directly measured, a Euclidean reconstruction is developed in this section to obtain the position and rotational error information by comparing multiple images acquired from the fixed, monocular vision system. Specifically, comparisons are made between the current image, the reference image obtained *a priori*, and the *a priori* known sequence of images that define the trajectory of $\mathcal{F}_d$. To facilitate the subsequent development, the normalized Euclidean coordinates of the points on π, π_d, and π^* can be respectively expressed in terms of $\mathcal{I}$ as $m_i(t)$, $m_{di}(t)$, $m_i^* \in \mathbb{R}^3$, as follows:

$$m_i \triangleq \frac{\bar{m}_i}{z_i} = \begin{bmatrix} \dfrac{x_i}{z_i} & \dfrac{y_i}{z_i} & 1 \end{bmatrix}^T \tag{10.5}$$

$$m_{di} \triangleq \frac{\bar{m}_{di}}{z_{di}} = \begin{bmatrix} \dfrac{x_{di}}{z_{di}} & \dfrac{y_{di}}{z_{di}} & 1 \end{bmatrix}^T \tag{10.6}$$

$$m_i^* \triangleq \frac{\bar{m}_i^*}{z_i^*} = \begin{bmatrix} \dfrac{x_i^*}{z_i^*} & \dfrac{y_i^*}{z_i^*} & 1 \end{bmatrix}^T. \tag{10.7}$$

From the expressions given in (10.3)–(10.7), the rotation and translation between the coordinate systems can now be related in terms of the normalized coordinates as follows:

$$m_i = \underbrace{\frac{z_i^*}{z_i}}_{\alpha_i} \underbrace{\left(\bar{R} + \bar{x}_h n^{*T}\right) m_i^*}_{H} \tag{10.8}$$

$$m_{di} = \underbrace{\frac{z_i^*}{z_{di}}}_{\alpha_{di}} \underbrace{\left(\bar{R}_d + \bar{x}_{hd} n^{*T}\right) m_i^*}_{H_d}, \tag{10.9}$$

where $\alpha_i(t)$, $\alpha_{di}(t) \in \mathbb{R}$ denote invertible depth ratios, $H(t)$, $H_d(t) \in \mathbb{R}^{3 \times 3}$ denote Euclidean homographies [66], and $\bar{x}_h(t)$, $\bar{x}_{hd}(t) \in \mathbb{R}^3$ denote scaled translation vectors that are defined as follows:

$$\bar{x}_h = \frac{\bar{x}_f}{d^*} \qquad \bar{x}_{hd} = \frac{\bar{x}_{fd}}{d^*} . \tag{10.10}$$

Each target point on π, π_d, and π^* will have a projected pixel coordinate expressed in terms of $\mathcal{I}$, denoted by $u_i(t)$, $v_i(t) \in \mathbb{R}$ for π, $u_{di}(t)$, $v_{di}(t) \in \mathbb{R}$ for π_d, and u_i^*, $v_i^* \in \mathbb{R}$ for π^*, that are defined as follows:

$$p_i \triangleq \begin{bmatrix} u_i & v_i & 1 \end{bmatrix}^T \qquad p_{di} \triangleq \begin{bmatrix} u_{di} & v_{di} & 1 \end{bmatrix}^T \qquad p_i^* \triangleq \begin{bmatrix} u_i^* & v_i^* & 1 \end{bmatrix}^T . \tag{10.11}$$

In (10.11), $p_i(t)$, $p_{di}(t)$, $p_i^* \in \mathbb{R}^3$ represent the image-space coordinates of the time-varying target points, the desired time-varying target point trajectory, and the constant reference target points, respectively.

$$p_i = \alpha_i \underbrace{\left(AHA^{-1} \right)}_{G} p_i^* \qquad p_{di} = \alpha_{di} \underbrace{\left(AH_dA^{-1} \right)}_{G_d} p_i^* , \tag{10.12}$$

where $G(t) = [g_{ij}(t)]$, $G_d(t) = [g_{dij}(t)]$ $\forall i, j = 1, 2, 3 \in \mathbb{R}^{3 \times 3}$ denote projective homographies, $A \in \mathbb{R}^{3 \times 3}$ is a known, constant, and invertible intrinsic camera calibration matrix. From the first relationship in (10.12), a set of 12 linearly independent equations given by the 4 target point pairs $(p_i^*, p_i(t))$ with 3 independent equations per target pair can be used to determine the projective homography up to a scalar multiple (i.e., the product $\alpha_i(t)G(t)$ can be determined). From the definition of $G(t)$ given in (10.12), various techniques can then be used (e.g., see [67,225]) to decompose the Euclidean homography, to obtain $\alpha_i(t)$, $G(t)$, $H(t)$, and the rotation and translation signals $\bar{R}(t)$ and $\bar{x}_h(t)$, and n^*. Likewise, by using the target point pairs $(p_i^*, p_{di}(t))$, the desired Euclidean homography can be decomposed to obtain $\alpha_{di}(t)$, $G_d(t)$, $H_d(t)$, and the desired rotation and translation signals $\bar{R}_d(t)$ and $\bar{x}_{hd}(t)$. The rotation matrices $R(t)$ and $R_d(t)$ can be computed from $\bar{R}(t)$ and $\bar{R}_d(t)$ by using (10.4) and the fact that R^* is assumed to be known. Hence, $R(t)$, $\bar{R}(t)$, $R_d(t)$, $\bar{R}_d(t)$, $\bar{x}_h(t)$, $\bar{x}_{hd}(t)$, and the depth ratios $\alpha_i(t)$ and $\alpha_{di}(t)$ are all known signals that can be used for control synthesis.

10.2.2 Control Development

10.2.2.1 Control Objective

The objective is to develop a visual servo controller that ensures that the trajectory of $\mathcal{F}$ tracks $\mathcal{F}_d$ (i.e., $\bar{m}_i(t)$ tracks $\bar{m}_{di}(t)$), where the trajectory of $\mathcal{F}_d$ is constructed relative to the reference camera position/orientation given by $\mathcal{F}^*$. To ensure that $\bar{m}_i(t)$ tracks $\bar{m}_{di}(t)$ from the Euclidean reconstruction given in (10.8) and (10.9), the tracking control objective can be stated as follows[2]: $\bar{R}(t) \rightarrow \bar{R}_d(t)$, $m_1(t) \rightarrow m_{d1}(t)$, and $z_1(t) \rightarrow z_{d1}(t)$ (and hence, $\bar{x}_h(t) \rightarrow \bar{x}_{hd}(t)$). The 3D control objective is complicated by the fact that only 2-D image information is measurable. That is, while the development of the homography provides a means to reconstruct some Euclidean information, the formulation of a controller is challenging due to the fact that the time varying signals $z_1(t)$ and $z_{d1}(t)$ are not measurable. In addition, it is desirable to servo on actual pixel information (in lieu of reconstructed Euclidean information) to improve robustness to intrinsic camera calibration parameters and to increase the likelihood that the object will stay in the field of view of the camera [152].

To reformulate the control objective in light of these issues, a hybrid translation tracking error, denoted by $e_v(t) \in \mathbb{R}^3$, is defined as follows:

$$e_v = p_e - p_{ed}, \tag{10.13}$$

where $p_e(t)$, $p_{ed}(t) \in \mathbb{R}^3$ are defined as follows:

$$p_e = \begin{bmatrix} u_1 & v_1 & -\ln(\alpha_1) \end{bmatrix}^T \qquad p_{ed} = \begin{bmatrix} u_{d1} & v_{d1} & -\ln(\alpha_{d1}) \end{bmatrix}^T, \tag{10.14}$$

and $\ln(\cdot)$ denotes the natural logarithm. A rotation tracking error, denoted by $e_\omega(t) \in \mathbb{R}^3$, is defined as follows:

$$e_\omega \triangleq \Theta - \Theta_d : \tag{10.15}$$

where $\Theta(t)$, $\Theta_d(t) \in \mathbb{R}^3$ denote the axis–angle representation of $\bar{R}(t)$ and $\bar{R}_d(t)$ as follows [197]

$$\Theta = u(t)\theta(t) \qquad \Theta_d = u_d(t)\theta_d(t). \tag{10.16}$$

For the representations in (10.16), $u(t)$, $u_d(t) \in \mathbb{R}^3$ represent unit rotation axes, and $\theta(t)$, $\theta_d(t) \in \mathbb{R}$ denote the respective rotation angles about $u(t)$ and $u_d(t)$ that are assumed to be confined to the following regions:

$$-\pi < \theta(t) < \pi \qquad -\pi < \theta_d(t) < \pi. \tag{10.17}$$

Based on the error system formulations in (10.13) and (10.15), the control objective can be stated as the desire to regulate the tracking error signals $e_v(t)$ and $e_\omega(t)$ to zero. If the tracking error signals $e_v(t)$ and $e_\omega(t)$ are regulated to zero then the object can be proven to be tracking the desired trajectory (see Appendix B.3 for further details).

---

[2]Any point O_i can be utilized in the subsequent development; however, to reduce the notational complexity, we have elected to select the image point O_1, and hence, the subscript 1 is utilized in lieu of i in the subsequent development.

Remark 10.2 As stated in [197], the axis–angle representation in (10.16) is not unique, in the sense that a rotation of $-\theta(t)$ about $-u(t)$ is equal to a rotation of $\theta(t)$ about $u(t)$. A particular solution for $\theta(t)$ and $u(t)$ can be determined as follows [197]:

$$\theta_p = \cos^{-1}\left(\frac{1}{2}(\text{tr}(\bar{R}) - 1)\right) \qquad [u_p]_\times = \frac{\bar{R} - \bar{R}^T}{2\sin(\theta_p)}, \qquad (10.18)$$

where the notation $\text{tr}(\cdot)$ denotes the trace of a matrix, and $[u_p]_\times$ denotes the 3×3 skew-symmetric expansion of $u_p(t)$. From (10.18), it is clear that

$$0 \leq \theta_p(t) \leq \pi. \qquad (10.19)$$

While (10.19) is confined to a smaller region than $\theta(t)$ in (10.17), it is not more restrictive in the sense that

$$u_p\theta_p = u\theta. \qquad (10.20)$$

The constraint in (10.19) is consistent with the computation of $[u(t)]_\times$ in (10.18) since a clockwise rotation (*i.e.*, $-\pi \leq \theta(t) \leq 0$) is equivalent to a counterclockwise rotation (*i.e.*, $0 \leq \theta(t) \leq \pi$) with the axis of rotation reversed. Hence, based on (10.20) and the functional structure of the object kinematics, the particular solutions $\theta_p(t)$ and $u_p(t)$ can be used in lieu of $\theta(t)$ and $u(t)$ without loss of generality and without confining $\theta(t)$ to a smaller region. Since, we do not distinguish between rotations that are off by multiples of 2π, all rotational possibilities are considered via the parameterization of (10.16) along with the computation of (10.18). Likewise, particular solutions can be found in the same manner for $\theta_d(t)$ and $u_d(t)$.

Remark 10.3 To develop a tracking control design, it is typical that the desired trajectory is used as a feedforward component in the control design. Hence, for a kinematic controller the desired trajectory is required to be at least first order differentiable and at least second order differentiable for a dynamic level controller. To this end, a sufficiently smooth function (e.g., a spline function) is used to fit the sequence of target points to generate the desired trajectory $p_{di}(t)$; hence, it is assumed that $p_{ed}(t)$ and $\dot{p}_{ed}(t)$ are bounded functions of time. From the projective homography introduced in (10.12), $p_{di}(t)$ can be expressed in terms of the *a priori* known, functions $\alpha_{di}(t)$, $H_d(t)$, $\bar{R}_d(t)$, and $\bar{x}_{hd}(t)$. Since these signals can be obtained from the pre-recorded sequence of images, sufficiently smooth functions can also be generated for these signals by fitting a sufficiently smooth spline function to the signals. Hence, in practice, the *a priori* developed smooth functions $\alpha_{di}(t)$, $\bar{R}_d(t)$, and $\bar{x}_{hd}(t)$ can be constructed as bounded functions with bounded time derivatives. Based on the assumption that $\bar{R}_d(t)$ is a bounded first order differentiable function with a bounded derivative, (10.18) can be used to conclude that $u_d(t)$ and $\theta_d(t)$ are bounded first order differentiable functions with a bounded derivative; hence, $\Theta_d(t)$ and $\dot{\Theta}_d(t)$ can be assumed to be bounded. In the subsequent tracking control development, the desired signals $\dot{p}_{ed}(t)$ and $\dot{\Theta}_d(t)$ will be used as a feedforward control term.

10.2.2.2 Open-Loop Error System

To develop the open-loop error system for $e_\omega(t)$, we take the time derivative of (10.15) and use (1.37) to obtain the following expression:

$$\dot{e}_\omega = L_\omega R \omega_e - \dot{\Theta}_d. \tag{10.21}$$

In (10.21), the Jacobian-like matrix $L_\omega(t) \in \mathbb{R}^{3 \times 3}$ is defined as:

$$L_\omega = I_3 - \frac{\theta}{2} [u]_\times + \left(1 - \frac{\text{sinc}(\theta)}{\text{sinc}^2\left(\dfrac{\theta}{2}\right)} \right) [u]_\times^2, \tag{10.22}$$

where

$$\text{sinc}(\theta(t)) \triangleq \frac{\sin \theta(t)}{\theta(t)},$$

and $\omega_e(t) \in \mathbb{R}^3$ denotes the angular velocity of the object expressed in $\mathcal{F}$. By exploiting the fact that $u(t)$ is a unit vector (i.e., $\|u\|^2 = 1$), the determinant of $L_\omega(t)$ can be calculated as [151]:

$$\det(L_\omega) = \frac{1}{\text{sinc}^2\left(\dfrac{\theta}{2}\right)}, \tag{10.23}$$

where $\det(\cdot)$ signifies the determinant operator. From (10.23), it is clear that $L_\omega(t)$ is only singular for multiples of 2π (i.e., out of the assumed workspace); therefore, $L_\omega(t)$ is invertible in the assumed workspace. To develop the open-loop error system for $e_v(t)$, we take the time derivative of (10.13) to obtain the following expression (see Appendix B.2 for further details):

$$z_1^* \dot{e}_v = \alpha_1 A_e L_v R \left[v_e + [\omega_e]_\times s_1 \right] - z_1^* \dot{p}_{ed}, \tag{10.24}$$

where $v_e(t) \in \mathbb{R}^3$ denotes the linear velocity of the object expressed in $\mathcal{F}$. In (10.24), $A_e \in \mathbb{R}^{3 \times 3}$ is defined as follows:

$$A_e = A - \begin{bmatrix} 0 & 0 & u_0 \\ 0 & 0 & v_0 \\ 0 & 0 & 0 \end{bmatrix}, \tag{10.25}$$

where $u_0, v_0 \in \mathbb{R}$ denote the pixel coordinates of the principal point[3], and the auxiliary Jacobian-like matrix $L_v(t) \in \mathbb{R}^{3 \times 3}$ is defined as:

$$L_v = \begin{bmatrix} 1 & 0 & -\dfrac{x_1}{z_1} \\ 0 & 1 & -\dfrac{y_1}{z_1} \\ 0 & 0 & 1 \end{bmatrix}. \tag{10.26}$$

---

[3]The principal point is the image center that is defined as the frame buffer coordinates of the intersection of the optical axis with the image plane.

Remark 10.4   It is easy to show that the product $A_e L_v$ is an invertible upper triangular matrix from (10.25) and (10.26).

10.2.2.3   Closed-Loop Error System

Based on the structure of the open-loop error systems and subsequent stability analysis, the angular and linear camera velocity control inputs for the object are defined as follows:

$$\omega_e = R^T L_\omega^{-1} (\dot{\Theta}_d - K_\omega e_\omega) \tag{10.27}$$

$$v_e = -\frac{1}{\alpha_1} R^T (A_e L_v)^{-1} (K_v e_v - \hat{z}_1^* \dot{p}_{ed}) - [\omega_e]_\times \hat{s}_1. \tag{10.28}$$

In (10.27) and (10.28), K_ω, $K_v \in \mathbb{R}^{3 \times 3}$ denote diagonal matrices of positive constant control gains, and $\hat{z}_1^*(t) \in \mathbb{R}$, $\hat{s}_1(t) \in \mathbb{R}^3$ denote parameter estimates that are generated according to the following adaptive update laws:

$$\dot{\hat{z}}_1^*(t) = -\gamma_1 e_v^T \dot{p}_{ed} \tag{10.29}$$

$$\dot{\hat{s}}_1 = -\alpha_1 \Gamma_2 [\omega_e]_\times R^T L_v^T A_e^T e_v, \tag{10.30}$$

where $\gamma_1 \in \mathbb{R}$ denotes a positive constant adaptation gain, and $\Gamma_2 \in \mathbb{R}^{3 \times 3}$ denotes a positive constant diagonal adaptation gain matrix. After substituting (10.27) into (10.21), the following closed-loop error dynamics can be obtained:

$$\dot{e}_\omega = -K_\omega e_\omega. \tag{10.31}$$

After substituting (10.28) into (10.24), the closed-loop translation error dynamics can be determined as follows:

$$z_1^* \dot{e}_v = -K_v e_v + \alpha_1 A_e L_v R [\omega_e]_\times \tilde{s}_1 - \tilde{z}_1^* \dot{p}_{ed}, \tag{10.32}$$

where the parameter estimation error signals $\tilde{z}_1^*(t) \in \mathbb{R}$ and $\tilde{s}_1(t) \in \mathbb{R}^3$ are defined as Follows:

$$\tilde{z}_1^* = z_1^* - \hat{z}_1^* \qquad \tilde{s}_1 = s_1 - \hat{s}_1. \tag{10.33}$$

From (10.31) it is clear that the angular velocity control input given in (10.27) is designed to yield an exponentially stable rotational error system. The linear velocity control input given in (10.28) and the adaptive update laws given in (10.29) and (10.30) are motivated to yield a negative feedback term in translational error system with additional terms included to cancel out cross-product terms involving the parameter estimation errors in the subsequent stability analysis.

10.2.2.4   Stability Analysis

Theorem 10.1
The control inputs designed in (10.27) and (10.28), along with the adaptive update laws defined in (10.29) and (10.30), ensure that $e_\omega(t)$ and $e_v(t)$ are asymptotically driven to zero in the sense that

$$\lim_{t \to \infty} \|e_\omega(t)\|, \|e_v(t)\| = 0. \tag{10.34}$$

Proof 10.1   To prove Theorem 10.1, a nonnegative function $V(t) \in \mathbb{R}$ is defined as follows:

$$V \triangleq \frac{1}{2}e_\omega^T e_\omega + \frac{z_1^*}{2}e_v^T e_v + \frac{1}{2\gamma_1}\tilde{z}_1^{*2} + \frac{1}{2}\tilde{s}_1^T \Gamma_2^{-1}\tilde{s}_1. \tag{10.35}$$

After taking the time derivative of (10.35) and then substituting for the closed-loop error systems developed in (10.31) and (10.32), the following expression can be obtained:

$$\dot{V} = -e_\omega^T K_\omega e_\omega + e_v^T \left(-K_v e_v + \alpha_1 A_e L_v R \left[\omega_e\right]_\times \tilde{s}_1 - \tilde{z}_1^* \dot{p}_{ed} \right)$$
$$- \frac{1}{\gamma_1}\tilde{z}_1^* \dot{\tilde{z}}_1^* - \tilde{s}_1^T \Gamma_2^{-1} \dot{\tilde{s}}_1 \tag{10.36}$$

where the time derivative of (10.33) was utilized. After substituting the adaptive update laws designed in (10.29) and (10.30) into (10.36), the following simplified expression can be obtained:

$$\dot{V} = -e_\omega^T K_\omega e_\omega - e_v^T K_v e_v, \tag{10.37}$$

where the fact that $\left[\omega_e\right]_\times^T = -\left[\omega_e\right]_\times$ was utilized. Based on (10.33), (10.35), and (10.37), it can be determined that $e_\omega(t), e_v(t), \tilde{z}_1^*(t), \hat{z}_1^*(t), \tilde{s}_1(t), \hat{s}_1(t) \in \mathcal{L}_\infty$ and that $e_\omega(t), e_v(t) \in \mathcal{L}_2$ [171]. Based on the assumption that $\dot{\Theta}_d(t)$ is designed as a bounded function, the expressions given in (10.15), (10.22), (10.23), and (10.27) can be used to conclude that $\omega_e(t) \in \mathcal{L}_\infty$. Since $e_v(t) \in \mathcal{L}_\infty$, (10.5), (10.11), (10.13), (10.14), and (10.26) can be used to prove that $m_1(t), L_v(t) \in \mathcal{L}_\infty$. Given that $\dot{p}_{ed}(t)$ is assumed to be bounded function, the expressions in (10.28)–(10.32) can be used to conclude that $\dot{\hat{z}}_1^*(t), \dot{\hat{s}}_1(t), v_e(t), \dot{e}_v(t), \dot{e}_\omega(t) \in \mathcal{L}_\infty$. Since $e_\omega(t), e_v(t) \in \mathcal{L}_2$ and $e_\omega(t), \dot{e}_\omega(t), e_v(t), \dot{e}_v(t) \in \mathcal{L}_\infty$, Barbalat's Lemma [192] can be used to prove the result given in (10.34).

Remark 10.5   The result in (10.34) is practically global in the sense that it is valid over the entire domain with the exception of the singularity introduced by the exponential parameterization of the rotation matrix (see (10.17)) and the physical restriction that $z_i(t), z_i^*(t)$, and $z_{di}(t)$ must remain positive. Although the result stated in Theorem 10.1 indicates asymptotic convergence for the rotation error $e_\omega(t)$, it is evident from (10.31) that

$$e_\omega(t) \leq e_\omega(0)\exp(-\lambda_{\min}(K_\omega)t)$$

where $\lambda_{\min}(K_\omega)$ denotes the minimum eigenvalue of the constant matrix K_ω. However, the fact that $e_\omega(t) \leq e_\omega(0)\exp(-\lambda_{\min}(K_\omega)t)$ does not simplify the control development or stability analysis and the overall resulting control objective of tracking a desired set of pre-recorded images is still asymptotically achieved.

10.3 Visual Tracking with Eye-in-Hand Configuration

10.3.1 Geometric Modeling

Based on the development provided for the fixed camera problem in the previous sections, a controller for the camera-in-hand problem can be developed in a similar manner. To formulate a controller for the camera-in-hand tracking problem consider the geometric relationships depicted in Figure 10.1. From the geometry between the coordinate frames, $\bar{m}_i^*$ can be related to $\bar{m}_i(t)$ and $\bar{m}_{di}(t)$ as follows:

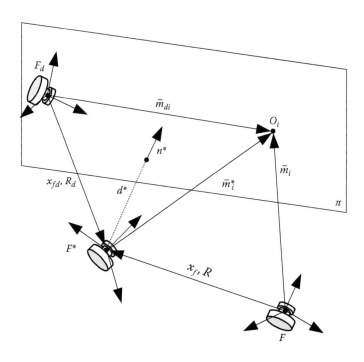

Figure 10.1: Coordinate frame relationships between the fixed feature point plane and the camera-in-hand at the current, desired, and reference position and orientation (i.e., $\mathcal{F}$, $\mathcal{F}_d$, and $\mathcal{F}^*$).

$$\bar{m}_i = x_f + R\bar{m}_i^*$$
$$\bar{m}_{di} = x_{fd} + R_d \bar{m}_i^*, \qquad (10.38)$$

where $\bar{m}_i(t)$, $\bar{m}_{di}(t)$, and $\bar{m}_i^*$ now denote the Euclidean coordinates of O_i expressed in $\mathcal{F}$, $\mathcal{F}_d$, and $\mathcal{F}^*$, respectively. By utilizing (10.2), (10.5)–(10.7), and a relationship similar to (10.10), the expressions in (10.38) can be written as follows:

$$m_i = \alpha_i \underbrace{\left(R + x_h n^{*T} \right)}_{H} m_i^* \qquad (10.39)$$

$$m_{di} = \alpha_{di} \underbrace{\left(R_d + x_{hd} n^{*T} \right)}_{H_d} m_i^*. \qquad (10.40)$$

In (10.39) and (10.40), $x_h(t)$, $x_{hd}(t) \in \mathbb{R}^3$ denote the following scaled translation vectors:

$$x_h = \frac{x_f}{d^*} \qquad x_{hd} = \frac{x_{fd}}{d^*}, \qquad (10.41)$$

$\alpha_i(t)$ and $\alpha_{di}(t)$ are introduced in (10.8) and (10.9), and $m_i(t)$, $m_{di}(t)$, and m_i^* now denote the normalized Euclidean coordinates of O_i expressed in $\mathcal{F}$, $\mathcal{F}_d$, and $\mathcal{F}^*$, respectively. Based on the development in (10.38)–(10.40), the Euclidean reconstruction and control formulation can be develop in the same manner as for the fixed camera problem. Specifically, the signals $R(t)$, $R_d(t)$, $x_h(t)$, $x_{hd}(t)$, and the depth ratios $\alpha_i(t)$ and $\alpha_{di}(t)$ can be computed.

10.3.2 Control Development

10.3.2.1 Open-Loop Error System

The error systems for the camera-in-hand problem are defined the same as for the fixed camera problem (i.e., see (10.13)–(10.16)); however, $u(t)$, $u_d(t)$, $\theta(t)$, and $\theta_d(t)$ are defined as in (10.18) in terms of $R(t)$ and $R_d(t)$, respectively, for the camera-in-hand problem. Based on this fact, the open-loop error dynamics for the rotation system can be derived as follows:

$$\dot{e}_\omega = -L_\omega \omega_c - \dot{\Theta}_d, \qquad (10.42)$$

where the fact that

$$[\omega_c]_\times = -\dot{R}R^T \qquad (10.43)$$

is used, and $\omega_c(t)$ denotes the camera angular velocity expressed in $\mathcal{F}$. After taking the time derivative of (10.38), the following expression for $\dot{\bar{m}}_1(t)$ can be derived for the camera-in-hand [63]:

$$\dot{\bar{m}}_1 = -v_c + [\bar{m}_1]_\times \omega_c, \qquad (10.44)$$

where $v_c(t)$ denotes the linear velocity of the camera expressed in terms of $\mathcal{F}$. After utilizing (10.44), the open-loop dynamics for $e_v(t)$ can be determined as follows:

$$z_1^* \dot{e}_v = -\alpha_1 A_e L_v v_c + \left(A_e L_v [m_1]_\times \omega_c - \dot{p}_{ed} \right) z_1^*, \tag{10.45}$$

where $e_v(t)$, $p_e(t)$, $p_{ed}(t)$ are defined in (10.13) and (10.14).

10.3.2.2 Controller Design

Based on the open-loop error systems in (10.42) and (10.45), the following control signals are designed:

$$\omega_c \triangleq L_\omega^{-1} \left(K_\omega e_\omega - \dot{\Theta}_d \right) \tag{10.46}$$

$$v_c \triangleq \frac{1}{\alpha_1} (A_e L_v)^{-1} \left(K_v e_v - \hat{z}_1^* \dot{p}_{ed} \right) + \frac{1}{\alpha_1} [m_1]_\times \omega_c \hat{z}_1^* \tag{10.47}$$

$$\dot{\hat{z}}_1^* \triangleq \gamma_1 e_v^T \left(A_e L_v [m_1]_\times \omega_c - \dot{p}_{ed} \right) \tag{10.48}$$

resulting in the following closed-loop error systems:

$$\dot{e}_\omega = -K_\omega e_\omega \tag{10.49}$$

$$z_1^* \dot{e}_v = -K_v e_v + \left(A_e L_v [m_1]_\times \omega_c - \dot{p}_{ed} \right) \tilde{z}_1^*. \tag{10.50}$$

The result in (10.34) can now be proven for the camera-in-hand problem using the same analysis techniques and the same nonnegative function as defined in (10.35) with the term containing $\tilde{s}_1(t)$ eliminated.

10.4 Simulation Results

Simulation studies were performed to illustrate the performance of the controller given in (10.27)–(10.30). For the simulation, the intrinsic camera calibration matrix is given as follows:

$$A = \begin{bmatrix} f k_u & -f k_u \cot \phi & u_0 \\ 0 & \dfrac{f k_v}{\sin \phi} & v_0 \\ 0 & 0 & 1 \end{bmatrix} \tag{10.51}$$

where $u_0 = 257$ (pixels), $v_0 = 253$ (pixels), $k_u = 101.4$ (pixels mm^{-1}) and $k_v = 101.4$ (pixels mm^{-1}) represent camera scaling factors, $\phi = 90$ deg is the angle between the camera axes, and $f = 12.5$ (mm) denotes the camera focal length. The control objective is defined in terms of tracking a desired image sequence. For the simulation, the desired image sequence was required to be arti-ficially generated. To generate an artificial image sequence for the simulation, the Euclidean coordinates of four target points were defined as follows:

$$s_1 = \begin{bmatrix} 0.1 \\ -0.1 \\ 0 \end{bmatrix} \quad s_2 = \begin{bmatrix} 0.1 \\ 0.1 \\ 0 \end{bmatrix} \quad s_3 = \begin{bmatrix} -0.1 \\ 0.1 \\ 0 \end{bmatrix} \quad s_4 = \begin{bmatrix} -0.1 \\ -0.1 \\ 0 \end{bmatrix} \quad (10.52)$$

and the initial translation and rotation between the current, desired, and reference image feature planes were defined as follows

$$x_f(0) = \begin{bmatrix} -0.3 \\ -0.1 \\ 3.7 \end{bmatrix} \quad x_{fd}(0) = \begin{bmatrix} 0.2 \\ 0.1 \\ 4 \end{bmatrix} \quad x_f^* = \begin{bmatrix} 0.2 \\ 0.1 \\ 4 \end{bmatrix} \quad (10.53)$$

$$R(0) = \begin{bmatrix} -0.4698 & -0.8660 & -0.1710 \\ -0.6477 & 0.4698 & -0.5997 \\ 0.5997 & -0.1710 & -0.7817 \end{bmatrix} \quad (10.54)$$

$$R_d(0) = \begin{bmatrix} 0.9568 & -0.2555 & -0.1386 \\ -0.2700 & -0.9578 & -0.0984 \\ -0.1077 & 0.1316 & -0.9854 \end{bmatrix} \quad (10.55)$$

$$R^* = \begin{bmatrix} 0.9865 & 0.0872 & -0.1386 \\ 0.0738 & -0.9924 & -0.0984 \\ -0.1462 & 0.0868 & -0.9854 \end{bmatrix}. \quad (10.56)$$

Based on (10.51)–(10.56), the initial pixel coordinates can be computed as follows:

$$p_1(0) = \begin{bmatrix} 170 & 182 & 1 \end{bmatrix}^T \quad p_2(0) = \begin{bmatrix} 110 & 213 & 1 \end{bmatrix}^T$$
$$p_3(0) = \begin{bmatrix} 138 & 257 & 1 \end{bmatrix}^T \quad p_4(0) = \begin{bmatrix} 199 & 224 & 1 \end{bmatrix}^T$$

$$p_{d1}(0) = \begin{bmatrix} 359 & 307 & 1 \end{bmatrix}^T \quad p_{d2}(0) = \begin{bmatrix} 343 & 246 & 1 \end{bmatrix}^T$$
$$p_{d3}(0) = \begin{bmatrix} 282 & 263 & 1 \end{bmatrix}^T \quad p_{d4}(0) = \begin{bmatrix} 298 & 324 & 1 \end{bmatrix}^T$$

$$p_1^* = \begin{bmatrix} 349 & 319 & 1 \end{bmatrix}^T \quad p_2^* = \begin{bmatrix} 355 & 256 & 1 \end{bmatrix}^T$$
$$p_3^* = \begin{bmatrix} 292 & 251 & 1 \end{bmatrix}^T \quad p_4^* = \begin{bmatrix} 286 & 314 & 1 \end{bmatrix}^T.$$

The time-varying desired image trajectory was then generated by the kinematics of the target plane where the desired linear and angular velocity were selected as follows:

$$v_{ed}(t) = \begin{bmatrix} 0.2\sin(t) & 0.3\sin(t) & 0 \end{bmatrix} (\text{m/s})$$
$$\omega_{ed}(t) = \begin{bmatrix} 0 & 0 & 0.52\sin(t) \end{bmatrix} (\text{rad/s}). \quad (10.57)$$

The desired translational trajectory is given in Figure 10.2, and the desired rotational trajectory is depicted in Figure 10.3. The generated desired image trajectory is a continuous function; however, in practice the image trajectory would be discretely represented by a sequence of pre-recorded images and would require a data interpolation scheme (i.e., a spline function) as described in Remark 10.3; hence, a spline function (i.e., the MATLAB spline routine) was utilized to generate a continuous curve to fit the desired image trajectory. For the top two subplots in Figure 10.2, the pixel values obtained from the pre-recorded image sequence are denoted by an asterisk (only select data points were included for clarity of illustration), and a cubic spline interpolation that was used to fit the data points is illustrated by a solid line. For the bottom subplot in Figure 10.2 and all the subplots in Figure 10.3, a plus sign denotes reconstructed Euclidean values computed using the pre-recorded pixel data, and the spline function is illustrated by a solid line.

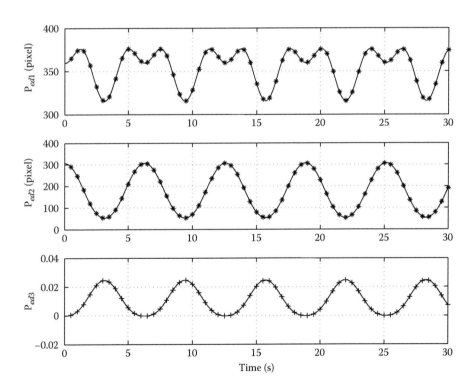

Figure 10.2: Desired translational trajectory of the manipulator end-effector generated by a spline function to fit pre-recorded image data.

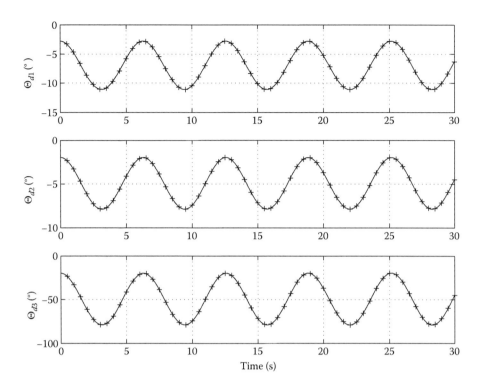

Figure 10.3: Desired rotational trajectory of the manipulator end-effector generated by a spline function to fit pre-recorded image data.

The control gains K_v and K_ω and the adaptation gains γ_1 and Γ_2 were adjusted through trial and error to the following values

$$K_v = diag\{6,8,5\} \quad K_\omega = diag\{0.6,0.8,0.7\}$$
$$\gamma_1 = 3 \times 10^{-6} \quad \Gamma_2 = 10^{-5} \times diag\{4.2,5.6,2.8\}. \tag{10.58}$$

The resulting errors between the actual relative translational and rotational of the target with respect to the reference target and the desired translational and rotational of the target with respect to thereference target are depicted in Figures 10.4 and 10.5, respectively. The parameter estimate signals are depicted in Figures 10.6 and 10.7. The angular and linear control input velocities (i.e., $\omega_e(t)$ and $v_e(t)$) defined in (10.27) and (10.28) are depicted in Figures 10.8 and 10.9.

While the results in Figures 10.4–10.9 provide an example of the performance of the tracking controller under ideal conditions, several issues must be considered for a practical implementation. For example, the performance of the tracking control algorithm is influenced by the accuracy of the image-space feedback signals and the accuracy of the reconstructed Euclidean information obtained from

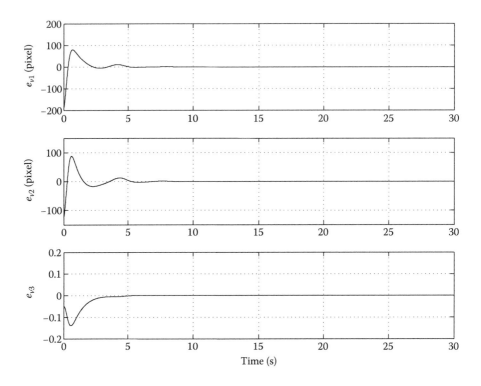

Figure 10.4: Error between the actual translation trajectory and the desired translation trajectory given in Figure 10.2 for the noise-free case.

constructing and decomposing the homography. That is, inaccuracies in determining the location of a feature from one frame to the next frame (i.e., feature tracking) will lead to errors in the construction and decomposition of the homography matrix, leading to errors in the feedback control signal. Inaccuracies in determining the feature point coordinates in an image is a similar problem faced in numerous sensor based feedback applications (e.g., noise associated with a force/torque sensor). Practically, errors related to sensor inaccuracies can often be addressed with an ad hoc filter scheme or other mechanisms (e.g., an intelligent image processing and feature tracking algorithm, redundant feature points and an optimal homography computation algorithm).

In light of these practical issues, another simulation was performed where random noise was injected with a standard deviation of 1 pixel (i.e., the measured feature coordinate was subject to ± 4 pixels of measurement error) as in [150]. As in any practical feedback control application in the presence of sensor noise, a filter was employed. Specifically, ad hoc third order butterworth low pass filters with a cutoff frequency of 10 rad/s were utilized to preprocess the corrupted

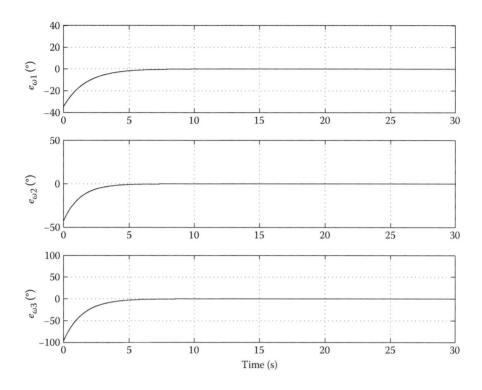

Figure 10.5: Error between the actual rotation trajectory and the desired rotation trajectory given in Figure 10.5 for the noise-free case.

image data. The control gains K_v and K_ω and the adaptation gains γ_1 and Γ_2 were tuned through trial and error to the following values:

$$K_v = diag\{17,11,9\} \quad K_\omega = diag\{0.4,0.4,0.4\}$$
$$\gamma_1 = 5 \times 10^{-7} \quad \Gamma_2 = 10^{-5} \times diag\{2.4,3.2,1.6\}. \tag{10.59}$$

The resulting translational and rotational errors of the target are depicted in Figures 10.10 and 10.11, respectively. The parameter estimate signals are depicted in Figures 10.12 and 10.13. The control input velocities $\omega_e(t)$ and $v_e(t)$ defined in (10.27) and (10.28) are depicted in Figures 10.14 and 10.15.

Another simulation was also performed to test the robustness of the controller with respect to the constant rotation matrix R^*. The constant rotation matrix R^* in (10.4) is coarsely calibrated as $diag\{1,-1,-1\}$. The resulting translational and rotational errors of the target are depicted in Figures 10.16 and 10.17, respectively.

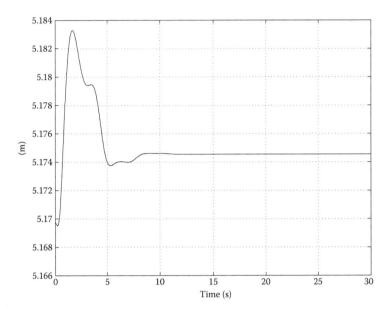

Figure 10.6: Parameter estimate for z_1^* for the noise-free case.

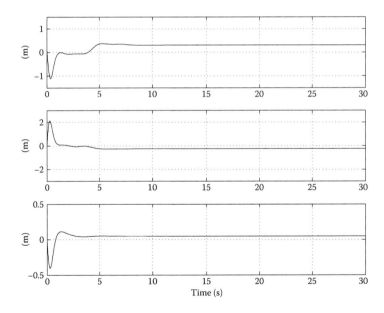

Figure 10.7: Parameter estimates for s_1 for the noise-free case.

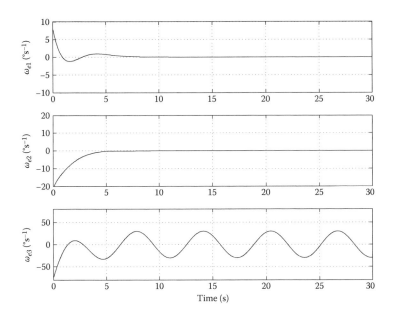

Figure 10.8: Angular velocity control input for the noise-free case.

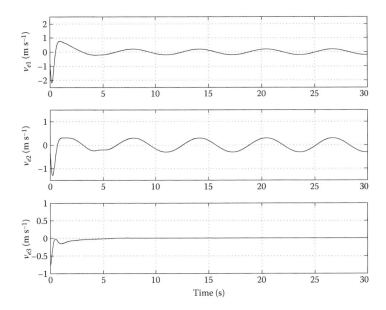

Figure 10.9: Linear velocity control input for the noise-free case.

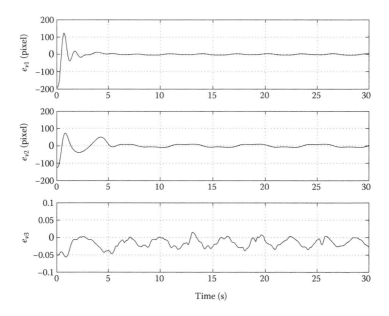

Figure 10.10: Error between the actual translation trajectory and the desired translation trajectory given in Figure 10.2 for the noise-injected case.

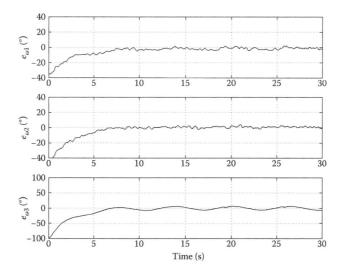

Figure 10.11: Error between the actual rotation trajectory and the desired rotation trajectory given in Figure 10.3 for the noise-injected case.

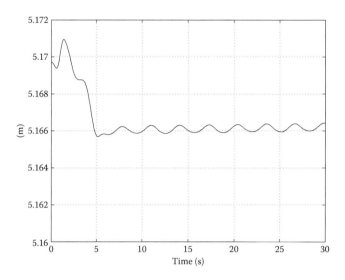

Figure 10.12: Parameter estimate for z_1^* for the noise-injected case.

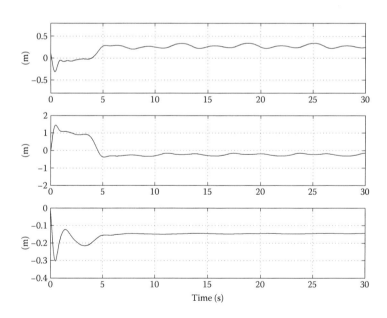

Figure 10.13: Parameter estimates for s_1 for the noise-injected case.

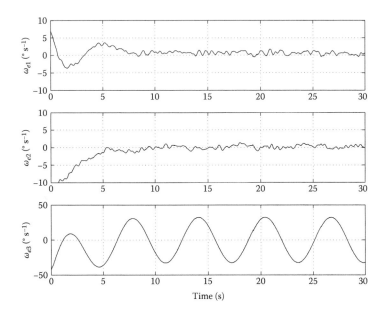

Figure 10.14: Angular velocity control input for the noise-injected case.

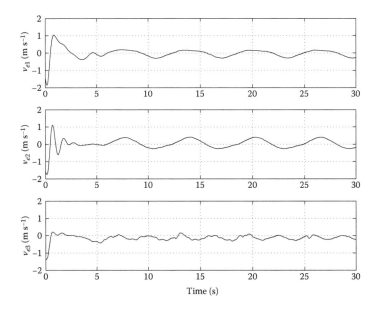

Figure 10.15: Linear velocity control input for the noise-injected case.

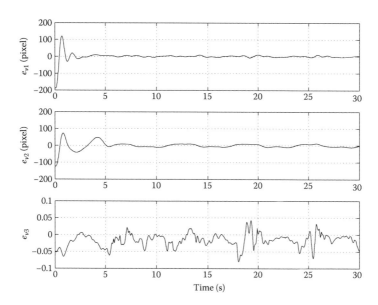

Figure 10.16: Error between the actual translation trajectory and the desired translation trajectory given in Figure 10.2 for the noise-injected case with a coarse calibration of R^*.

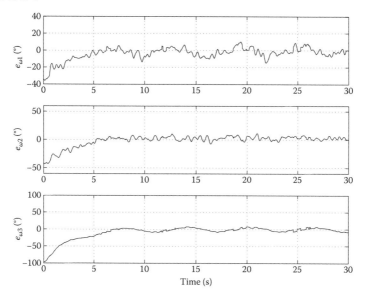

Figure 10.17: Error between the actual rotation trajectory and the desired rotation trajectory given in Figure 10.3 for the noise-injected case with a coarse calibration of R^*.

10.5 Conclusion

In this chapter, an adaptive visual servo controller is developed for the fixed camera configuration to enable the end-effector of a robot manipulator to track a desired trajectory determined by an *a priori* available sequence of images. The controller is formulated using a hybrid composition of image-space pixel information and reconstructed Euclidean information that is obtained via projective homography relationships between the actual image, a reference image, and the desired image. To achieve the objective, a Lyapunov-based adaptive control strategy is employed to actively compensate for the lack of unknown depth measurements and unknown object model parameters. Based on the development for the fixed camera controller, an extension is provided to enable a camera held by a robot end-effector to track a desired trajectory determined from a sequence of images (i.e., camera-in-hand tracking). Simulation results were provided to demonstrate the performance of the controller for the fixed camera problem.

Chapter 11

Robust Moving Object Tracking Control

11.1 Introduction

This chapter proposes a homography-based visual servo controller for a rigid body to track a moving object in 3-D space with a fixed relative pose (i.e., position and orientation) using visual feedback. Specifically, a monocular camera is mounted on the rigid body, and a reference image is pre-recorded to express the desired relative pose. By leveraging the previous work [27], the homography between the current and reference images is developed to construct the rotational and translational system errors without using the information of the object model. Since the leading object's velocities and the distance information are unknown, system uncertainties exist. Based on the RISE methodology [168,217], a continuous robust nonlinear controller is developed to achieve asymptotic tracking in the presence of the system uncertainties. To facilitate the stability analysis, the system uncertainties are divided into the error-unrelated and the error-related system uncertainties, then the composited system errors are utilized to derive the upper bounds of the error-related system uncertainties. Rigorous stability analysis is conducted on the basis of the Lyapunov theory and the derived upper bounds. Simulations are carried out based on the Virtual Robot Experimentation Platform (V-REP) [175], and the results are provided and analyzed for the performance evaluation of the proposed approach.

The objective is to develop a visual servo controller that ensures the following rigid body (i.e., camera), denoted by $\mathcal{F}$, tracking the leading object, denoted by

$\mathcal{F}_l$, with a fixed relative rotation R^* and a fixed relative translation x_f^* in the sense that

$$R(t) \to R^*, \; x_f(t) \to x_f^* \qquad \text{as} \qquad t \to \infty.$$

11.2 Vision System Model

11.2.1 Camera Geometry

Denote the coplanar and non-collinear feature points by $O_i \; \forall i = 1, 2, \ldots, N$ ($N \geq 4$). Let the 3-D coordinate of O_i on the plane π be defined as $\bar{m}_i(t) \triangleq \left[x_i(t) \; y_i(t) \; z_i(t)\right]^T \in \mathbb{R}^3$ expressed in terms of $\mathcal{F}$ and $\bar{m}_i^* \triangleq \left[x_i^* \; y_i^* \; z_i^*\right]^T \in \mathbb{R}^3$ expressed in terms of $\mathcal{F}^*$. From the geometry between the coordinate frames, the following relationships can be determined:

$$\bar{m}_i = \bar{x}_f + \bar{R}\bar{m}_i^*, \tag{11.1}$$

where $\bar{R}(t) = R(R^*)^T \in SO(3)$ and $\bar{x}_f(t) = x_f - \bar{R}x_f^* \in \mathbb{R}^3$ denote the rotation and translation between $\mathcal{F}^*$ and $\mathcal{F}$ expressed in terms of $\mathcal{F}$, respectively. The normalized Euclidean coordinates of the feature points on π can be expressed in $\mathcal{F}$ and $\mathcal{F}^*$, respectively, as follows:

$$m_i \triangleq \frac{\bar{m}_i}{z_i} \qquad m_i^* \triangleq \frac{\bar{m}_i^*}{z_i^*} \tag{11.2}$$

under the standard assumption that the depth $z_i(t)$, $z_i^* > \varepsilon$ where ε denotes an arbitrarily small positive constant [29]. Each feature point on π has a pixel coordinate expressed in the image coordinate frame for the current image and the desired image, denoted by $p_i(t) \triangleq \left[u_i(t) \; v_i(t) \; 1\right]^T$, $p_i^* \triangleq \left[u_i^* \; v_i^* \; 1\right]^T \in \mathbb{R}^3$, respectively. The pixel coordinates of the feature points are related to $m_i(t)$, and m_i^* by the following pinhole lens models [66]

$$p_i = A m_i \qquad p_i^* = A m_i^*, \tag{11.3}$$

where $A \in \mathbb{R}^{3 \times 3}$ is a known, constant, and invertible intrinsic camera calibration matrix. Since the Euclidean position of $\mathcal{F}_l$ and $\mathcal{F}^*$ cannot be directly measured, a Euclidean reconstruction needs to be developed to obtain the rotational and translational errors between the current image and the reference image.

11.2.2 Euclidean Reconstruction

In Figure 11.1, $n^* \in \mathbb{R}^3$ denotes the constant unit normal of the plane π expressed in the frame of $\mathcal{F}^*$, and the unknown constant distance from the origin of $\mathcal{F}^*$ to π

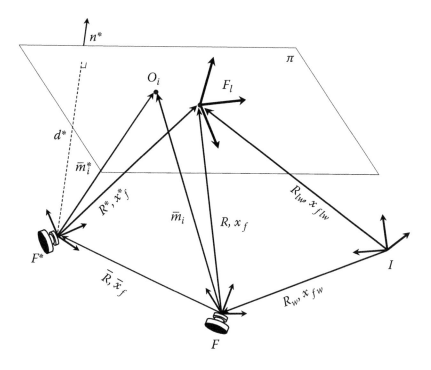

Figure 11.1: Coordinate frame relationships between a camera viewing a moving object at different poses.

is denoted by $d^* = n^{*T} \bar{m}_i^* \in \mathbb{R}$. From (11.1)–(11.3) and the expression of d^*, the relationship between the image coordinates of the corresponding feature points in $\mathcal{F}$ and $\mathcal{F}^*$ can be developed as:

$$
p_i = \underbrace{\frac{z_i^*}{z_i}}_{\alpha_i} \underbrace{A\left(\bar{R} + e_t n^{*T}\right) A^{-1} p_i^*}_{G} , \tag{11.4}
$$

where $\alpha_i(t) \in \mathbb{R}$ denotes the invertible depth ratio, $G(t) \in \mathbb{R}^{3 \times 3}$ denotes the projective homography, and $e_t(t) \in \mathbb{R}^3$ denotes a scaled translational vector that is defined as follows:

$$
e_t \triangleq \frac{\bar{x}_f}{d^*}. \tag{11.5}
$$

The feature point pairs $(p_i(t), p_i^*)$ can be utilized to estimate the projective homography $G(t)$, and then $\alpha_i(t)$, $\bar{R}(t)$, and $e_t(t)$ can be decomposed from $G(t)$.

11.3   Control Development

11.3.1   Open-Loop Error System

To quantify the rotation of $\mathcal{F}$ relative to $\mathcal{F}^*$ (i.e., $\bar{R}(t)$), a rotational error, denoted by $e_r(t) \in \mathbb{R}^3$, is defined as follows:

$$e_r \triangleq \phi\mu, \tag{11.6}$$

where $\phi(t) \in \mathbb{R}, \mu(t) \in \mathbb{R}^3$ denote the axis-angle representation of $\bar{R}(t)$. Given the rotation matrix $\bar{R}(t)$, a particular solution for $\phi(t)$ and $\mu(t)$ can be calculated [86]. Considering the dynamics of the following rigid body, it is assumed that $\phi(t)$ is confined as:

$$-\pi < \phi < \pi. \tag{11.7}$$

Based on the defined error system (11.5) and (11.6), the control objective can be stated as to regulate the error signals $e_r(t)$ and $e_t(t)$ to zero. If $e_r(t)$ is regulated to zero, $\bar{R}(t)$ will converge to I_3 [86], where I_i denotes an $i \times i$ identity matrix. Furthermore, since $\bar{R}(t) = R(R^*)^T$, it can be concluded that $R(t) \to R^*$ as $e_r(t) \to 0$. From (11.5) and the fact that $\bar{x}_f(t) = x_f - \bar{R}x_f^*$, it can be obtained that $x_f(t) \to x_f^*$ as $e_r(t) \to 0$ and $e_t(t) \to 0$.

After taking the time derivative of (11.6), the open-loop dynamics for $e_r(t)$ can be obtained (see [27] for further details)

$$\dot{e}_r = -L_\omega \omega + L_\omega R \omega_l, \tag{11.8}$$

where $\omega(t) \in \mathbb{R}^3$ denotes the angular velocity of the following rigid body expressed in $\mathcal{F}$, and $\omega_l(t) \in \mathbb{R}^3$ denotes the unknown angular velocity of the leading object expressed in $\mathcal{F}_l$. In (11.8), the Jacobian-like matrix $L_\omega(t) \in \mathbb{R}^{3\times3}$ is measurable and invertible in the assumed region of $\phi(t)$ [29].

Taking the time derivative of (11.5), the open-loop error system of $e_t(t)$ can be expressed as follows [27]:

$$d^*\dot{e}_t = -v + d^*[e_t]_\times \omega + Rv_l - [R\omega_l]_\times \bar{R}x_f^*. \tag{11.9}$$

In (11.9), $v(t) \in \mathbb{R}^3$ is the linear velocity of the following rigid body expressed in $\mathcal{F}$, and $v_l(t) \in \mathbb{R}^3$ is the unknown linear velocity of the leading object in $\mathcal{F}_l$. For any vector $x \in \mathbb{R}^3$, $[x]_\times$ denotes the 3×3 skew-symmetric expansion of x.

Assumption 11.1 *The unknown angular and linear velocities $\omega_l(t)$, $v_l(t)$ are continuously differentiable and bounded up to their second-order time derivative, i.e.,*

$$\|\omega_l\| \leq \bar{\omega}_l \qquad \|\dot{\omega}_l\| \leq \bar{\dot{\omega}}_l \qquad \|\ddot{\omega}_l\| \leq \bar{\ddot{\omega}}_l$$
$$\|v_l\| \leq \bar{v}_l \qquad \|\dot{v}_l\| \leq \bar{\dot{v}}_l \qquad \|\ddot{v}_l\| \leq \bar{\ddot{v}}_l$$

where $\|\cdot\|$ denotes the Euclidean norm and $\bar{\omega}_l$, $\bar{v}_l$, $\bar{\dot{\omega}}_l$, $\bar{\dot{v}}_l$, $\bar{\ddot{\omega}}_l$, $\bar{\ddot{v}}_l \in \mathbb{R}$ are positive constants.

Remark 11.1   To aid the subsequent design and analysis for the controller, vector functions $\text{Sgn}(\cdot)$, $\text{Tanh}(\cdot)$, and $\text{Lncosh}(\cdot) \in \mathbb{R}^n$ are defined as:

$$\text{Sgn}(x) \triangleq \begin{bmatrix} \text{sgn}(x_1) & \text{sgn}(x_2) & \cdots & \text{sgn}(x_n) \end{bmatrix}^T$$

$$\text{Tanh}(x) \triangleq \begin{bmatrix} \tanh(x_1) & \tanh(x_2) & \cdots & \tanh(x_n) \end{bmatrix}^T$$

$$\text{Lncosh}(x) \triangleq \begin{bmatrix} \ln(\cosh(x_1)) & \ln(\cosh(x_2)) & \cdots & \ln(\cosh(x_n)) \end{bmatrix}^T$$

where $x = \begin{bmatrix} x_1 & x_2 & \cdots & x_n \end{bmatrix}^T \in \mathbb{R}^n$, and $\text{sgn}(\cdot)$, $\tanh(\cdot)$, $\ln(\cdot)$, and $\cosh(\cdot)$ are the standard signum, hyperbolic tangent, natural logarithm, and hyperbolic cosine functions, respectively. Throughout this chapter, $\lambda_{\min}(\cdot)$ and $\lambda_{\max}(\cdot)$ denote the minimum and maximum eigenvalues of a matrix, respectively.

11.3.2   Control Design

Based on the structure of the open-loop error system (11.8) and (11.9), we design the angular and linear velocity control inputs for the following rigid body as follows:

$$\omega \triangleq L_\omega^{-1} \omega_r$$
$$v \triangleq v_t + \hat{d}^* [e_t]_\times \omega \tag{11.10}$$

where $\hat{d}^* \in \mathbb{R}$ is the best-guess estimate of d^*, and $\omega_r(t)$, $v_t(t) \in \mathbb{R}^3$ are the auxiliary variables. Inspired by [168,217], $\omega_r(t)$ and $v_t(t)$ are designed using a RISE feedback structure as follows:

$$\omega_r \triangleq (K_{pr} + I_3) \left(e_r(t) - e_r(0) + \int_0^t (K_{Ir} e_r(\tau)) d\tau \right)$$

$$+ \int_0^t (K_{sr} \text{Sgn}(e_r(\tau)) + K_{tr} \text{Tanh}(e_r(\tau))) d\tau$$

$$v_t \triangleq (K_{pt} + I_3) \left(e_t(t) - e_t(0) + \int_0^t (K_{It} e_t(\tau)) d\tau \right)$$

$$+ \int_0^t (K_{st} \text{Sgn}(e_t(\tau)) + K_{tt} \text{Tanh}(e_t(\tau))) d\tau. \tag{11.11}$$

In (11.11), $K_{pr}, K_{pt}, K_{Ir}, K_{It}, K_{sr}, K_{st}, K_{tr}, K_{tt} \in \mathbb{R}^{3 \times 3}$ are positive-definite, diagonal, control gain matrices. Especially, $K_{pr}, K_{pt}, K_{sr}, K_{st}$ are selected with appropriate values to guarantee asymptotic tracking, which will be discussed further in the subsequent stability analysis. Note that the terms $-e_r(0)$, $-e_t(0)$ in (11.11) are introduced to drive the initial values of $\omega_r(t)$, $v_t(t)$ to $\omega_r(0) = 0$, $v_t(0) = 0$ (i.e., $\omega(0) = 0$, $v(0) = 0$). In general, small initial control inputs can avoid physical damage to the system. However, the introduction of $-e_r(0)$, $-e_t(0)$ could influence the system response rate.

Remark 11.2 Different control techniques can be utilized to stabilize the dynamic system with uncertainties or disturbances. Discontinuous control techniques, such as variable structure or sliding mode control, can achieve asymptotic convergence, but they suffer from chattering or infinite control bandwidth. Classical high-gain feedback control methods are continuous, but only achieve uniformly ultimately bounded convergence [70]. Since the RISE-based controller designed in (11.10), (11.11) utilizes the integral of the signum function, it is continuous and an asymptotic tracking result can be achieved. For more detailed differences between the RISE-based control and other control methods, please refer to [168,217,218].

11.3.3 Closed-Loop Error System

By substituting (11.10) into (11.8) and (11.9), the closed-loop error system can be obtained.

$$\dot{e}_r = -\omega_r + \delta_r \qquad d^* \dot{e}_t = -v_t + \delta_t, \tag{11.12}$$

where $\omega_r(t)$, $v_t(t)$ are designed as in (11.11), and $\delta_r(t)$, $\delta_t(t) \in \mathbb{R}^3$ can be confirmed as

$$\delta_r \triangleq L_\omega R \omega_l \qquad \delta_t \triangleq \tilde{d}^* [e_t]_\times \omega + R v_l - [R \omega_l]_\times \bar{R} x_f^*. \tag{11.13}$$

In (11.13), $\tilde{d}^* \triangleq d^* - \hat{d}^* \in \mathbb{R}$ is the constant estimate error of distance from the origin of $\mathcal{F}^*$ to π.

To facilitate the subsequent development, two unmeasurable filtered tracking errors $\eta_r(t)$, $\eta_t(t) \in \mathbb{R}^3$ are defined as follows:

$$\eta_r \triangleq \dot{e}_r + K_{Ir} e_r \qquad \eta_t \triangleq \dot{e}_t + K_{It} e_t. \tag{11.14}$$

After substituting (11.11) and (11.12) into the time derivatives of (11.14), it can be obtained that

$$\begin{aligned} \dot{\eta}_r &= -e_r - (K_{pr} + I_3)\eta_r - K_{sr}\mathrm{Sgn}(e_r) - K_{tr}\mathrm{Tanh}(e_r) + N_r \\ d^* \dot{\eta}_t &= -d^* e_t - (K_{pt} + I_3)\eta_t - K_{st}\mathrm{Sgn}(e_t) - K_{tt}\mathrm{Tanh}(e_t) + N_t, \end{aligned} \tag{11.15}$$

where $N_r(e_r, \dot{e}_r, t)$, $N_t(e_t, \dot{e}_t, t) \in \mathbb{R}^3$ are the system uncertainty terms defined as:

$$\begin{aligned} N_r(e_r, \dot{e}_r, t) &\triangleq e_r + K_{Ir}\dot{e}_r + \dot{\delta}_r \\ N_t(e_t, \dot{e}_t, t) &\triangleq d^* e_t + d^* K_{It}\dot{e}_t + \dot{\delta}_t. \end{aligned} \tag{11.16}$$

Before analyzing the stability of the closed-loop error systems, $N_r(e_r, \dot{e}_r, t)$, $N_t(e_t, \dot{e}_t, t)$ are divided into the error-unrelated system uncertainties $N_{rd}(t)$, $N_{td}(t) \in \mathbb{R}^3$ and the error-related system uncertainties $\tilde{N}_r(t), \tilde{N}_t(t) \in \mathbb{R}^3$ as follows

$$\begin{aligned} N_{rd} &\triangleq N_r(0,0,t) & \tilde{N}_r &\triangleq N_r(e_r, \dot{e}_r, t) - N_{rd} \\ N_{td} &\triangleq N_t(0,0,t) & \tilde{N}_t &\triangleq N_t(e_t, \dot{e}_t, t) - N_{td}. \end{aligned} \tag{11.17}$$

Based on (11.17), (11.15) can be rewritten as:

$$\dot{\eta}_r = -e_r - (K_{pr} + I_3)\eta_r - K_{sr}\text{Sgn}(e_r) - K_{tr}\text{Tanh}(e_r) + N_{rd} + \tilde{N}_r$$
$$d^*\dot{\eta}_t = -d^* e_t - (K_{pt} + I_3)\eta_t - K_{st}\text{Sgn}(e_t) - K_{tt}\text{Tanh}(e_t) + N_{td} + \tilde{N}_t. \tag{11.18}$$

In (11.18), the differential equations are discontinuous due to the signum functions. Utilizing Filippov's theory of differential inclusions [69], the existence of solutions to (11.18) can be established for $\dot{\eta}_r(t)$ and $\dot{\eta}_t(t)$ (see Appendix C.1 for further details). Moreover, the structure of $\eta_r(t)$, $\eta_t(t)$, $\dot{\eta}_r(t)$, and $\dot{\eta}_t(t)$ shown in (11.14), (11.18) is motivated by the need to inject and cancel terms in the subsequent stability analysis.

11.4 Stability Analysis

11.4.1 Convergence of the Rotational Error

Property 11.1 *The error-related system uncertainty $\tilde{N}_r(t)$ defined in (11.17) can be upper bounded as:*

$$\|\tilde{N}_r\| \leq \rho_r \|z_r\|, \tag{11.19}$$

where $\rho_r \in \mathbb{R}$ is a positive constant, and the composited system error $z_r(t) \in \mathbb{R}^6$ is defined as:

$$z_r \triangleq \begin{bmatrix} e_r^T & \eta_r^T \end{bmatrix}^T. \tag{11.20}$$

Proof 11.1   See Appendix C.2.

Existing RISE-based control strategies usually apply the mean value theorem to upper bound the error-related system uncertainty $N_r(t)$ [70,168,185,217], and semi-global asymptotic tracking can be obtained if the expression of ρ_r contains the composited system error $z_r(t)$. In this chapter, based on the system characteristics, we prove that ρ_r in (11.19) is irrelevant to the system error (see Appendix C.2 for further details), which leads to global asymptotic tracking.

Lemma 11.1
Define an auxiliary function $\Lambda_r(t) \in \mathbb{R}$ and a positive constant $\xi_r \in \mathbb{R}$ as follows

$$\Lambda_r \triangleq \eta_r^T (N_{rd} - K_{sr}Sgn(e_r) - K_{tr}Tanh(e_r))$$
$$\xi_r \triangleq a^T K_{tr}Lncosh(e_r(0)) - e_r^T(0)(N_{rd}(0) - K_{sr}Sgn(e_r(0))), \tag{11.21}$$

where $a = \begin{bmatrix} 1 & 1 & 1 \end{bmatrix}^T$. If K_{sr} is selected with sufficiently large diagonal elements to satisfy the following condition:

$$\lambda_{\min}(K_{sr}) \geq \sup_{t \geq 0, \|e_r\| \neq 0}$$

$$\times \max \left\{ \|N_{rd}\|, \frac{\lambda_{\max}(K_{Ir})\|N_{rd}\| + \|\dot{N}_{rd}\|}{\lambda_{\min}(K_{Ir})} - 0.5\lambda_{\min}(K_{tr})\|Tanh(e_r)\| \right\};$$

(11.22)

then,

$$\xi_r - \int_0^t \Lambda_r(\tau) d\tau \geq 0.$$

(11.23)

Proof 11.2   See Appendix C.3.

In this chapter, hyperbolic functions are introduced to reduce the magnitude of K_{sr}, since a large control gain may intensify the chattering phenomenon and deteriorate the system performance [125,220]. Specifically, if the control inputs designed in (11.10) and (11.11) do not contain the nonlinear integral term $Tanh(\cdot)$, K_{sr} should be chosen to satisfy $\lambda_{\min}(K_{sr}) \geq \sup_{t \geq 0} \frac{\lambda_{\max}(K_{Ir})\|N_{rd}\| + \|\dot{N}_{rd}\|}{\lambda_{\min}(K_{Ir})}$, which is similar to the classical RISE-based control strategies [70,168,217]. Note that $\sup_{t \geq 0} \frac{\lambda_{\max}(K_{Ir})\|N_{rd}\| + \|\dot{N}_{rd}\|}{\lambda_{\min}(K_{Ir})}$ is larger than the right-hand side of (11.22). Therefore, the magnitude of K_{sr} can be reduced with the introduction of the hyperbolic functions.

Theorem 11.1
The control inputs designed in (11.10) and (11.11) ensure that the rotational error $e_r(t)$ asymptotically converges to zero in the sense that

$$\lim_{t \to \infty} e_r(t) = 0$$

(11.24)

on condition that the control gain matrix K_{sr} is adjusted according to (11.22), and the control gain matrix K_{pr} is selected sufficiently large to satisfy

$$\lambda_{\min}(K_{pr}) > \frac{\rho_r^2}{4\lambda_r},$$

(11.25)

where $\lambda_r \triangleq \min\{\lambda_{\min}(K_{Ir}), 1\}$ and ρ_r is given in Property 11.1.

Proof 11.3   To prove Theorem 11.1, a non-negative function $V_r(e_r, \eta_r, t) : \mathbb{R}^6 \times [0, \infty) \to \mathbb{R}$ is defined as:

$$V_r \triangleq \frac{1}{2}e_r^T e_r + \frac{1}{2}\eta_r^T \eta_r + \Delta_r,$$

(11.26)

where $\Delta_r(t) \in \mathbb{R}$ is an auxiliary function given by:

$$\Delta_r \triangleq \xi_r - \int_0^t \Lambda_r(\tau)d\tau. \tag{11.27}$$

According to Lemma 11.1, it is easy to see that $\Delta_r(t) \geq 0$ if (11.22) is satisfied. Taking the time derivative of (11.27), $\dot{\Delta}_r(t) = -\Lambda_r$ can be obtained. Based on (11.18) and (11.21), it is clear that $\dot{\eta}_r(t)$ and $\dot{\Delta}_r(t)$ are discontinuous due to the signum functions. The Filippov solutions [69,165] (see Definitions 11.1 and 11.2 in Appendix C.1) can be established for $\dot{\eta}_r(t)$, $\dot{\Delta}_r(t)$, and it can be concluded that $\eta_r(t)$ and $\Delta_r(t)$ are absolutely continuous and for almost all $t \in [0,\infty)$

$$\begin{aligned} \dot{\eta}_r &\in -e_r - (K_{pr}+I_3)\eta_r - K_{sr}\mathrm{SGN}_1(e_r) - K_{tr}\mathrm{Tanh}(e_r) + N_{rd} + \tilde{N}_r \\ \dot{\Delta}_r &\in -\eta_r^T(N_{rd} - K_{sr}\mathrm{SGN}_2(e_r) - K_{tr}\mathrm{Tanh}(e_r)) \end{aligned}, \tag{11.28}$$

where $\mathrm{SGN}_i(\cdot)$ $(i = 1,2)$ is a set-valued function [165] such that for $x = \begin{bmatrix} x_1 & x_2 & \cdots & x_n \end{bmatrix}^T \in \mathbb{R}^n$,

$$\mathrm{SGN}_i(x_j) = \begin{cases} 1, & x_j > 0 \\ [-1,1], & x_j = 0, \quad \forall j = 1,2,\ldots,n. \\ -1, & x_j < 0 \end{cases}$$

Taking the time derivative of $V_r(e_r, \eta_r, t)$ and substituting (11.14), it can be obtained that

$$\dot{V}_r = \begin{bmatrix} e_r^T & \eta_r^T & \dot{\Delta}_r \end{bmatrix} \begin{bmatrix} \dot{e}_r \\ \dot{\eta}_r \\ 1 \end{bmatrix} = \begin{bmatrix} e_r^T & \eta_r^T & \dot{\Delta}_r \end{bmatrix} \begin{bmatrix} \eta_r - K_{Ir}e_r \\ \dot{\eta}_r \\ 1 \end{bmatrix}. \tag{11.29}$$

According to [41,71] (see Definition 11.3 in Appendix C.1), (11.29), and the set-valued map of $\dot{\Delta}_r(t)$ given in (11.28), the generalized gradient of $V_r(e_r, \eta_r, t)$ at (e_r, η_r, t), denoted by $\partial V_r(t) \in \mathbb{R}^7$, can be defined as:

$$\partial V_r \triangleq \begin{bmatrix} e_r^T & \eta_r^T & -\eta_r^T(N_{rd} - K_{sr}\mathrm{SGN}_2(e_r) - K_{tr}\mathrm{Tanh}(e_r)) \end{bmatrix}. \tag{11.30}$$

Furthermore, by utilizing the chain rule [71,165] (see Lemma 11.1 in Appendix C.1), (11.29), and the set-valued map of $\dot{\eta}_r(t)$ given in (11.28), the following expression can be obtained:

$$\dot{V}_r \overset{a.e.}{\in} \dot{\tilde{V}}_r = \bigcap_{V_d \in \partial V_r} V_d^T \begin{bmatrix} \eta_r - K_{Ir}e_r \\ -e_r - (K_{pr}+I_3)\eta_r - K_{sr}\mathrm{SGN}_1(e_r) - K_{tr}\mathrm{Tanh}(e_r) + N_{rd} + \tilde{N}_r \\ 1 \end{bmatrix}. \tag{11.31}$$

For all $\upsilon_2 \in \mathrm{SGN}_2(e_r)$, it can be concluded from (11.30) and (11.31) that

$$\dot{\tilde{V}}_r = \bigcap_{\upsilon_2 \in \mathrm{SGN}_2(e_r)} \left(-e_r^T K_{Ir}e_r - \eta_r^T(K_{pr}+I_3)\eta_r + \eta_r^T \tilde{N}_r + \eta_r^T K_{sr}(\upsilon_2 - \mathrm{SGN}_1(e_r)) \right).$$

$$\tag{11.32}$$

If $e_r \neq 0$, $v_2 - \mathrm{SGN}_1(e_r) = 0$, and if $e_r = 0$, the following equation holds:

$$\bigcap_{v_2 \in \mathrm{SGN}_2(e_r)} (v_2 - \mathrm{SGN}_1(e_r)) = \bigcap_{v_2 \in [-1,1]} (v_2 - [-1,1]) = \bigcap_{v_2 \in [-1,1]} ([v_2 - 1, v_2 + 1]) = 0.$$

(11.33)

Thus, it can be determined that

$$\dot{V}_r = -e_r^T K_{Ir} e_r - \eta_r^T (K_{pr} + I_3) \eta_r + \eta_r^T \tilde{N}_r.$$

(11.34)

Based on Property 11.1 (i.e., (11.19), (11.20)) and Young's Inequality, the right-hand side of (11.34) can be upper bounded as:

$$\dot{V}_r \leq -\lambda_{\min}(K_{Ir}) \|e_r\|^2 - \|\eta_r\|^2 - \lambda_{\min}(K_{pr}) \|\eta_r\|^2 + \rho_r \|\eta_r\| \|z_r\|$$
$$\leq -\lambda_r \|z_r\|^2 + \frac{\rho_r^2}{4\lambda_{\min}(K_{pr})} \|z_r\|^2 ,$$

(11.35)

where $\lambda_r \triangleq \min\{\lambda_{\min}(K_{Ir}), 1\}$. If K_{pr} is selected to satisfy the condition (11.25), then it can be stated that

$$\dot{V}_r \leq -\gamma_r \|z_r\|^2 \leq 0,$$

(11.36)

where $\gamma_r \in \mathbb{R}$ is some positive constant and can be bounded as $0 < \gamma_r \leq \lambda_r - \frac{\rho_r^2}{4\lambda_{\min}(K_{pr})}$. Since it can be concluded from (C.13) in Appendix C.2 that ρ_r is bounded and does not contain K_{pr}, the existence of K_{pr} that satisfies the condition of (11.25) is guaranteed.

Based on (11.26), (11.36), and the fact that $\dot{V}_r \overset{a.e.}{\in} \dot{\tilde{V}}_r$, it can be concluded that $V_r(t) \in \mathcal{L}_\infty$; hence, $z_r(t) \in \mathcal{L}_\infty \cap \mathcal{L}_2$ [51]. Since $z_r(t) = \begin{bmatrix} e_r^T & \eta_r^T \end{bmatrix}^T$, it can be derived that $e_r(t), \eta_r(t) \in \mathcal{L}_\infty \cap \mathcal{L}_2$. Utilizing (11.14) and linear analysis techniques [53], $\dot{e}_r(t) \in \mathcal{L}_\infty$ can be obtained. From (11.8), it can be determined that $\omega(t) = -L_\omega^{-1} \dot{e}_r + R\omega_l$. Since $\dot{e}_r(t), \omega_l(t) \in \mathcal{L}_\infty$ and $L_\omega(t), R(t)$ are invertible matrices, it can be confirmed that $\omega(t) \in \mathcal{L}_\infty$. Besides, it can be inferred from (11.19) and $z_r(t) \in \mathcal{L}_\infty$ that $\|\tilde{N}_r\| \in \mathcal{L}_\infty$, and $\|N_{rd}\| \in \mathcal{L}_\infty$ can be obtained from (C.15) in Appendix C.3. Based on the first equation of (11.18) and $e_r(t), \eta_r(t), \|\tilde{N}_r\|$, $\|N_{rd}\| \in \mathcal{L}_\infty$, it can be concluded that $\dot{\eta}_r(t) \in \mathcal{L}_\infty$.

Define $W_r(t) \in \mathbb{R}$ as $W_r(t) \triangleq \int_0^t \frac{1}{2}(e_r(\tau)^T e_r(\tau) + \eta_r(\tau)^T \eta_r(\tau)) d\tau$. Then, it is easy to obtain that $\dot{W}_r(t) = \frac{1}{2}(e_r^T e_r + \eta_r^T \eta_r)$ and $\ddot{W}_r(t) = e_r^T \dot{e}_r + \eta_r^T \dot{\eta}_r$. Since $e_r(t), \eta_r(t) \in \mathcal{L}_2$, $W_r(t)$ is bounded. Based on $\dot{e}_r(t), \dot{\eta}_r(t), e_r(t), \eta_r(t) \in \mathcal{L}_\infty$, it can be inferred that $\ddot{W}_r(t) \in \mathcal{L}_\infty$, which is a sufficient condition for $\dot{W}_r(t)$ being uniformly continuous. Therefore, by Barbalat's lemma [115], $\dot{W}_r(t) \to 0$ as $t \to \infty$, i.e., $e_r(t), \eta_r(t) \to 0$ as $t \to \infty$.

11.4.2 Convergence of the Translational Error

Property 11.2 *The error-related system uncertainty $\tilde{N}_t(t)$ defined in (11.17) can be upper bounded as*

$$\|\tilde{N}_t\| \leq \rho_t \|z_t\|,$$

(11.37)

where $\rho_t \in \mathbb{R}$ is a positive constant, and the composited system error $z_t(t) \in \mathbb{R}^6$ is defined as

$$z_t \triangleq \begin{bmatrix} e_t^T & \eta_t^T \end{bmatrix}^T. \tag{11.38}$$

Proof 11.4   See Appendix C.4.

Lemma 11.2
Define an auxiliary function $\Lambda_t(t) \in \mathbb{R}$ as follows:

$$\Lambda_t \triangleq \eta_t^T (N_{td} - K_{st} Sgn(e_t) - K_{tt} Tanh(e_t)). \tag{11.39}$$

If K_{st} is selected to satisfy the following condition

$$\lambda_{\min}(K_{st}) \geq \sup_{t \geq 0, \|e_t\| \neq 0}$$

$$\times \max \left\{ \|N_{td}\|, \frac{\lambda_{\max}(K_{It}) \|N_{td}\| + \|\dot{N}_{td}\|}{\lambda_{\min}(K_{It})} - 0.5\lambda_{\min}(K_{tt}) \|Tanh(e_t)\| \right\}; \tag{11.40}$$

then,

$$\xi_t - \int_0^t \Lambda_t(\tau) d\tau \geq 0, \tag{11.41}$$

where $\xi_t \in \mathbb{R}$ is a constant defined as

$$\xi_t \triangleq a^T K_{tt} Lncosh(e_t(0)) - e_t^T(0)(N_{td}(0) - K_{st} Sgn(e_t(0))). \tag{11.42}$$

Proof 11.5   The proof is similar to Lemma 11.1, but it should be noted that $\dot{e}_r(t)$, $\dot{\eta}_r(t) \in \mathcal{L}_\infty$ need to be used to infer that $\|N_{td}\|, \|\dot{N}_{td}\| \in \mathcal{L}_\infty$.

Theorem 11.2
The control inputs designed in (11.10) and (11.11) ensure that the translational error $e_t(t)$ asymptotically converges to zero in the sense that

$$\lim_{t \to \infty} e_t(t) = 0 \tag{11.43}$$

on condition that the control gain matrix K_{st} is adjusted according to (11.40), and the control gain matrix K_{pt} is selected sufficiently large to satisfy

$$\lambda_{\min}(K_{pt}) > \frac{\rho_t^2}{4\lambda_t} \tag{11.44}$$

where $\lambda_t \triangleq \min\{d^* \lambda_{\min}(K_{It}), 1\}$ and ρ_t is given in Property 11.2.

Proof 11.6   Most of the proof is the same as the proof of Theorem 11.1, and the remaining part is trivial.

Remark 11.3 In (11.10), the term $\hat{d}^* [e_t]_\times \omega$ is introduced into the control design to reduce the magnitude of $\lambda_{\min}(K_{pt})$. Specifically, if the controller designed in (11.10) does not contain the term $\hat{d}^* [e_t]_\times \omega$, it is easy to conclude that $|\tilde{d}^*|$ existing in the expression of ρ_t (see (C.23) in Appendix C.4) should be replaced by $|d^*|$. Since $\tilde{d}^* = d^* - \hat{d}^*$ and $\hat{d}^*$ is the best-guess estimate of d^*, the magnitude of ρ_t will become larger if the term $\hat{d}^* [e_t]_\times \omega$ is dropped. Then, based on (11.44), the magnitude of $\lambda_{\min}(K_{pt})$ will become larger as well.

11.5 Simulation Results

11.5.1 Simulation Configuration

In this section, simulation studies are performed to illustrate the performance of the proposed controller with the robot simulator V-REP [175]. As shown in Figure 11.2a, the experimental platform consists of two six degrees of freedom (DOF) robotic manipulators. The simulated perspective camera is held by the end effector of the following robot and a textured plane is attached at the leading robot as shown in Figure 11.2b. The feature points are extracted from the textured plane using speeded-up robust feature [11], and then corresponding feature points between the desired image and the current image are matched. To reduce the influence of image noise existing in the simulated environment, homography is estimated using RANSAC based method [72]. The control system is implemented in MATLAB which then communicates with the simulator via the remote API of V-REP. Images captured by the simulated camera are sent to the image processing unit and the controller in MATLAB, and the control signals generated by the controller are sent back to the following robot in the simulator.

(a)   (b)

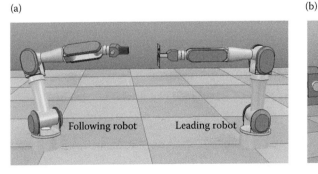 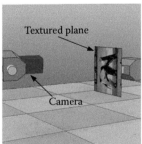

Figure 11.2: Simulation configuration for moving object tracking. (a) Robot platform and (b) setup details.

The simulation is conducted in the following steps: first, a reference image of the textured plane is captured by the camera to express the desired relative position and orientation between the leading object and the following robot. Second, two robots are placed at a initial relative pose where the textured plane is in the field of the camera. Then, the leading robot is actuated by a pre-set signal and the textured plane moves in the workspace. Meanwhile, the relevant controller is implemented to drive the following robot to track the leading robot with a fixed relative pose using visual feedback.

In the simulation, the total running time is 20 s and the control cycle is 50 ms. The intrinsic camera calibration matrix is given as follows:

$$A = \begin{bmatrix} 443.4050 & 0 & 256 \\ 0 & 443.4050 & 256 \\ 0 & 0 & 1 \end{bmatrix}$$

and the resolution of gray scale images captured by the camera is 512×512 pixels. The initial translation and rotation between $\mathcal{F}_l$ and $\mathcal{I}$, $\mathcal{F}$ and $\mathcal{I}$ are set as:

$$x_{flw}(0) = \begin{bmatrix} 0.5450 \\ 0 \\ 0.4279 \end{bmatrix} (\text{m}) \qquad R_{lw}(0) = \begin{bmatrix} 0 & 0 & -1 \\ 0 & 1 & 0 \\ 1 & 0 & 1 \end{bmatrix}$$

$$x_{fw}(0) = \begin{bmatrix} 0.3723 \\ 0.0497 \\ 0.4468 \end{bmatrix} (\text{m}) \qquad R_w(0) = \begin{bmatrix} 0.1716 & -0.0291 & 0.9847 \\ 0.9697 & -0.1714 & -0.1741 \\ 0.1738 & 0.9848 & -0.0012 \end{bmatrix}.$$

The velocities of moving object expressed in $\mathcal{I}$ are selected as follows:

$$v_{lw}(t) = \begin{bmatrix} 0.01 \sin\left(\frac{\pi t}{10}\right) & -0.045 \cos\left(\frac{\pi t}{10}\right) & 0.045 \cos\left(\frac{\pi t}{10}\right) \end{bmatrix} (\text{m/s})$$

$$\omega_{lw}(t) = \begin{bmatrix} \frac{\pi^2}{180} \cos\left(\frac{\pi t}{10}\right) & 0 & 0 \end{bmatrix} (\text{rad/s}).$$

11.5.2 Simulation Results and Discussion

To compare and validate the performance of the proposed method, three control algorithms are implemented on the robot simulator V-REP.

■ *Tracking with Proposed Controller:* The controller proposed in (11.10), (11.11) are employed in the simulation. Considering the sufficient conditions in (11.22), (11.25), (11.40), and (11.44), K_{sr}, K_{pr}, K_{st}, and K_{pt} can be first estimated approximately and then selected sufficiently large.

After the tuning process, the control gain matrices are set to the following values:

$$
\begin{aligned}
K_{pr} &= 0.1 diag \begin{bmatrix} 1 & 3 & 6 \end{bmatrix} & K_{pt} &= 0.1 diag \begin{bmatrix} 8 & 8 & 8 \end{bmatrix} \\
K_{Ir} &= 0.1 diag \begin{bmatrix} 1 & 3 & 6 \end{bmatrix} & K_{It} &= 0.01 diag \begin{bmatrix} 1 & 1 & 1 \end{bmatrix} \\
K_{sr} &= 0.001 diag \begin{bmatrix} 5 & 15 & 30 \end{bmatrix} & K_{st} &= 0.01 diag \begin{bmatrix} 5 & 6 & 6 \end{bmatrix} \\
K_{tr} &= 0.001 diag \begin{bmatrix} 6 & 18 & 36 \end{bmatrix} & K_{tt} &= 0.1 diag \begin{bmatrix} 4 & 5 & 5 \end{bmatrix}
\end{aligned}
\tag{11.45}
$$

and the best-guess estimate of d^* is chosen as $\hat{d}^* = 0.1 \,(m)$.

■ *Tracking with a PI-like Controller:* To validate the effectiveness of the nonlinear integral terms in (11.11), a PI-like controller is implemented as

$$
\begin{aligned}
\omega &= L_\omega^{-1} (K_{pr} + I_3) \left(e_r(t) - e_r(0) + \int_0^t (K_{Ir} e_r(\tau)) d\tau \right) \\
v &= (K_{pt} + I_3) \left(e_t(t) - e_t(0) + \int_0^t (K_{It} e_t(\tau)) d\tau \right) + \hat{d}^* [e_t]_\times \omega,
\end{aligned}
$$

where $\hat{d}^* = 0.1 \,(m)$, and K_{pr}, K_{pt}, K_{Ir}, and K_{It} are control gains selected as (11.45). It is easy to find that by removing the nonlinear integral terms from (11.11), the PI-like controller can be obtained.

■ *Tracking with a Super-twisting Controller:* Sliding mode techniques have been widely used to compensate for the system uncertainties. Therefore, to compare with the proposed method, a super-twisting controller [9] is also implemented as:

$$
\omega = L_\omega^{-1} \left(K_{pr} \begin{bmatrix} \sqrt{|e_{r,1}|} \, \text{sgn}(e_{r,1}) \\ \sqrt{|e_{r,2}|} \, \text{sgn}(e_{r,2}) \\ \sqrt{|e_{r,3}|} \, \text{sgn}(e_{r,3}) \end{bmatrix} + \int_0^t K_{sr} \text{Sgn}(e_r(\tau)) d\tau \right)
$$

$$
v = K_{pt} \begin{bmatrix} \sqrt{|e_{t,1}|} \, \text{sgn}(e_{t,1}) \\ \sqrt{|e_{t,2}|} \, \text{sgn}(e_{t,2}) \\ \sqrt{|e_{t,3}|} \, \text{sgn}(e_{t,3}) \end{bmatrix} + \int_0^t K_{st} \text{Sgn}(e_t(\tau)) d\tau
$$

where K_{pr}, K_{pt}, K_{sr}, and K_{st} are control gains adjusted as $K_{pr} = 0.2 diag \begin{bmatrix} 1 & 1 & 6 \end{bmatrix}$, $K_{pt} = 0.15 diag \begin{bmatrix} 1 & 1 & 1 \end{bmatrix}$, $K_{sr} = 0.18 diag \begin{bmatrix} 1 & 1 & 6 \end{bmatrix}$, and $K_{st} = 0.12 diag \begin{bmatrix} 1 & 1 & 1 \end{bmatrix}$.

The simulation results of the proposed controller are depicted in Figures 11.3, 11.4, and 11.5. Specifically, Figure 11.3a shows the reference image and Figure 11.3b–i are images captured by the simulated perspective camera during the control process at different moments. It can be concluded that the images captured after 8 s are very close to the reference one. Figure 11.4 depicts the partial trajectories of the leading moving object and the following rigid body expressed in

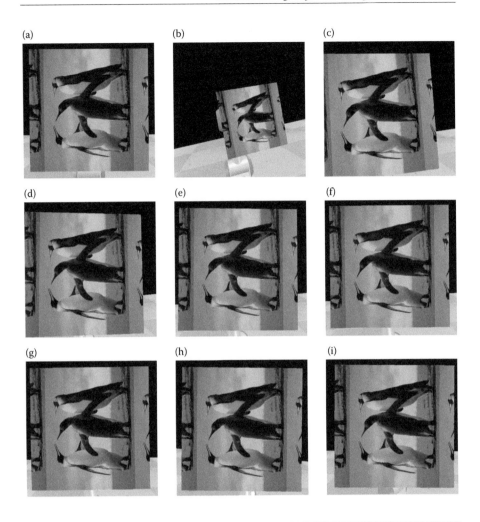

Figure 11.3: Images of the textured plane. (a) Reference, (b) 0 s, (c) 2 s, (d) 4 s, (e) 6 s, (f) 8 s, (g) 10 s, (h) 15 s, and (i) 20 s.

terms of the inertial coordinate frame $\mathcal{I}$. The resulting rotational and translational errors are shown in Figure 11.5a and b, respectively. It is clear that the rotational and translational errors all converge to small values that are very close to zero in the presence of image noise. The control inputs $\omega(t)$ and $v(t)$ defined in (11.10) are depicted in Figure 11.5c and d, respectively.

To validate the effectiveness of the nonlinear integral terms in (11.11), a PI-like controller is implemented for the same simulation scene. The corresponding results are shown in Figure 11.6. Comparing Figure 11.5 with Figure 11.6, it is obvious that the tracking performance of the PI-like controller is worse than the proposed controller. Furthermore, we can conclude that the nonlinear integral

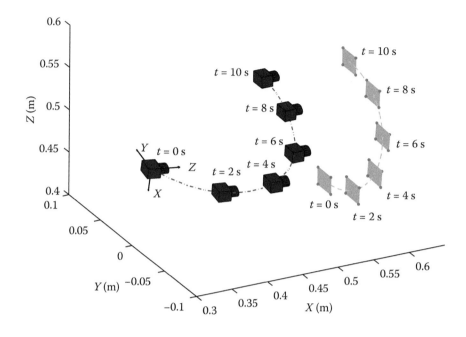

Figure 11.4: Trajectories in Euclidean space.

terms, which eliminate the influence of the system uncertainties, play a significant role in the convergence of system errors.

Table 11.1 presents the comparison with respect to the steady state root mean square (RMS) convergence errors ($t = 16 \sim 20$ s) and the maximum norm of control inputs for different methods. Results of the controller designed in [27] are also included. Note that the proposed method is different from the work [27]. The proportional controller developed in [27] only achieves uniformly ultimately bounded tracking by increasing the control gains to reduce the influence of the system uncertainty terms. In this chapter, however, the proposed controller can fully eliminate the influence of the system uncertainty terms to obtain asymptotic tracking. As seen from Table 11.1, the steady state RMS convergence errors of the proposed method are the smallest of the four implemented controllers. Moreover, due to the introduction of $e_r(0)$ and $e_t(0)$ in (11.10) and (11.11), the proposed controller can guarantee the convergence performance with smaller control inputs than the controller in [27] and the super-twisting controller. Figure 11.7 presents the results of the super-twisting controller. Based on Figures 11.5c, d and 11.7c, d, it is clear that the control inputs of the proposed method are smoother than the super-twisting controller. From the above, it can be concluded that the proposed controller shows better performance than the other controllers. The video of the simulation can be seen at https://youtu.be/JOId8RZBdLw, in

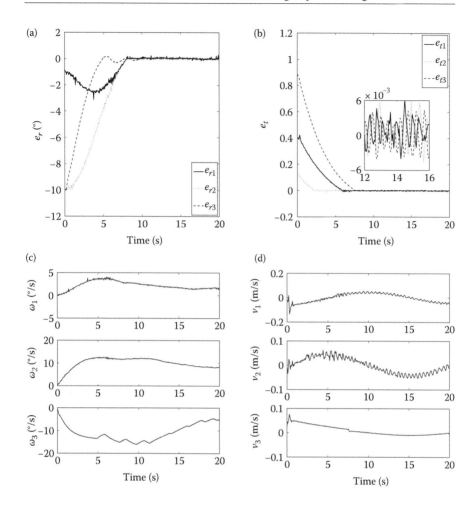

Figure 11.5: Results of proposed controller. (a) Rotational error, (b) translational error, (c) angular velocity input, and (d) linear velocity input.

which the simulation configuration, tracking process, and the simulation results of the proposed method are shown.

11.6   Conclusion

This chapter proposes a visual servo control strategy for a rigid body equipped with a monocular camera to follow a moving object in 3-D space with a fixed relative pose. The only required *a priori* knowledge is a reference image of the leading object recorded from the following rigid body at a desired relative pose.

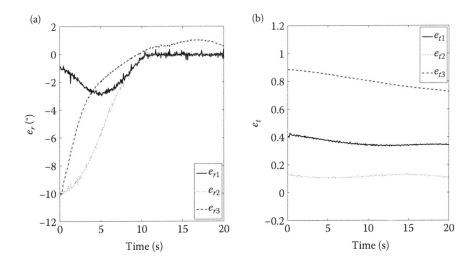

Figure 11.6: Results of PI-like controller. (a) Rotational error and (b) translational error.

Table 11.1   Comparison between the Proposed Controller and Other Controllers

	Proposed	*PI-like*	*[27]*	*Super-Twisting*
Steady state RMS of e_r	0.1093	0.9043	0.5043	0.3345
Steady state RMS of e_t	0.0038	0.8230	0.0307	0.0093
Maximum value of $\|\omega\|$	20.6766	17.3080	47.4954	107.9325
Maximum value of $\|v\|$	0.1553	0.0521	0.9803	0.2434

Based on homography techniques, orientation and scaled position information can be obtained for control formulation. A robust nonlinear controller is then developed using the robust integral of the signum of the error (RISE) methodology. The proposed control strategy eliminates the requirement of knowing the moving object's angular and linear velocities as well as the distance information. The asymptotic convergence of the proposed control algorithm is guaranteed via Lyapunov-based stability analysis. Simulation studies comparing the proposed method with a PI-like and a super-twisting controller are provided, which demonstrate the performance and effectiveness of the proposed strategy.

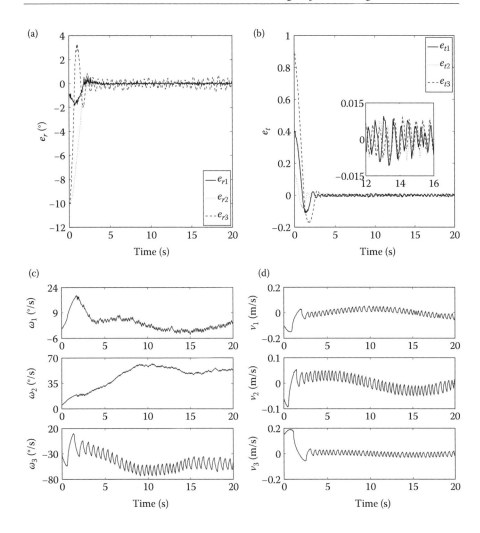

Figure 11.7: Results of super-twisting controller. (a) Rotational error and (b) translational error, (c) angular velocity input, and (d) linear velocity input.

Chapter 12

Visual Control with Field-of-View Constraints

12.1 Introduction

Motivated by the image space navigation function developed in [46], an off-line desired image trajectory generator is proposed based on a new image Jacobian-like matrix for the monocular, camera-in-hand problem. This approach generates a desired camera pose trajectory that moves the camera from the initial camera pose to a goal camera pose while ensuring that all the feature points of the object remain visible under certain technical restrictions. To develop a desired camera pose trajectory that ensures all feature points remain visible, a unique relationship is formulated between the desired image feature vector and the desired camera pose. The resulting image Jacobian-like matrix is related to the camera pose, rather than the camera velocity as in other approaches [23]. Motivation for the development of this relationship is that the resulting image Jacobian-like matrix is measurable, and hence, does not suffer from the lack of robustness associated with estimation based methods. Furthermore, the desired image generated with this image Jacobian-like matrix satisfies rigid body constraints (the terminology, rigid body constraints, this chapter is utilized to denote the image feature vector constraints in which feature points have a fixed relative position to each other in Euclidean space). Building on our recent research in [29], an adaptive homography-based visual tracking controller is then developed to ensure that the actual camera pose tracks the desired camera pose trajectory (i.e., the actual features track the desired feature point trajectory) despite the fact that time-varying depth from the camera to the reference image plane is not measurable from the monocular camera system. Based on the analysis of the homography-based controller, bounds are developed that can be used to ensure that the actual image features also remain visible under certain technical restrictions. A Lyapunov-based analysis is provided to support the claims for the path planner and to analyze the stability of the adaptive tracking controller. Simulation results are provided to illustrate the performance of the proposed approach.

12.2 Geometric Modeling

12.2.1 Euclidean Homography

Four feature points, denoted by O_i $\forall i = 1, 2, 3, 4$, are assumed to be located on a reference plane π (see Figure 12.1), and are considered to be coplanar[1] and not colinear. The reference plane can be related to the coordinate frames $\mathcal{F}$, $\mathcal{F}_d$, and $\mathcal{F}^*$ depicted in Figure 12.1 that denote the actual, desired, and goal pose of the camera, respectively.

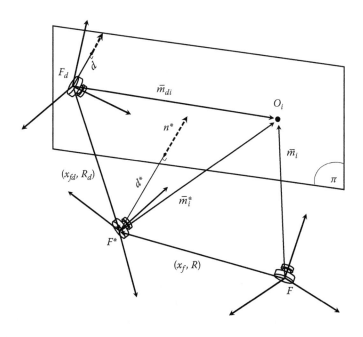

Figure 12.1: Coordinate frame relationships of the camera-in-hand system.

---

[1]It should be noted that if four coplanar target points are not available then the subsequent development can exploit the classic eight-points algorithm [150] with no four of the eight target points being coplanar.

Specifically, the following relationships can be developed from the geometry between the coordinate frames and the feature points located on π:

$$\bar{m}_i = x_f + R\bar{m}_i^* \\ \bar{m}_{di} = x_{fd} + R_d\bar{m}_i^* ,$$

(12.1)

where $\bar{m}_i(t) \triangleq [\ x_i(t) \quad y_i(t) \quad z_i(t)\ ]^T, \bar{m}_{di}(t) \triangleq [\ x_{di}(t) \quad y_{di}(t) \quad z_{di}(t)\ ]^T$, and $\bar{m}_i^* \triangleq [\ x_i^* \quad y_i^* \quad z_i^*\ ]^T$ denote the Euclidean coordinates of O_i expressed in $\mathcal{F}$, $\mathcal{F}_d$, and $\mathcal{F}^*$, respectively. In (12.1), $R(t), R_d(t) \in SO(3)$ denote the rotation between $\mathcal{F}$ and $\mathcal{F}^*$ and between $\mathcal{F}_d$ and $\mathcal{F}^*$, respectively, and $x_f(t), x_{fd}(t) \in \mathbb{R}^3$ denote translation vectors from $\mathcal{F}$ to $\mathcal{F}^*$ and $\mathcal{F}_d$ to $\mathcal{F}^*$ expressed in the coordinates of $\mathcal{F}$ and $\mathcal{F}_d$, respectively. The constant unknown distance from the origin of $\mathcal{F}^*$ to π is denoted by $d^* \in \mathbb{R}$ and is defined as follows:

$$d^* = n^{*T}\bar{m}_i^* ,$$

(12.2)

where $n^* \in \mathbb{R}^3$ denotes the constant unit normal to the plane π expressed in the coordinates of $\mathcal{F}^*$. Also from Figure 12.1, the unknown, time-varying distance from the origin of $\mathcal{F}_d$ to π, denoted by $d(t) \in \mathbb{R}$, can be expressed as follows:

$$d = n^{*T}R_d^T\bar{m}_{di}.$$

(12.3)

Since the Euclidean position of $\mathcal{F}, \mathcal{F}_d$, and $\mathcal{F}^*$ cannot be directly measured, the expressions in (12.1) need to be related to the measurable image-space coordinates. To this end, the normalized Euclidean coordinates of O_i expressed in terms of $\mathcal{F}, \mathcal{F}_d$, and $\mathcal{F}^*$ as $m_i(t), m_{di}(t), m_i^* \in \mathbb{R}^3$, respectively, are defined as follows:

$$m_i \triangleq \frac{\bar{m}_i}{z_i} \qquad m_{di} \triangleq \frac{\bar{m}_{di}}{z_{di}} \qquad m_i^* \triangleq \frac{\bar{m}_i^*}{z_i^*}$$

(12.4)

under the standard assumption that $z_i(t), z_{di}(t), z_i^* > \varepsilon$ where ε denotes an arbitrarily small positive constant. Based on (12.2) and (12.4), the expression in (12.1) can be rewritten as follows:

$$m_i = \underbrace{\frac{z_i^*}{z_i}}_{\alpha_i} \underbrace{\left(R + \frac{x_f}{d^*}n^{*T}\right)}_{H} m_i^*$$

(12.5)

$$m_{di} = \underbrace{\frac{z_i^*}{z_{di}}}_{\alpha_{di}} \underbrace{\left(R_d + \frac{x_{fd}}{d^*}n^{*T}\right)}_{H_d} m_i^* ,$$

(12.6)

where $\alpha_i(t)$, $\alpha_{di}(t) \in \mathbb{R}$ denote invertible depth ratios, $H(t)$, $H_d(t) \in \mathbb{R}^{3 \times 3}$ denote Euclidean homographies [66].

12.2.2 Projective Homography

Each feature point on π has a projected pixel coordinate denoted by $u_i(t)$, $v_i(t) \in \mathbb{R}$ in $\mathcal{F}$, $u_{di}(t)$, $v_{di}(t) \in \mathbb{R}$ in $\mathcal{F}_d$, and u_i^*, $v_i^* \in \mathbb{R}$ in $\mathcal{F}^*$, that are defined as follows:

$$
p_i \triangleq \begin{bmatrix} u_i & v_i & 1 \end{bmatrix}^T \qquad p_{di} \triangleq \begin{bmatrix} u_{di} & v_{di} & 1 \end{bmatrix}^T
$$
$$
p_i^* \triangleq \begin{bmatrix} u_i^* & v_i^* & 1 \end{bmatrix}^T. \tag{12.7}
$$

In (12.7), $p_i(t)$, $p_{di}(t)$, $p_i^* \in \mathbb{R}^3$ represent the image-space coordinates of the time-varying feature points, the desired time-varying feature point trajectory, and the constant reference feature points, respectively. To calculate the Euclidean homography given in (12.5) and (12.6) from pixel information, the projected pixel coordinates of the target points are related to $m_i(t)$, $m_{di}(t)$, and m_i^* by the following pinhole lens models [66]

$$
p_i = A m_i \qquad p_{di} = A m_{di} \qquad p_i^* = A m_i^*, \tag{12.8}
$$

where $A \in \mathbb{R}^{3 \times 3}$ is a known, constant, and invertible intrinsic camera calibration matrix with the following form:

$$
A = \begin{bmatrix} a_1 & a_2 & a_4 \\ 0 & a_3 & a_5 \\ 0 & 0 & 1 \end{bmatrix}, \tag{12.9}
$$

where $a_i \in \mathbb{R}$ $\forall i = 1, 2, ..., 5$, denote known, constant calibration parameters. After substituting (12.8) into (12.5) and (12.6), the following relationships can be developed

$$
p_i = \alpha_i \underbrace{\left(A H A^{-1} \right)}_{G} p_i^* \qquad p_{di} = \alpha_{di} \underbrace{\left(A H_d A^{-1} \right)}_{G_d} p_i^*, \tag{12.10}
$$

where $G(t)$, $G_d(t) \in \mathbb{R}^{3 \times 3}$ denote projective homographies. Given the images of the four feature points on π expressed in $\mathcal{F}$, $\mathcal{F}_d$, and $\mathcal{F}^*$, a linear system of equations can be developed from (12.10). From the linear system of equations, a decomposition algorithm (e.g., the Faugeras decomposition algorithm in [66]) can be used to compute $\alpha_i(t)$, $\alpha_{di}(t)$, n^*, $R(t)$, and $R_d(t)$ (see [29] for details)[2]. Hence, $\alpha_i(t)$, $\alpha_{di}(t)$, n^*, $R(t)$, and $R_d(t)$ are known signals that can be used in the subsequent development.

---

[2]The initial best-guess of n^* can be utilized to resolve the decomposition ambiguity. See [31] for details.

12.2.3 Kinematic Model of Vision System

The camera pose, denoted by $\Upsilon(t) \in \mathbb{R}^6$, can be expressed in terms of a hybrid of pixel and reconstructed Euclidean information as follows:

$$\Upsilon(t) \triangleq \begin{bmatrix} p_{e1}^T & \Theta^T \end{bmatrix}^T, \tag{12.11}$$

where the extended pixel coordinate $p_{e1}(t) \in \mathbb{R}^3$ is defined as follows:

$$p_{e1} = \begin{bmatrix} u_1 & v_1 & -\ln(\alpha_1) \end{bmatrix}^T, \tag{12.12}$$

and $\Theta(t) \in \mathbb{R}^3$ denotes the following axis–angle representation of $R(t)$ (see [29] for details):

$$\Theta = \mu(t)\theta(t). \tag{12.13}$$

In (12.12), $\ln(\cdot)$ denotes the natural logarithm, and $\alpha_1(t)$ is introduced in (12.5). In (12.13), $\mu(t) \in \mathbb{R}^3$ represents the unit axis of rotation, and $\theta(t)$ denotes the rotation angle about that axis. Based on the development in Appendix D.1, the open-loop dynamics for $\Upsilon(t)$ can be expressed as follows:

$$\dot{\Upsilon} = \begin{bmatrix} \dot{p}_{e1} \\ \dot{\Theta} \end{bmatrix} = \begin{bmatrix} -\frac{1}{z_1} A_{e1} & A_{e1}[m_1]_\times \\ 0 & -L_\omega \end{bmatrix} \begin{bmatrix} v_c \\ \omega_c \end{bmatrix}, \tag{12.14}$$

where $v_c(t) \in \mathbb{R}^3$ and $\omega_c(t) \in \mathbb{R}^3$ denote the linear and angular velocity of the camera expressed in terms of $\mathcal{F}$, $A_{ei}(u_i, v_i) \in \mathbb{R}^{3 \times 3}$ is a known, invertible matrix defined as follows:

$$A_{ei} = A - \begin{bmatrix} 0 & 0 & u_i \\ 0 & 0 & v_i \\ 0 & 0 & 0 \end{bmatrix} \qquad i = 1, 2, 3, 4, \tag{12.15}$$

and the invertible Jacobian-like matrix $L_\omega(\theta, \mu) \in \mathbb{R}^{3 \times 3}$ is defined as [152]

$$L_\omega = I_3 - \frac{\theta}{2}[\mu]_\times + \left(1 - \frac{\text{sinc}(\theta)}{\text{sinc}^2\left(\frac{\theta}{2}\right)}\right)[\mu]_\times^2, \tag{12.16}$$

where $I_n \in \mathbb{R}^{n \times n}$ denotes the $n \times n$ identity matrix, $[\mu_p]_\times$ denotes the 3×3 skew-symmetric expansion of $\mu_p(t)$, and

$$\text{sinc}(\theta(t)) \triangleq \frac{\sin\theta(t)}{\theta(t)}.$$

Remark 12.1 As stated in [197], the axis–angle representation of (12.13) is not unique, in the sense that a rotation of $-\theta(t)$ about $-\mu(t)$ is equal to a rotation of $\theta(t)$ about $\mu(t)$. A particular solution for $\theta(t)$ and $\mu(t)$ can be determined as follows [197]

$$\theta_p = \cos^{-1}\left(\frac{1}{2}\left(\text{tr}(R) - 1\right)\right) \qquad [\mu_p]_\times = \frac{R - R^T}{2\sin(\theta_p)}, \qquad (12.17)$$

where the notation $tr(\cdot)$ denotes the trace of a matrix. From (12.17), it is clear that

$$0 \le \theta_p(t) \le \pi. \qquad (12.18)$$

12.3 Image-Based Path Planning

The path planning objective involves regulating the pose of a camera held by the end-effector of a robot manipulator to a desired camera pose along an image-space trajectory while ensuring the target points remain visible. To achieve this objective, a desired camera pose trajectory is constructed in this section so that the desired image feature vector, denoted by $\bar{p}_d(t) \triangleq \begin{bmatrix} u_{d1}(t) & v_{d1}(t) \cdots u_{d4}(t) & v_{d4}(t) \end{bmatrix}^T \in \mathbb{R}^8$, remains in a set, denoted by $\mathcal{D} \subset \mathbb{R}^8$, where all four feature points of the target remain visible for a valid camera pose. The constant, goal image feature vector $\bar{p}^* \triangleq \begin{bmatrix} u_1^* & v_1^* \cdots u_4^* & v_4^* \end{bmatrix}^T \in \mathbb{R}^8$ is assumed be in the interior of $\mathcal{D}$. To generate the desired camera pose trajectory such that $\bar{p}_d(t) \in \mathcal{D}$, the special artificial potential function coined a navigation function in [117], can be used. Specifically, the navigation functions used in this chapter are defined as follows [174]:

Definition 12.1 A map $\varphi(\bar{p}_d) : \mathcal{D} \to [0, 1]$, is a NF if:
   P 1) Analytic on $\mathcal{D}$ (at least the first and second partial derivatives exist and are bounded on $\mathcal{D}$);
   P 2) a unique minimum exists at $\bar{p}^*$;
   P 3) it obtains a maximum value on the boundary of $\mathcal{D}$ (i.e., admissible on $\mathcal{D}$);
   P 4) it is a Morse function (i.e., the matrix of second partial derivatives, the Hessian, evaluated at its critical points is nonsingular (and has bounded elements based on the smoothness property in P 1)).

12.3.1 Pose Space to Image Space Relationship

To develop a desired camera pose trajectory that ensures $\bar{p}_d(t) \in \mathcal{D}$, the desired image feature vector is related to the desired camera pose, denoted by $\Upsilon_d(t) \in \mathbb{R}^6$, through the following relationship

$$\bar{p}_d = \Pi(\Upsilon_d) \qquad (12.19)$$

where $\Pi(\cdot) : \mathbb{R}^6 \rightarrow \mathcal{D}$ denotes an unknown function that maps the camera pose to the image feature vector[3]. In (12.19), the desired camera pose is defined as follows:

$$\Upsilon_d(t) \triangleq \begin{bmatrix} p_{ed1}^T & \Theta_d^T \end{bmatrix}^T, \tag{12.20}$$

where $p_{ed1}(t) \in \mathbb{R}^3$ denotes the desired extended pixel coordinates defined as follows:

$$p_{ed1} = \begin{bmatrix} u_{d1} & v_{d1} & -\ln(\alpha_{d1}) \end{bmatrix}^T, \tag{12.21}$$

where $\alpha_{d1}(t)$ is introduced in (12.6), and $\Theta_d(t) \in \mathbb{R}^3$ denotes the axis–angle representation of $R_d(t)$ as follows:

$$\Theta_d = \mu_d(t)\theta_d(t) \tag{12.22}$$

where $\mu_d(t) \in \mathbb{R}^3$ and $\theta_d(t) \in \mathbb{R}$ are defined in the same manner as $\mu(t)$ and $\theta(t)$ in (12.13) with respect to $R_d(t)$.

12.3.2 Desired Image Trajectory Planning

After taking the time derivative of (12.19), the following expression can be obtained:

$$\dot{p}_d = L_{\Upsilon_d}\dot{\Upsilon}_d, \tag{12.23}$$

where $L_{\Upsilon_d}(\bar{p}_d) \triangleq \dfrac{\partial \bar{p}_d}{\partial \Upsilon_d} \in \mathbb{R}^{8 \times 6}$ denotes an image Jacobian-like matrix. Based on the development in Appendix D.2, a measurable expression for $L_{\Upsilon_d}(t)$ can be developed as follows:

$$L_{\Upsilon_d} = \bar{I}T, \tag{12.24}$$

where $\bar{I} \in \mathbb{R}^{8 \times 12}$ denotes a constant, row-delete matrix defined as follows:

$$\bar{I} = \begin{bmatrix} I_2 & 0^2 & 0_2 & 0^2 & 0_2 & 0^2 & 0_2 & 0^2 \\ 0_2 & 0^2 & I_2 & 0^2 & 0_2 & 0^2 & 0_2 & 0^2 \\ 0_2 & 0^2 & 0_2 & 0^2 & I_2 & 0^2 & 0_2 & 0^2 \\ 0_2 & 0^2 & 0_2 & 0^2 & 0_2 & 0^2 & I_2 & 0^2 \end{bmatrix},$$

---

[3]The reason we choose four feature points to construct the image feature vector is that the same image of three points can be seen from four different camera poses [94]. A unique camera pose can theoretically be obtained by using at least four points [23]. Therefore, the map $\Pi(\cdot)$ is a unique mapping with the image feature vector corresponding to a valid camera pose.

where $0_n \in \mathbb{R}^{n \times n}$ denotes an $n \times n$ matrix of zeros, $0^n \in \mathbb{R}^n$ denotes an $n \times 1$ column of zeros, and $T(t) \in \mathbb{R}^{12 \times 6}$ is a measurable auxiliary matrix defined as follows:

$$
T = \begin{bmatrix} I_3 & 0_3 \\ \frac{\beta_1}{\beta_2} A_{ed2} A_{ed1}^{-1} & A_{ed2} \left[\frac{\beta_1}{\beta_2} m_{d1} - m_{d2} \right]_\times L_{\omega d}^{-1} \\ \frac{\beta_1}{\beta_3} A_{ed3} A_{ed1}^{-1} & A_{ed3} \left[\frac{\beta_1}{\beta_3} m_{d1} - m_{d3} \right]_\times L_{\omega d}^{-1} \\ \frac{\beta_1}{\beta_4} A_{ed4} A_{ed1}^{-1} & A_{ed4} \left[\frac{\beta_1}{\beta_4} m_{d1} - m_{d4} \right]_\times L_{\omega d}^{-1} \end{bmatrix}. \tag{12.25}
$$

In (12.25), $A_{edi}(u_{di}, v_{di}) \in \mathbb{R}^{3 \times 3}$ and the Jacobian-like matrix $L_{\omega d}(\theta_d, \mu_d) \in \mathbb{R}^{3 \times 3}$ are defined as in (12.15) and (12.16) with respect to $u_{di}(t), v_{di}(t), \mu_d(t)$, and $\theta_d(t)$, respectively. The auxiliary variable $\beta_i(t) \in \mathbb{R}$ in (12.25) is defined as follows:

$$
\beta_i \triangleq \frac{z_{di}}{d} \quad i = 1, 2, 3, 4. \tag{12.26}
$$

Based on (12.3), (12.4), and (12.8), $\beta_i(t)$ can be rewritten in terms of computed and measurable terms as follows:

$$
\beta_i = \frac{1}{n^{*T} R_d^T A^{-1} p_{di}}. \tag{12.27}
$$

Motivated by (12.23) and the definition of the navigation function in Definition 12.1, the desired camera pose trajectory is designed as follows:

$$
\dot{\Upsilon}_d = -K(\Upsilon_d) L_{\Upsilon_d}^T \nabla \varphi, \tag{12.28}
$$

where $K(\Upsilon_d) \triangleq k_1 \left(L_{\Upsilon_d}^T L_{\Upsilon_d} \right)^{-1}$, $k_1 \in \mathbb{R}$ denotes a positive constant, and $\nabla \varphi(\bar{p}_d) \triangleq \left(\frac{\partial \varphi(\bar{p}_d)}{\partial \bar{p}_d} \right)^T \in \mathbb{R}^8$ denotes the gradient vector of $\varphi(\bar{p}_d)$. The development of a particular image space NF and its gradient are provided in Appendix D.3. After substituting (12.28) into (12.23), the desired image trajectory can be expressed as follows:

$$
\dot{\bar{p}}_d = -L_{\Upsilon_d} K(\Upsilon_d) L_{\Upsilon_d}^T \nabla \varphi, \tag{12.29}
$$

where it is assumed that $\nabla \varphi(\bar{p}_d)$ is not a member of the null space of $L_{\Upsilon_d}^T(\bar{p}_d)$. Based on (12.23) and (12.28), it is clear that the desired image trajectory generated by (12.29) will satisfy the rigid body constraints.

Remark 12.2 Based on comments in [23] and the current development, it seems that a remaining open problem is to develop a rigorous, theoretical and general approach to ensure that $\nabla \varphi(\bar{p}_d)$ is not a member of the null space of $L_{\Upsilon_d}^T(\bar{p}_d)$ (i.e., $\nabla \varphi(\bar{p}_d) \notin NS(L_{\Upsilon_d}^T(\bar{p}_d))$ where $NS(\cdot)$ denotes the null space operator). However, since the approach in this chapter is developed in terms of the desired image-space trajectory (and hence, is an off-line approach), a particular desired image trajectory can be chosen (e.g., by trial and error) *a priori* to ensure that $\nabla\varphi(\bar{p}_d) \notin NS(L_{\Upsilon_d}^T(\bar{p}_d))$. Similar comments are provided in [23] and [158] that indicate that in practice this assumption can be readily satisfied for particular cases. Likewise, a particular desired image trajectory is also assumed to be *a priori* selected to ensure that $\Upsilon_d(t), \dot{\Upsilon}_d(t) \in \mathcal{L}_\infty$ if $\bar{p}_d(t) \in \mathcal{D}$. Based on the structure of (12.20) and (12.21), the assumption that $\Upsilon_d(t), \dot{\Upsilon}_d(t) \in \mathcal{L}_\infty$ if $\bar{p}_d(t) \in \mathcal{D}$ is considered mild in the sense that the only possible alternative case is if the camera could somehow be positioned at an infinite distance from the target while all four feature points remain visible.

Remark 12.3　　It is clear that $K(\Upsilon_d)$ is positive definite if $L_{\Upsilon_d}(\bar{p}_d)$ is full rank. Similar to the statement in Remark 12.2, this assumption is readily satisfied for the proposed off-line path planner approach. Based on this assumption, $K(\Upsilon_d)$ satisfies the following inequalities:

$$\underline{k}\|\xi\|^2 \leqslant \xi^T K(\Upsilon_d)\,\xi \leqslant \bar{k}(\cdot)\|\xi\|^2 \quad \forall \xi \in \mathbb{R}^6, \tag{12.30}$$

where $\underline{k} \in \mathbb{R}$ denotes a positive constant, and $\bar{k}(\cdot)$ denotes a positive, non-decreasing function.

12.3.3　Path Planner Analysis

Theorem 12.1
Provided the desired feature points can be a priori selected to ensure that $\bar{p}_d(0) \in \mathcal{D}$ and that $\nabla\varphi(\bar{p}_d) \notin NS(L_{\Upsilon_d}^T(\bar{p}_d))$, then the desired image trajectory generated by (12.29) ensures that $\bar{p}_d(t) \in \mathcal{D}$ and (12.29) has the asymptotically stable equilibrium point $\bar{p}^*$.

Proof 12.1　　Let $V_1(\bar{p}_d) : \mathcal{D} \to \mathbb{R}$ denote a nonnegative function defined as follows:

$$V_1(\bar{p}_d) \triangleq \varphi(\bar{p}_d). \tag{12.31}$$

After taking the time derivative of (12.31), the following expression can be obtained

$$\dot{V}_1(\bar{p}_d(t)) = (\nabla\varphi)^T \dot{\bar{p}}_d. \tag{12.32}$$

After substituting (12.29) into (12.32), the following expression can be obtained

$$\dot{V}_1(\bar{p}_d(t)) = -\left(L_{\Upsilon_d}^T \nabla\varphi\right)^T K(\Upsilon_d) L_{\Upsilon_d}^T \nabla\varphi. \tag{12.33}$$

Based on (12.30), $\dot{V}_1(\bar{p}_d(t))$ can be upper bounded as follows

$$\dot{V}_1(\bar{p}_d(t)) \leqslant -\underline{k}\left\|L_{\Upsilon_d}^T \nabla\varphi\right\|^2, \tag{12.34}$$

which clearly shows that $V_1(\bar{p}_d(t))$ is a non-increasing function in the sense that

$$V_1(\bar{p}_d(t)) \leq V_1(\bar{p}_d(0)). \tag{12.35}$$

From (12.31), (12.35), and the development in Appendix D.3, it is clear that for any initial condition $\bar{p}_d(0) \in \mathcal{D}$, that $\bar{p}_d(t) \in \mathcal{D} \; \forall t > 0$; therefore, $\mathcal{D}$ is a positively invariant set [115]. Let $E_1 \subset \mathcal{D}$ denote the following set $E_1 \triangleq \{\bar{p}_d(t) \mid \dot{V}_1(\bar{p}_d) = 0\}$. Based on (12.33), it is clear that $\left\| L_{\Upsilon_d}^T(\bar{p}_d) \triangledown \varphi(\bar{p}_d) \right\| = 0$ in E_1; hence, from (12.28) and (12.29), it can be determined that $\|\dot{\Upsilon}_d(t)\| = \left\| \dot{\bar{p}}_d(t) \right\| = 0$ in E_1, and that E_1 is the largest invariant set. By invoking LaSalle's Theorem [115], it can be determined that every solution $\bar{p}_d(t) \in \mathcal{D}$ approaches E_1 as $t \to \infty$, and hence, $\left\| L_{\Upsilon_d}^T(\bar{p}_d) \triangledown \varphi(\bar{p}_d) \right\| \to 0$. Since $\bar{p}_d(t)$ are chosen *a priori* via the off-line path planning routine in (12.29), the four feature points can be *a priori* selected to ensure that $\triangledown \varphi(\bar{p}_d) \notin NS(L_{\Upsilon_d}^T(\bar{p}_d))$. Provided $\triangledown \varphi(\bar{p}_d) \notin NS(L_{\Upsilon_d}^T(\bar{p}_d))$, then $\left\| L_{\Upsilon_d}^T(\bar{p}_d) \triangledown \varphi(\bar{p}_d) \right\| = 0$ implies that $\|\triangledown \varphi(\bar{p}_d)\| = 0$. Based on development given in Appendix D.3, we can now show that $\triangledown \varphi(\bar{p}_d(t)) \to 0$, and hence, that $\bar{p}_d(t) \to \bar{p}^*$.

12.4   Tracking Control Development

Based on Theorem 12.1, the desired camera pose trajectory can be generated from (12.28) to ensure that the camera moves along a path generated in the image space such that the desired object features remain visible (i.e., $\bar{p}_d(t) \in \mathcal{D}$). The objective in this section is to develop a controller so that the actual camera pose $\Upsilon(t)$ tracks the desired camera pose $\Upsilon_d(t)$ generated by (12.28), while also ensuring that the object features remain visible (i.e., $\bar{p}(t) \triangleq \begin{bmatrix} u_1(t) & v_1(t) & \dots & u_4(t) & v_4(t) \end{bmatrix}^T \in \mathcal{D}$). To quantify this objective, a rotational tracking error, denoted by $e_\omega(t) \in \mathbb{R}^3$, is defined as:

$$e_\omega \triangleq \Theta - \Theta_d, \tag{12.36}$$

and a translational tracking error, denoted by $e_v(t) \in \mathbb{R}^3$, is defined as follows:

$$e_v = p_{e1} - p_{ed1}. \tag{12.37}$$

12.4.1   Control Development

After taking the time derivative of (12.36) and (12.37), the open-loop dynamics for $e_\omega(t)$ and $e_v(t)$ can be obtained as follows:

$$\dot{e}_\omega = -L_\omega \omega_c - \dot{\Theta}_d \tag{12.38}$$

$$\dot{e}_v = -\frac{1}{z_1} A_{e1} v_c + A_{e1} [m_1]_\times \omega_c - \dot{p}_{ed1}, \tag{12.39}$$

where (12.14) was utilized. Based on the open-loop error systems in (12.38) and (12.39), $v_c(t)$ and $\omega_c(t)$ are designed as follows:

$$\omega_c \triangleq L_\omega^{-1} \left(K_\omega e_\omega - \dot{\Theta}_d \right) \tag{12.40}$$

$$v_c \triangleq \frac{1}{\alpha_1} A_{e1}^{-1} \left(K_v e_v - \hat{z}_1^* \dot{p}_{ed1} \right) + \frac{1}{\alpha_1} [m_1]_\times \omega_c \hat{z}_1^*, \tag{12.41}$$

where $K_\omega, K_v \in \mathbb{R}^{3 \times 3}$ denote diagonal matrices of positive constant control gains, and $\hat{z}_1^*(t) \in \mathbb{R}$ denotes a parameter estimate for z_1^* that is designed as follows:

$$\dot{\hat{z}}_1^* \triangleq k_2 e_v^T \left(A_{e1} [m_1]_\times \omega_c - \dot{p}_{ed1} \right), \tag{12.42}$$

where $k_2 \in \mathbb{R}$ denotes a positive constant adaptation gain. After substituting (12.40) and (12.41) into (12.38) and (12.39), the following closed-loop error systems can be developed:

$$\dot{e}_\omega = -K_\omega e_\omega \tag{12.43}$$

$$z_1^* \dot{e}_v = -K_v e_v + \left(A_{e1} [m_1]_\times \omega_c - \dot{p}_{ed1} \right) \tilde{z}_1^*, \tag{12.44}$$

where the parameter estimation error signal $\tilde{z}_1^*(t) \in \mathbb{R}$ is defined as follows:

$$\tilde{z}_1^* = z_1^* - \hat{z}_1^*. \tag{12.45}$$

12.4.2 Controller Analysis

Theorem 12.2
The controller introduced in (12.40) and (12.41), along with the adaptive update law defined in (12.42), ensure that the actual camera pose tracks the desired camera pose trajectory in the sense that

$$\|e_\omega(t)\| \to 0 \quad \|e_v(t)\| \to 0 \text{ as } t \to \infty. \tag{12.46}$$

Proof 12.2   Let $V_2(t) \in \mathbb{R}$ denote a nonnegative function defined as follows:

$$V_2 \triangleq \frac{1}{2} e_\omega^T e_\omega + \frac{z_1^*}{2} e_v^T e_v + \frac{1}{2k_2} \tilde{z}_1^{*2}. \tag{12.47}$$

After taking the time derivative of (12.47) and then substituting for the closed-loop error systems developed in (12.43) and (12.44), the following expression can be obtained:

$$\dot{V}_2 = -e_\omega^T K_\omega e_\omega - e_v^T K_v e_v \\ + e_v^T \left(A_{e1} [m_1]_\times \omega_c - \dot{p}_{ed1} \right) \tilde{z}_1^* - \frac{1}{k_2} \tilde{z}_1^* \dot{\hat{z}}_1^*, \tag{12.48}$$

where the time derivative of (12.45) was utilized. After substituting the adaptive update law designed in (12.42) into (12.48), the following expression can be obtained:

$$\dot{V}_2 = -e_\omega^T K_\omega e_\omega - e_v^T K_v e_v. \tag{12.49}$$

Based on (12.45), (12.47), and (12.49), it can be determined that $e_\omega(t)$, $e_v(t)$, $\tilde{z}_1^*(t)$, $\hat{z}_1^*(t) \in \mathcal{L}_\infty$ and that $e_\omega(t)$, $e_v(t) \in \mathcal{L}_2$. Based on the assumption that $\dot{\Theta}_d(t)$ is bounded (see Remark 12.2), the expressions given in (12.36), (12.40), and $L_\omega(t)$ in (12.16) can be used to conclude that $\omega_c(t) \in \mathcal{L}_\infty$. Since $e_v(t) \in \mathcal{L}_\infty$, (12.37), (12.12), (12.8), and $A_{e1}(t)$ in (12.15) can be used to prove that $u_1(t), v_1(t), \alpha_1(t), m_1(t), A_{e1}(t) \in \mathcal{L}_\infty$. Based on the assumption that $\dot{p}_{ed1}(t)$ is bounded (see Remark 12.2), the expressions in (12.41), (12.42), and (12.44) can be used to conclude that $v_c(t), \dot{\hat{z}}_1^*(t), \dot{e}_v(t) \in \mathcal{L}_\infty$. Since $e_\omega(t) \in \mathcal{L}_\infty$, it is clear from (12.43) that $\dot{e}_\omega(t) \in \mathcal{L}_\infty$. Since $e_\omega(t), e_v(t) \in \mathcal{L}_2$ and $e_\omega(t), \dot{e}_\omega(t), e_v(t), \dot{e}_v(t) \in \mathcal{L}_\infty$, Barbalat's Lemma [192] can be used to prove the result given in (12.46).

Remark 12.4   Based on the result provided in (12.46), it can be proven from the Euclidean reconstruction given in (12.5) and (12.6) that $R(t) \to R_d(t)$, $m_1(t) \to m_{d1}(t)$, and $z_1(t) \to z_{d1}(t)$ (and hence, $x_f(t) \to x_{fd}(t)$). Based on these results, (12.1) can be used to also prove that $\bar{m}_i(t) \to \bar{m}_{di}(t)$. Since $\Pi(\cdot)$ is a unique mapping, we can conclude that the desired camera pose converges to the goal camera pose based on the previous result $\bar{p}_d(t) \to \bar{p}^*$ from Theorem 12.1. Based on the above analysis, $\bar{m}_i(t) \to \bar{m}^*$.

Remark 12.5   Based on (12.47) and (12.49), the following inequality can be obtained:

$$e_\omega^T e_\omega + e_v^T e_v \; \leqslant \; 2\max\left\{1, \frac{1}{z_1^*}\right\} V_2(t) \tag{12.50}$$

$$\leqslant \; 2\max\left\{1, \frac{1}{z_1^*}\right\} V_2(0),$$

where

$$V_2(0) = \frac{1}{2} e_\omega^T(0) e_\omega(0) + \frac{z_1^*}{2} e_v^T(0) e_v(0) + \frac{1}{2k_2} \tilde{z}_1^{*2}(0).$$

From (12.11), (12.20), (12.36), (12.37), and the inequality in (12.50), the following inequality can be developed:

$$\|\Upsilon - \Upsilon_d\| \leqslant \sqrt{2\max\left\{1, \frac{1}{z_1^*}\right\} V_2(0)}. \tag{12.51}$$

Based on (12.19), the following expression can be developed:

$$\bar{p} = \Pi(\Upsilon) - \Pi(\Upsilon_d) + \bar{p}_d. \tag{12.52}$$

After applying the mean-value theorem to (12.52), the following inequality can be obtained:

$$\|\bar{p}\| \leqslant \|L_{\Upsilon_d}\| \|\Upsilon - \Upsilon_d\| + \|\bar{p}_d\|. \tag{12.53}$$

Since all signals are bounded, it can be shown that $L_{\Upsilon_d}^T(\bar{p}_d) \in \mathcal{L}_\infty$; hence, the following inequality can be developed from (12.51) and (12.53):

$$\|\bar{p}\| \leqslant \zeta_b \sqrt{V_2(0)} + \|\bar{p}_d\|, \tag{12.54}$$

for some positive constant $\zeta_b \in \mathbb{R}$, where $\bar{p}_d(t) \in \mathcal{D}$ based on Theorem 12.1. To ensure that $\bar{p}(t) \in \mathcal{D}$, the image space needs to be sized to account for the effects of $\zeta_b V_2(0)$. Based on (12.47), $V_2(0)$ can be made arbitrarily small by increasing k_2 and initializing $\bar{p}_d(0)$ close or equal to $\bar{p}(0)$.

12.5   Simulation Results

To solve the self-occlusion problem (the terminology, self-occlusion, in this chapter is utilized to denote the case when the center of the camera is in the plane determined by the feature points) from a practical point of view, we define a distance ratio $\gamma(t) \in \mathbb{R}$ as follows:

$$\gamma(t) = \frac{d}{d^*}. \tag{12.55}$$

From [152], $\gamma(t)$ is measurable. The idea to avoid the self-occlusion is to plan a desired image trajectory without self-occlusion. Based on (12.54), we can assume that the actual trajectory is close enough to the desired trajectory such that no self-occlusion occurred for the actual trajectory.

To illustrate the performance of the path planner given in (12.29) and the controller given in (12.40)–(12.42), numerical simulations will be performed for four standard visual servo tasks, which are believed to represent the most interesting tasks encountered by a visual servo system [77]:

- Task 1: Optical axis rotation, a pure rotation about the optic axis.

- Task 2: Optical axis translation, a pure translation along the optic axis.

- Task 3: Camera y-axis rotation, a pure rotation of the camera about the y-axis of the camera coordinate frame.

- Task 4: General camera motion, a transformation that includes a translation and rotation about an arbitrary axis.

For the simulation, the intrinsic camera calibration matrix is given as follows:

$$A = \begin{bmatrix} fk_u & -fk_u \cot\phi & u_0 \\ 0 & \dfrac{fk_v}{\sin\phi} & v_0 \\ 0 & 0 & 1 \end{bmatrix}, \tag{12.56}$$

where $u_0 = 257$ (pixels), $v_0 = 253$ (pixels) represent the pixel coordinates of the principal point, $k_u = 101.4$ (pixels mm^{-1}) and $k_v = 101.4$ (pixels mm^{-1}) represent camera scaling factors, $\phi = 90$ deg is the angle between the camera axes, and $f = 12.5$ (mm) denotes the camera focal length. For all simulations, we select $p_i(0) = p_{di}(0) \ \forall i = 1, 2, 3, 4, \kappa = 8$.

12.5.1 Optical Axis Rotation

The desired and actual image trajectories of the feature points are depicted in Figures 12.2 and 12.3, respectively. The translational and rotational tracking errors of the target are depicted in Figures 12.4 and 12.5, respectively, and the parameter estimate signal is depicted in Figure 12.6. For the resulting Figures 12.2–12.8, the control parameters were selected as follows:

$$K_v = I_3 \quad K_\omega = 0.3I_3 \quad k_1 = 400000 \quad k_2 = 0.04$$

$$K = diag\{10, 10, 10, 18, 13, 15, 10, 10\}.$$

The control input velocities $\omega_c(t)$ and $v_c(t)$ defined in (12.40) and (12.41) are depicted in Figures 12.7 and 12.8. From Figures 12.2 and 12.3, it is clear that

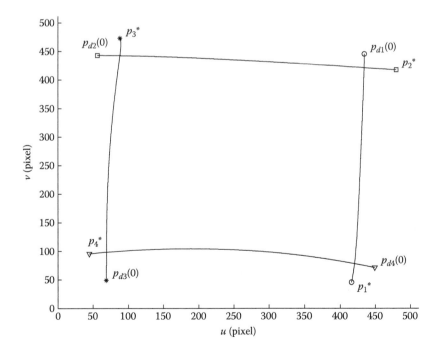

Figure 12.2: Desired image trajectory of Task 1.

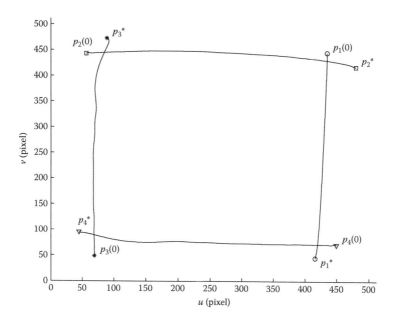

Figure 12.3: Actual image trajectory of Task 1.

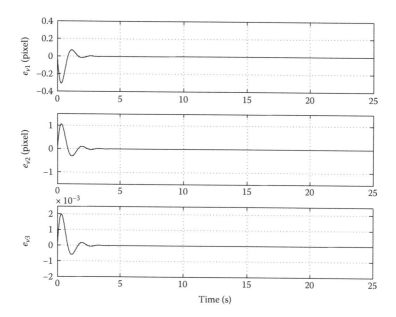

Figure 12.4: Translational tracking error of Task 1.

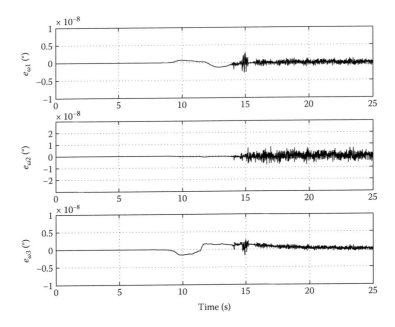

Figure 12.5: Rotational tracking error of Task 1.

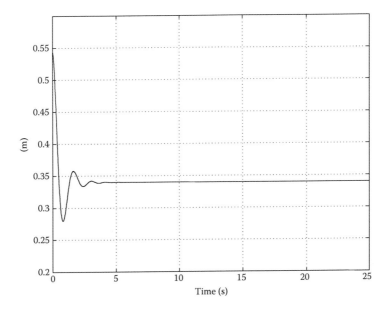

Figure 12.6: Estimate of z_1^* of Task 1.

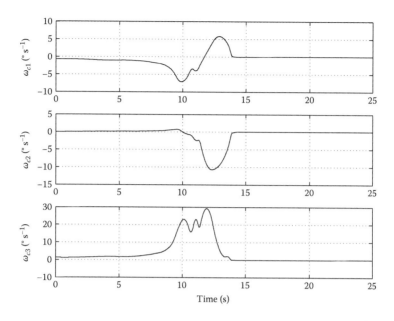

Figure 12.7: Angular velocity of Task 1.

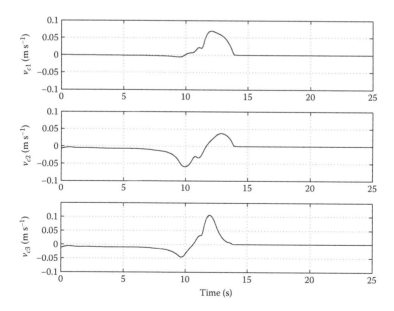

Figure 12.8: Linear velocity input of Task 1.

the desired feature points and actual feature points remain in the camera field of view and converge to the goal feature points. Figures 12.4 and 12.5 show that the tracking errors go to zero as $t \to \infty$.

12.5.2 Optical Axis Translation

The control parameters were selected as follows:

$$K_v = I_3 \quad K_\omega = 0.3I_3 \quad k_1 = 10000 \quad k_2 = 0.0004$$

$$K = diag\{30, 20, 10, 28, 33, 25, 10, 40\}.$$

The actual image trajectories of the feature points are depicted in Figure 12.9. The control input velocities $\omega_c(t)$ and $v_c(t)$ defined in (12.40) and (12.41) are depicted in Figure 12.10 and 12.11. From Figure 12.9, it is clear that the actual feature points remain in the camera field of view and converge to the goal feature points. To reduce the length of the chapter, we only provide figures for the actual image trajectory, angular velocity, and linear velocity for this task (also for Tasks 3 and 4). For more simulation results, see [25].

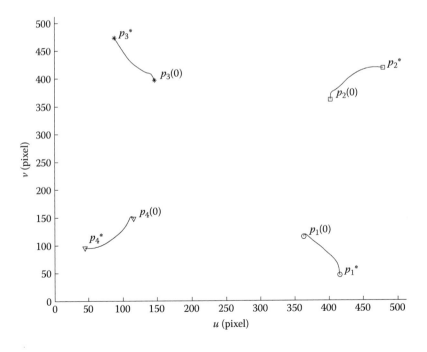

Figure 12.9: Actual image trajectory of Task 2.

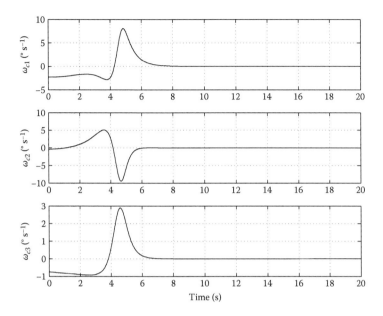

Figure 12.10: Angular velocity input of Task 2.

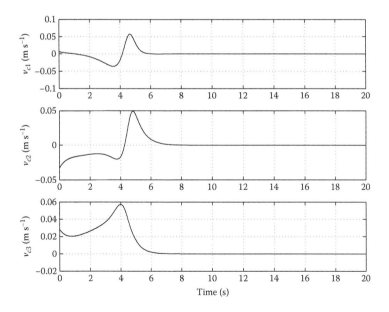

Figure 12.11: Linear velocity input of Task 2.

12.5.3 Camera y-Axis Rotation

The control parameters were selected as follows

$$K_v = 5I_3 \quad K_\omega = 0.3I_3 \quad k_1 = 1000000 \quad k_2 = 0.04$$

$$K = diag\{30, 20, 10, 28, 33, 25, 10, 40\}.$$

The actual image trajectories of the feature points are depicted in Figure 12.12. The control input velocities $\omega_c(t)$ and $v_c(t)$ defined in (12.40) and (12.41) are depicted in Figure 12.13 and 12.14. From Figure 12.12, it is clear that the actual feature points remain in the camera field of view and converge to the goal feature points.

12.5.4 General Camera Motion

The control parameters were selected as follows:

$$K_v = I_3 \quad K_\omega = 0.3I_3 \quad k_1 = 200000 \quad k_2 = 0.004$$

$$K = diag\{10, 10, 10, 18, 13, 15, 10, 10\}.$$

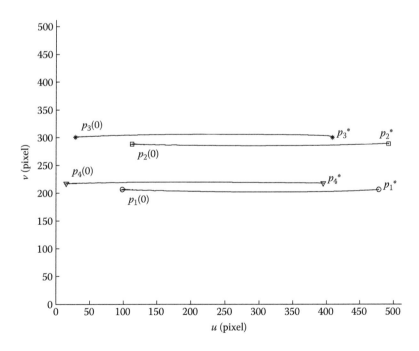

Figure 12.12: Actual image trajectory of Task 3.

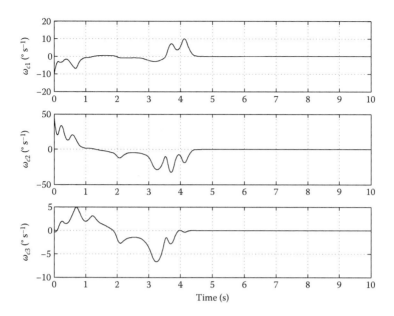

Figure 12.13: Angular velocity input of Task 3.

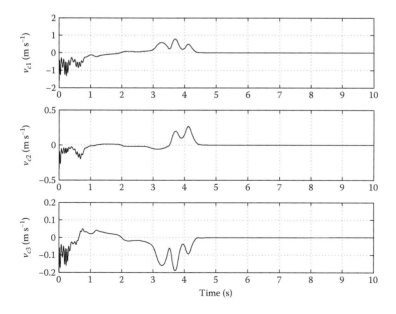

Figure 12.14: Linear velocity input of Task 3.

The actual image trajectories of the feature points are depicted in Figure 12.15. The control input velocities $\omega_c(t)$ and $v_c(t)$ defined in (12.40) and (12.41) are depicted in Figure 12.16 and 12.17. From Figure 12.15, it is clear that the actual feature points remain in the camera field of view and converge to the goal feature points.

Unlike position-based visual servoing and 2.5 D visual servoing, our visual servo algorithm applies an image-based navigation function to ensure all feature points remain in the camera field of view. From Figures 12.3, 12.9, 12.12, and 12.15, it is clear that all actual image trajectories are very close to a straight lines connecting the initial images and the goal images which is the main advantage of image-based visual servoing with regard to keeping the feature points in the camera field of view. Since our desired image trajectories are generated by (12.29), they satisfy the rigid body constraints. The common image-based visual servoing algorithms do not have mechanisms to satisfy rigid body constraints which cause image-based visual servo robot systems to behave irregularly when the initial error is large. By monitoring the distance ratio $\gamma(t)$, all the desired trajectories in our simulations avoid the self-occlusion problem. By comparing the desired image trajectories and the actual image trajectories in our simulations, it is clear that they are very close to each other. During closed-loop operation,

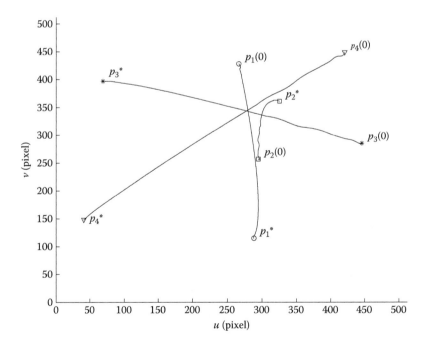

Figure 12.15: Actual image trajectory of Task 4.

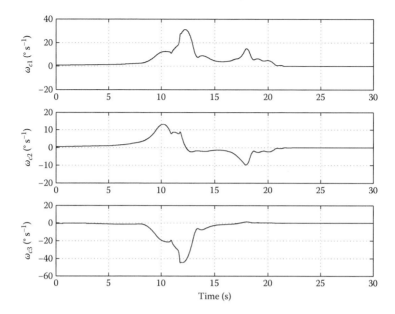

Figure 12.16: Angular velocity input of Task 4.

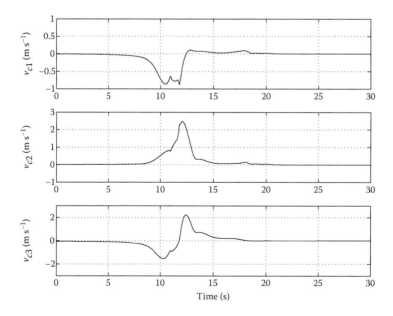

Figure 12.17: Linear velocity input of Task 4.

we had verified that there is no self-occlusion for the actual robot system in the simulations. To the best of our knowledge, no other result can guarantee all feature points remain in the camera field of view without irregular behavior while avoiding self-occlusion for full six degrees-of-freedom (DOF) robot systems.

12.6 Conclusions

A path planner is developed based on an image-space NF that ensures the desired image trajectory converges to the goal position while also ensuring the desired image features remain in a visibility set under certain technical restrictions. An adaptive, homography-based visual servo tracking controller is then developed to navigate the camera-in-hand pose along the desired trajectory despite the lack of depth information from a monocular camera system. The path planner and the tracking controller are analyzed through a Lyapunov-based analysis. Simulation results are provided to illustrate the performance of the proposed approach. Further experimental verification of the performance of our proposed algorithm will be provided in the future when relevant facilities are available.

Chapter 13

Visual Control of Mobile Robots

13.1 Introduction

In this chapter, visual trajectory tracking tasks for wheeled mobile robots are considered including both eye-to-hand and eye-in-hand configurations. Like that in Chapter 10, the camera can be installed at some fixed position in the workspace or on the mobile robot. Due to the nonholonomic constraints of wheeled mobile robots, the error system development, control design, and stability analysis in this effort requires fundamentally new control approaches in comparison to general robotic systems, a typical feedback linearizing controller developed for holonomic systems cannot be used to solve the problem.

For the eye-to-hand case, an adaptive visual servo controller is developed for the off-board, monocular camera configuration that ensures global tracking for a mobile robot. New development is presented to address the open-loop error system resulting from the geometric relationship related to the camera configuration. The controller in this chapter does not require the camera to be mounted so that the optical axis is perpendicular to the wheeled mobile robot (WMR) plane of motion (i.e., the depth to the WMR is allowed to be time varying). To address the fact that the distance from the camera to the WMR is an unknown time-varying signal, a Lyapunov-based control strategy is employed, which provides a framework for the construction of an adaptive update law to actively compensate for an unknown depth-related scaling constant. In lieu of tracking a pre-recorded set of images, the problem could also be formulated in terms of tracking the desired linear and angular velocities of a reference robot. An extension is provided to illustrate how a visual servo regulation controller can also be designed for the fixed monocular camera configuration. Simulation results are provided to demonstrate the performance of the tracking control design.

For the eye-in-hand case, a homography-based visual servo control strategy is used to force the Euclidean position/orientation of a camera mounted on a mobile robot (i.e., the camera-in-hand problem) to track a desired time-varying trajectory defined by a pre-recorded sequence of images. By comparing the feature points of an object from a reference image to feature points of an object in the current image and the pre-recorded sequence of images, projective geometric relationships are exploited to enable the reconstruction of the Euclidean coordinates of the target points with respect to the WMR coordinate frame. The tracking control objective is naturally defined in terms of the Euclidean space; however, the translation error is unmeasurable. That is, the Euclidean reconstruction is scaled by an unknown distance from the camera/WMR to the target, and while the scaled position is measurable through the homography, the unscaled position error is unmeasurable. To overcome this obstacle, a Lyapunov-based control strategy is employed, which provides a framework for the construction of an adaptive update law to actively compensate for the unknown depth-related scaling constant.

13.2 Goemetric Reconstruction

13.2.1 Eye-to-Hand Configuration

To facilitate the subsequent development, the WMR is assumed to be identifiable by four feature points denoted by $O_i \; \forall i = 1, 2, 3, 4$ that are considered to be coplanar[1] and not colinear. It is also assumed that a feature point is located at the origin of the coordinate frame attached to the WMR. Each feature point on $\mathcal{F}$, $\mathcal{F}_d$, and $\mathcal{F}^*$ has a 3D Euclidean coordinate that can be expressed in terms of $\mathcal{I}$, as $\bar{m}_i(t)$, $\bar{m}_{di}(t)$, $\bar{m}_i^* \in \mathbb{R}^3$, respectively. Based on the geometric relationships between the coordinate frames (see Figure 9.3a), the following expressions[2] can be developed to relate the feature points on $\mathcal{F}$ and $\mathcal{F}_d$ to the feature points on $\mathcal{F}^*$

$$\bar{m}_i = \left(\bar{R} + \bar{x}_h n^{*T} \right) \bar{m}_i^* \qquad \bar{m}_{di} = \left(\bar{R}_d + \bar{x}_{hd} n^{*T} \right) \bar{m}_i^*. \tag{13.1}$$

---

[1] It should be noted that if four coplanar target points are not available then the subsequent development can exploit the classic eight-points algorithm [150] with no four of the eight target points being coplanar.

[2] See [28] for details regarding the development of the following relationships.

In (13.1), $n^* \in \mathbb{R}^3$ denotes the constant unit normally expressed in $\mathcal{I}$ that is perpendicular to the WMR plane of motion, and $\bar{R}(t)$, $\bar{R}_d(t) \in SO(3)$ and $\bar{x}_h(t)$, $\bar{x}_{hd}(t) \in \mathbb{R}^3$ are defined as follows:

$$\bar{R} \triangleq R(R^*)^T \qquad \bar{R}_d \triangleq R_d(R^*)^T$$
$$\bar{x}_h \triangleq x_h - \bar{R}x_h^* \qquad \bar{x}_{hd} \triangleq x_h - \bar{R}_d x_h^*, \tag{13.2}$$

where $R(t)$, $R_d(t)$, $R^* \in SO(3)$ denote the rotation from $\mathcal{F}$ to $\mathcal{I}$, $\mathcal{F}_d$ to $\mathcal{I}$, and $\mathcal{F}^*$ to $\mathcal{I}$, respectively, and $x_h(t)$, $x_{hd}(t)$, $x_h^* \in \mathbb{R}^3$ denote the respective scaled translation vectors expressed in $\mathcal{I}$ (See Figure 9.3a). Also in Figure 9.3a, $s_i \in \mathbb{R}^3$ denotes the constant coordinates of the i-th feature point, and $d^* \in \mathbb{R}$ denotes the constant distance from the origin of $\mathcal{I}$ to the WMR plane of motion along the normal n^*. From the geometry illustrated in Figure 9.3a, the following relationships can be determined:

$$d^* = n^{*T}\bar{m}_i = n^{*T}\bar{m}_{di} \qquad n^* = R\begin{bmatrix} 0 & 0 & -1 \end{bmatrix}^T = R_d\begin{bmatrix} 0 & 0 & -1 \end{bmatrix}^T. \tag{13.3}$$

The relationship given by (13.1) provides a means to quantify a translation and rotation error between $\mathcal{F}$ and $\mathcal{F}^*$ and between $\mathcal{F}_d$ and $\mathcal{F}^*$. Since the Euclidean position of $\mathcal{F}$, $\mathcal{F}_d$, and $\mathcal{F}^*$ cannot be directly measured, further development is required to obtain the position and rotational error information in terms of measurable image-space coordinates. To this end, the normalized Euclidean coordinates of the points on $\mathcal{F}$, $\mathcal{F}_d$, and $\mathcal{F}^*$ expressed in terms of $\mathcal{I}$ as $m_i(t)$, $m_{di}(t)$, $m_i^* \in \mathbb{R}^3$, are defined as follows:

$$m_i \triangleq \frac{\bar{m}_i}{z_i} \qquad m_{di} \triangleq \frac{\bar{m}_{di}}{z_{di}} \qquad m_i^* \triangleq \frac{\bar{m}_i^*}{z_i^*} \tag{13.4}$$

under the standard assumption that the distances from the origin of $\mathcal{I}$ to the feature points along the focal axis remains positive (i.e., $z_i(t)$, $z_{di}(t)$, $z_i^* > \varepsilon$ where ε denotes an arbitrarily small positive constant). Based on (13.4), the expression in (13.1) can be rewritten as follows:

$$m_i = \underbrace{\frac{z_i^*}{z_i}}_{\alpha_i} \underbrace{\left(\bar{R} + \bar{x}_h n^{*T}\right) m_i^*}_{H} \qquad m_{di} = \underbrace{\frac{z_i^*}{z_{di}}}_{\alpha_{di}} \underbrace{\left(\bar{R}_d + \bar{x}_{hd} n^{*T}\right) m_i^*}_{H_d,} \tag{13.5}$$

where $\alpha_i(t)$, $\alpha_{di}(t) \in \mathbb{R}$ denote invertible depth ratios, and $H(t)$, $H_d(t) \in \mathbb{R}^{3\times 3}$ denote Euclidean homographies.

Remark 13.1 The subsequent development requires that the constant rotation matrix R^* be known. This is a mild assumption since the constant rotation matrix R^* can be obtained *a priori* using various methods (e.g., a second camera or an additional well-calibrated WMR pose, Euclidean measurements).

Each feature point on $\mathcal{F}$, $\mathcal{F}_d$, and $\mathcal{F}^*$ will also have a projected pixel coordinate expressed in terms of $\mathcal{I}$, denoted by $u_i(t)$, $v_i(t) \in \mathbb{R}$ for $\mathcal{F}$, $u_{di}(t)$, $v_{di}(t) \in \mathbb{R}$ for $\mathcal{F}_d$, and u_i^*, $v_i^* \in \mathbb{R}$ for $\mathcal{F}^*$, that are defined as elements of $p_i(t)$ (i.e., the actual time-varying feature points), $p_{di}(t)$ (i.e., the desired time-varying feature point trajectory), and p_i^* (i.e., the constant reference feature points), respectively. To calculate the Euclidean homography given in (13.5) from pixel information, the projected 2D pixel coordinates of the feature points are related to $m_i(t)$, $m_{di}(t)$, and m_i^* by the following pinhole lens models:

$$p_i = Am_i \qquad p_{di} = Am_{di} \qquad p_i^* = Am_i^*, \qquad (13.6)$$

where $A \in \mathbb{R}^{3 \times 3}$ is a known, constant, and invertible intrinsic camera calibration matrix. After substituting (13.6) into (13.5), the following relationships can be developed

$$p_i = \alpha_i \underbrace{\left(AHA^{-1}\right)}_{G} p_i^* \qquad p_{di} = \alpha_{di} \underbrace{\left(AH_d A^{-1}\right)}_{G_d,} p_i^* \qquad (13.7)$$

where $G(t) = [g_{ij}(t)]$, $G_d(t) = [g_{dij}(t)] \quad \forall i,j = 1,2,3 \in \mathbb{R}^{3 \times 3}$ denote projective homographies.

From the first relationship in (13.7), a set of 12 linearly independent equations given by the 4 feature point pairs $(p_i^*, p_i(t))$ with 3 independent equations per feature point pair can be used to determine the projective homography up to a scalar multiple (i.e., the product $\alpha_i(t)G(t)$ can be determined). From the definition of $G(t)$ given in (13.7), various techniques can then be used (e.g., see [67], [225]) to decompose the Euclidean homography, to obtain $\alpha_i(t)$, $G(t)$, $H(t)$, and the rotation signal $\bar{R}(t)$. Likewise, by using the feature point pairs $(p_i^*, p_{di}(t))$, the desired Euclidean homography can be decomposed to obtain $\alpha_{di}(t)$, $G_d(t)$, $H_d(t)$, and the desired rotation signal $\bar{R}_d(t)$. The rotation matrices $R(t)$ and $R_d(t)$ can be computed from $\bar{R}(t)$ and $\bar{R}_d(t)$ by using (13.2) and the fact that R^* is assumed to be known. The normalized coordinates $m_i(t)$ and $m_{di}(t)$ can be computed from $p_i(t)$ and $p_{di}(t)$ by using (13.6). Hence, $R(t)$, $\bar{R}(t)$, $R_d(t)$, $\bar{R}_d(t)$, $m_i(t)$, $m_{di}(t)$, and the depth ratios $\alpha_i(t)$ and $\alpha_{di}(t)$ are all known signals that can be used for control synthesis.

Based on the definitions for $R(t)$, $R_d(t)$, and R^* provided in the previous development, the rotation from $\mathcal{F}$ to $\mathcal{F}^*$ and from $\mathcal{F}_d$ to $\mathcal{F}^*$, denoted by $R_1(t) \triangleq R^{*T} R$ and $R_{d1}(t) \triangleq R^{*T} R_d$, can be expressed as follows:

$$R_1 = \begin{bmatrix} \cos\theta & -\sin\theta & 0 \\ \sin\theta & \cos\theta & 0 \\ 0 & 0 & 1 \end{bmatrix} \qquad R_{d1} = \begin{bmatrix} \cos\theta_d & -\sin\theta_d & 0 \\ \sin\theta_d & \cos\theta_d & 0 \\ 0 & 0 & 1 \end{bmatrix}, \qquad (13.8)$$

where $\theta(t) \in \mathbb{R}$ denotes the right-handed rotation angle about the z-axis that aligns $\mathcal{F}$ with $\mathcal{F}^*$, and $\theta_d(t) \in \mathbb{R}$ denotes the right-handed rotation angle about the z-axis that aligns $\mathcal{F}_d$ with $\mathcal{F}^*$. From Figure 9.3a and the definitions of $\theta(t)$ and $\theta_d(t)$, it is clear that

$$\dot{\theta} = \omega_c \qquad \dot{\theta}_d = \omega_{cd}. \tag{13.9}$$

Based on the fact that $R(t)$, $R_d(t)$, and R^* are known, it is clear from (13.8) to (13.9) that $\theta(t)$, $\theta_d(t)$, and $\omega_{cd}(t)$ are known signals that can be used in the subsequent control development. Note that $v_{cd}(t)$ is not a measurable signal. To facilitate the subsequent development, $\theta(t)$ and $\theta_d(t)$ are assumed to be confined to $-\pi < \theta(t) \leqslant \pi$ and $-\pi < \theta_d(t) \leqslant \pi$.

13.2.2 Eye-in-Hand Configuration

13.2.2.1 Geometric Modeling

In this section, geometric relationships are developed between the coordinate systems $\mathcal{F}$, $\mathcal{F}_d$, and $\mathcal{F}^*$, and a reference plane π that is defined by three target points O_i $\forall i = 1, 2, 3$ that are not collinear. The 3D Euclidean coordinates of O_i expressed in terms of $\mathcal{F}$, $\mathcal{F}_d$, and $\mathcal{F}^*$ as $\bar{m}_i(t)$, $\bar{m}_{di}(t)$, $\bar{m}_i^* \in \mathbb{R}^3$, respectively, are defined as follows (see Figure 10.1)

$$
\begin{aligned}
\bar{m}_i(t) &\triangleq \begin{bmatrix} x_i(t) & y_i(t) & z_i(t) \end{bmatrix}^T \\
\bar{m}_{di}(t) &\triangleq \begin{bmatrix} x_{di}(t) & y_{di}(t) & z_{di}(t) \end{bmatrix}^T \\
\bar{m}_i^* &\triangleq \begin{bmatrix} x_i^* & y_i^* & z_i^* \end{bmatrix}^T
\end{aligned}
\tag{13.10}
$$

under the standard assumption that the distances from the origin of the respective coordinate frames to the targets along the focal axis remains positive (i.e., $x_i(t)$, $x_{di}(t)$, $x_i^* \geq \varepsilon > 0$ where ε is an arbitrarily small positive constant). The rotation from $\mathcal{F}^*$ to $\mathcal{F}$ is denoted by $R(t) \in SO(3)$, and the translation from $\mathcal{F}$ to $\mathcal{F}^*$ is denoted by $x_f(t) \in \mathbb{R}^3$, where $x_f(t)$ is expressed in $\mathcal{F}$. Similarly, $R_d(t) \in SO(3)$ denotes the desired time-varying rotation from $\mathcal{F}^*$ to $\mathcal{F}_d$, and $x_{fd}(t) \in \mathbb{R}^3$ denotes the desired translation from $\mathcal{F}_d$ to $\mathcal{F}^*$, where $x_{fd}(t)$ is expressed in $\mathcal{F}_d$. Since the motion of the WMR is constrained to the xy-plane, $x_f(t)$ and $x_{fd}(t)$ are defined as follows:

$$
\begin{aligned}
x_f(t) &\triangleq \begin{bmatrix} x_{f1} & x_{f2} & 0 \end{bmatrix}^T \\
x_{fd}(t) &\triangleq \begin{bmatrix} x_{fd1} & x_{fd2} & 0 \end{bmatrix}^T.
\end{aligned}
\tag{13.11}
$$

From the geometry between the coordinate frames depicted in Figure 10.1, $\bar{m}_i^*$ can be related to $\bar{m}_i(t)$ and $\bar{m}_{di}(t)$ as follows:

$$\bar{m}_i = x_f + R\bar{m}_i^* \qquad \bar{m}_{di} = x_{fd} + R_d\bar{m}_i^*. \tag{13.12}$$

In (13.12), $R(t)$ and $R_d(t)$ are defined as follows:

$$R \triangleq \begin{bmatrix} \cos\theta & -\sin\theta & 0 \\ \sin\theta & \cos\theta & 0 \\ 0 & 0 & 1 \end{bmatrix}$$

$$R_d \triangleq \begin{bmatrix} \cos\theta_d & -\sin\theta_d & 0 \\ \sin\theta_d & \cos\theta_d & 0 \\ 0 & 0 & 1, \end{bmatrix}$$

(13.13)

where $\theta(t) \in \mathbb{R}$ denotes the right-handed rotation angle about $z_i(t)$ that aligns the rotation of $\mathcal{F}$ with $\mathcal{F}^*$, and $\theta_d(t) \in \mathbb{R}$ denotes the right-handed rotation angle about $z_{di}(t)$ that aligns the rotation of $\mathcal{F}_d$ with $\mathcal{F}^*$. From Figures 9.3b and (13.13), it is clear that

$$\dot\theta = -\omega_c \qquad \dot\theta_d = -\omega_{cd} \tag{13.14}$$

where $\omega_{cd}(t) \in \mathbb{R}$ denotes the desired angular velocity of the WMR expressed in $\mathcal{F}_d$. The rotation angles are assumed to be confined to the following regions:

$$-\pi < \theta(t) < \pi \qquad -\pi < \theta_d(t) < \pi. \tag{13.15}$$

From the geometry given in Figure 10.1, the distance $d^* \in \mathbb{R}$ from $\mathcal{F}^*$ to π along the unit normal of π is given by:

$$d^* = n^{*T} \bar{m}_i^*, \tag{13.16}$$

where $n^* = \begin{bmatrix} n_x^* & n_y^* & n_z^* \end{bmatrix}^T \in \mathbb{R}^3$ denotes the constant unit normal to π. Based on the definition of d^* in (13.16) and the fact that n^* and $\bar{m}^*$ do not change, it is clear that d^* is a constant. From (13.16), the relationships in (13.12) can be expressed as follows:

$$\bar{m}_i = \left(R + \frac{x_f}{d^*} n^{*T} \right) \bar{m}_i^*$$

$$\bar{m}_{di} = \left(R_d + \frac{x_{fd}}{d^*} n^{*T} \right) \bar{m}_i^*. \tag{13.17}$$

13.2.2.2 Euclidean Reconstruction

The relationship given in (13.12) provides a means to quantify the translational and rotational error between $\mathcal{F}$ and $\mathcal{F}^*$ and between $\mathcal{F}_d$ and $\mathcal{F}^*$. Since the position of $\mathcal{F}$, $\mathcal{F}_d$, and $\mathcal{F}^*$ cannot be directly measured, this section illustrates how the normalized Euclidean coordinates of the target points can be reconstructed by relating multiple images. Specifically, comparisons are made between an image acquired from the camera attached to $\mathcal{F}$, the reference image, and the pre-recorded sequence of images that define the trajectory of $\mathcal{F}_d$. To facilitate the subsequent development, the normalized Euclidean coordinates of O_i expressed in terms of $\mathcal{F}$, $\mathcal{F}_d$, and $\mathcal{F}^*$ are denoted by $m_i(t)$, $m_{di}(t)$, $m_i^* \in \mathbb{R}^3$, respectively, are explicitly defined as follows:

$$m_i \triangleq \begin{bmatrix} 1 & m_{iy} & m_{iz} \end{bmatrix}^T = \frac{\bar{m}_i}{x_i}$$

$$m_{di} \triangleq \begin{bmatrix} 1 & m_{diy} & m_{diz} \end{bmatrix}^T = \frac{\bar{m}_{di}}{x_{di}} \quad (13.18)$$

$$m_i^* \triangleq \begin{bmatrix} 1 & m_{iy}^* & m_{iz}^* \end{bmatrix}^T = \frac{\bar{m}_i^*}{x_i^*},$$

where $\bar{m}_i(t)$, $\bar{m}_{di}(t)$, and $\bar{m}_i^*$ were introduced in (13.10). In addition to having a Euclidean coordinate, each target point O_i will also have a projected pixel coordinate denoted by $u_i(t)$, $v_i(t) \in \mathbb{R}$ for $\mathcal{F}$, u_i^*, $v_i^* \in \mathbb{R}$ for $\mathcal{F}^*$, and $u_{di}(t)$, $v_{di}(t) \in \mathbb{R}$ for $\mathcal{F}_d$, that are defined as elements of $p_i(t) \in \mathbb{R}^3$ (i.e., the actual time-varying image points), $p_{di}(t) \in \mathbb{R}^3$ (i.e., the desired image point trajectory), and $p_i^* \in \mathbb{R}^3$ (i.e., the constant reference image points), respectively, as follows:

$$p_i \triangleq \begin{bmatrix} 1 & v_i & u_i \end{bmatrix}^T \qquad p_{di} \triangleq \begin{bmatrix} 1 & v_{di} & u_{di} \end{bmatrix}^T$$

$$p_i^* \triangleq \begin{bmatrix} 1 & v_i^* & u_i^* \end{bmatrix}^T. \quad (13.19)$$

The normalized Euclidean coordinates of the target points are related to the image data through the following pinhole lens models:

$$p_i = A m_i \qquad p_{di} = A m_{di} \qquad p_i^* = A m_i^*, \quad (13.20)$$

where $A \in \mathbb{R}^{3 \times 3}$ is a known, constant, and invertible intrinsic camera calibration matrix.

Given that $m_i(t)$, $m_{di}(t)$, and m_i^* can be obtained from (13.20), the rotation and translation between the coordinate systems can now be related in terms of the normalized Euclidean coordinates as follows:

$$m_i = \underbrace{\frac{x_i^*}{x_i}}_{\alpha_i} \underbrace{\left(R + x_h n^{*T} \right)}_{H} m_i^* \quad (13.21)$$

$$m_{di} = \underbrace{\frac{x_i^*}{x_{di}}}_{\alpha_{di}} \underbrace{\left(R_d + x_{hd} n^{*T} \right)}_{H_d,} m_i^* \quad (13.22)$$

where $\alpha_i(t)$, $\alpha_{di}(t) \in \mathbb{R}$ denote the depth ratios, $H(t)$, $H_d(t) \in \mathbb{R}^{3 \times 3}$ denote Euclidean homographies, and $x_h(t)$, $x_{hd}(t) \in \mathbb{R}^3$ denote scaled translation vectors that are defined as follows:

$$x_h \triangleq \begin{bmatrix} x_{h1} & x_{h2} & 0 \end{bmatrix}^T = \frac{x_f}{d^*}$$

$$x_{hd} \triangleq \begin{bmatrix} x_{hd1} & x_{hd2} & 0 \end{bmatrix}^T = \frac{x_{fd}}{d^*}. \quad (13.23)$$

By using (13.13) and (13.23), the Euclidean homography in (13.21) can be rewritten as follows:

$$H = [H_{jk}]$$

$$= \begin{bmatrix} \cos\theta + x_{h1}n_x^* & -\sin\theta + x_{h1}n_y^* & x_{h1}n_z^* \\ \sin\theta + x_{h2}n_x^* & \cos\theta + x_{h2}n_y^* & x_{h2}n_z^* \\ 0 & 0 & 1 \end{bmatrix}. \tag{13.24}$$

By examining the terms in (13.24), it is clear that $H(t)$ contains signals that are not directly measurable (e.g., $\theta(t)$, $x_h(t)$, and n^*). By expanding $H_{jk}(t)$ $\forall j = 1,2$, $k = 1,2,3$, the following expressions can be obtained from (13.18), (13.21), and (13.24):

$$1 = \alpha_i \left(H_{11} + H_{12}m_{iy}^* + H_{13}m_{iz}^* \right) \tag{13.25}$$

$$m_{iy} = \alpha_i \left(H_{21} + H_{22}m_{iy}^* + H_{23}m_{iz}^* \right) \tag{13.26}$$

$$m_{iz} = \alpha_i m_{iz}^*. \tag{13.27}$$

From (13.25) to (13.27), it is clear that three independent equations with nine unknowns (i.e., $H_{jk}(t)$ $\forall j = 1,2$, $k = 1,2,3$ and $\alpha_i(t)$ $\forall i = 1,2,3$) can be generated for each target point. Hence, by determining the normalized Euclidean coordinate of three target points in $\mathcal{F}$ and $\mathcal{F}^*$ from the image data and (13.20), the unknown elements of $H(t)$ and the unknown ratio $\alpha_i(t)$ can be determined. Likewise, for the same three target points in $\mathcal{F}_d$ and $\mathcal{F}^*$, the unknown elements of $H_d(t)$ and the unknown ratio $\alpha_{di}(t)$ can be determined. Once the elements of $H(t)$ and $H_d(t)$ are determined, various techniques (e.g., see [67], [225]) can be used to decompose the Euclidean homographies to obtain the rotation and translation components. There are, in general, four solutions generated by the decomposition of $H(t)$ (and likewise for $H_d(t)$) depending on the multiplicity of the singular values. As stated in [67], some additional information (e.g., provided by the physical nature of the problem) must be used to determine the unique solution. For example, physical insight can be used to determine the unique solution among the four possible solutions for n^* (see Section 5.2 for details regarding one method to resolve the decomposition ambiguity). Hence, $R(t)$, $R_d(t)$, $x_h(t)$, and $x_{hd}(t)$ can all be computed and used for the subsequent control synthesis. Since $R(t)$ and $R_d(t)$ are known matrices, then (13.13) can be used to determine $\theta(t)$ and $\theta_d(t)$.

Remark 13.2   Motivation for using a homography-based approach is the desire to craft the error systems in a manner to facilitate the development of an adaptive update law to compensate for the unmeasurable depth parameter d^*, rather than infer depth from an object model. With the proposed approach, the mismatch between the estimated and actual depth information can be explicitly included in the stability analysis. Moreover, the resulting error systems do not depend on an image-Jacobian that could introduce singularities in the controller, which is an endemic problem with pure image-based visual servo control strategies.

Remark 13.3 To develop a tracking controller, it is typical that the desired trajectory is used as a feedforward component in the control design. Hence, for a kinematic controller the desired trajectory is required to be at least first order differentiable and at least second order differentiable for a dynamic level controller. From the Euclidean homography introduced in (13.22), $m_d(t)$ can be expressed in terms of the *a priori* known, functions $\alpha_{di}(t)$, $H_d(t)$, $R_d(t)$, and $x_{hd}(t)$. Since these signals can be obtained from the pre-recorded sequence of images, sufficiently smooth functions can be generated for these signals by fitting a sufficiently smooth spline function to the signals. Hence, in practice, the *a priori* developed smooth functions $\alpha_{di}(t)$, $R_d(t)$, and $x_{hd}(t)$ can be constructed as bounded functions with sufficiently bounded time derivatives. Given $\theta_d(t)$ and the time derivative of $R_d(t)$, $\dot{\theta}_d(t)$ can be determined. In the subsequent tracking control development, $\dot{x}_{hd1}(t)$ and $\dot{\theta}_d(t)$ will be used in feedforward control terms.

13.3 Control Development for Eye-to-Hand Configuration

13.3.1 Control Objective

The control objective in this chapter is to ensure that the coordinate frame $\mathcal{F}$ tracks the time-varying trajectory of $\mathcal{F}_d$ (i.e., the Euclidean coordinates of the WMR feature points track the desired time-varying trajectory in the sense that $\bar{m}_i(t) \to \bar{m}_{di}(t)$). To quantify the control objective, the translation and rotation tracking error, denoted by $e(t) \triangleq [e_1(t), e_2(t), e_3(t)]^T \in \mathbb{R}^3$, is defined as follows:

$$e_1 \triangleq \eta_1 - \eta_{d1} \qquad e_2 \triangleq \eta_2 - \eta_{d2} \qquad e_3 \triangleq \theta - \theta_d, \tag{13.28}$$

where $\theta(t)$ and $\theta_d(t)$ are introduced in (13.8), respectively, and the auxiliary signals $\eta(t) \triangleq [\eta_1(t), \eta_2(t), \eta_3]^T$, $\eta_d(t) \triangleq [\eta_{d1}(t), \eta_{d2}(t), \eta_{d3}]^T \in \mathbb{R}^3$ are defined as follows[3]

$$\eta(t) \triangleq \frac{1}{z_1^*} R^T \bar{m}_1 \qquad \eta_d(t) \triangleq \frac{1}{z_1^*} R_d^T \bar{m}_{d1}. \tag{13.29}$$

From (13.3) and (13.29), it can be determined that

$$\eta_3 = \eta_{d3} = \frac{-d^*}{z_1^*}. \tag{13.30}$$

---

[3] Any point O_i can be utilized in the subsequent development; however, to reduce the notational complexity, we have elected to select the image point O_1, and hence, the subscript 1 is utilized in lieu of i in the subsequent development.

The expression in (13.4)–(13.6) can be used to rewrite $\eta(t)$ and $\eta_d(t)$ in terms of the measurable signals $\alpha_1(t)$, $\alpha_{d1}(t)$, $R(t)$, $R_d(t)$, $p_1(t)$, and $p_{d1}(t)$ as follows:

$$\eta(t) = \frac{1}{\alpha_1} R^T A^{-1} p_1 \qquad \eta_d(t) = \frac{1}{\alpha_{d1}} R_d^T A^{-1} p_{d1}. \tag{13.31}$$

Based on (13.28), (13.31), and the fact that $\theta(t)$ and $\theta_d(t)$ are measurable, it is clear that $e(t)$ is measurable. By examining (13.28)–(13.30), it can be shown that the control objective is achieved if $\|e(t)\| \to 0$. Specifically, if $e_3(t) \to 0$, then it is clear from (13.8) and (13.28) that $R(t) \to R_d(t)$. If $e_1(t) \to 0$ and $e_2(t) \to 0$, then from (13.28) and (13.30) it is clear that $\eta(t) \to \eta_d(t)$. Given that $R(t) \to R_d(t)$ and that $\eta(t) \to \eta_d(t)$, then (13.29) can be used to conclude that $\bar{m}_1(t) \to \bar{m}_{d1}(t)$. If $\bar{m}_1(t) \to \bar{m}_{d1}(t)$ and $R(t) \to R_d(t)$, then (13.1) can be used to prove that $\bar{m}_i(t) \to \bar{m}_{di}(t)$.

Remark 13.4   To develop a tracking controller, it is typical for the desired trajectory to be used as a feedforward component in the control design. Hence, for a kinematic controller the desired trajectory is required to be at least first order differentiable and at least second order differentiable for a dynamic level controller. From the Euclidean homography introduced in (13.5), $m_d(t)$ can be expressed in terms of the *a priori* known functions $\alpha_{di}(t)$, $H_d(t)$, $R_d(t)$, and $x_{hd}(t)$. Since these signals can be obtained from the pre-recorded sequence of images, sufficiently smooth functions can be generated for these signals by fitting a sufficiently smooth spline function to the signals. Hence, in practice, the *a priori* developed smooth functions $\alpha_{di}(t)$, $R_d(t)$, and $\eta_d(t)$ can be constructed as bounded functions with sufficiently bounded time derivatives. Given $\theta_d(t)$ and the time derivative of $R_d(t)$, $\dot{\theta}_d(t)$ can be determined. In the subsequent tracking control development, $\dot{\eta}_{d1}(t)$ and $\dot{\theta}_d(t)$ will be used as feedforward control terms.

13.3.2   Open-Loop Error System

To facilitate the development of the open-loop tracking error system, we take the time derivative of (13.29) as follows:

$$\dot{\eta} = \frac{v}{z_1^*} + \left[\eta - \frac{s_1}{z_1^*}\right]_\times \omega, \tag{13.32}$$

where the following relationships were utilized [28]

$$\dot{\bar{m}}_1 = Rv + R[\omega]_\times s_1 \qquad \dot{R} = R[\omega]_\times$$

and $v(t)$, $\omega(t) \in \mathbb{R}^3$ denote the respective linear and angular velocity of the WMR expressed in $\mathcal{F}$ as:

$$v \triangleq \begin{bmatrix} v_c & 0 & 0 \end{bmatrix}^T \qquad \omega \triangleq \begin{bmatrix} 0 & 0 & \omega_c \end{bmatrix}^T. \tag{13.33}$$

In (13.32), the notation $[\cdot]_\times$ denotes the 3×3 skew-symmetric matrix form of the vector argument.

Without loss of generality, we assume that the feature point O_1 is the feature point assumed to be located at the origin of the coordinate frame attached to the WMR, so that $s_1 = [0, 0, 0]^T$. Based on (13.33) and the assumption that $s_1 = [0, 0, 0]^T$, we can rewrite (13.32) as follows:

$$\dot{\eta}_1 = \frac{v_c}{z_1^*} + \eta_2 \omega_c \qquad \dot{\eta}_2 = -\eta_1 \omega_c. \tag{13.34}$$

Since the desired trajectory is assumed to be generated in accordance with the WMR motion constraints, a similar expression to (13.34) can be developed as follows:

$$\dot{\eta}_{d1} = \frac{v_{cd}}{z_1^*} + \eta_{d2} \omega_{cd} \qquad \dot{\eta}_{d2} = -\eta_{d1} \omega_{cd}. \tag{13.35}$$

After taking the time derivative of (13.28) and utilizing (13.14) and (13.34), the following open-loop error system can be obtained

$$z_1^* \dot{e}_1 = v_c + z_1^* (\eta_2 \omega_c - \dot{\eta}_{d1}) \qquad \dot{e}_2 = -\eta_1 \omega_c + \eta_{d1} \dot{\theta}_d \qquad \dot{e}_3 = \omega_c - \dot{\theta}_d. \tag{13.36}$$

To facilitate the subsequent development, the auxiliary variable $\bar{e}_2(t) \in \mathbb{R}$ is defined as:

$$\bar{e}_2 \overset{\triangle}{=} e_2 + \eta_{d1} e_3. \tag{13.37}$$

After taking the time derivative of (13.37) and utilizing (13.36), the following expression is obtained:

$$\dot{\bar{e}}_2 = -e_1 \omega_c + \dot{\eta}_{d1} e_3. \tag{13.38}$$

Based on (13.37), it is clear that if $\bar{e}_2(t)$, $e_3(t) \to 0$, then $e_2(t) \to 0$. Based on this observation and the open-loop dynamics given in (13.38), the following control development is based on the desire to show that $e_1(t), \bar{e}_2(t), e_3(t)$ are asymptotically driven to zero.

13.3.3 Closed-Loop Error System

Based on the open-loop error systems in (13.36) and (13.38), the linear and angular velocity control inputs for the WMR are designed as follows:

$$v_c \overset{\triangle}{=} -k_v e_1 + \bar{e}_2 \omega_c - \hat{z}_1^* (\eta_2 \omega_c - \dot{\eta}_{d1}) \qquad \omega_c \overset{\triangle}{=} -k_\omega e_3 + \dot{\theta}_d - \dot{\eta}_{d1} \bar{e}_2, \tag{13.39}$$

where k_v, $k_\omega \in \mathbb{R}$ denote positive, constant control gains, and $\dot{\theta}_d(t)$ and $\dot{\eta}_{d1}(t)$ are generated as described in Remark 13.4. The parameter update law $\hat{z}_1^*(t) \in \mathbb{R}$ is generated by the following differential equation:

$$\dot{\hat{z}}_1^* = \gamma_1 e_1 (\eta_2 \omega_c - \dot{\eta}_{d1}), \tag{13.40}$$

where $\gamma_1 \in \mathbb{R}$ is a positive, constant adaptation gain. After substituting the kinematic control signals designed in (13.39) into (13.36), the following closed-loop error systems are obtained:

$$z_1^* \dot{e}_1 = -k_v e_1 + \bar{e}_2 \omega_c + \tilde{z}_1^* (\eta_2 \omega_c - \dot{\eta}_{d1})$$
$$\dot{e}_2 = -e_1 \omega_c + \dot{\eta}_{d1} e_3 \qquad (13.41)$$
$$\dot{e}_3 = -k_\omega e_3 - \dot{\eta}_{d1} \bar{e}_2,$$

where (13.38) was utilized, and the depth-related parameter estimation error, denoted by $\tilde{z}_1^*(t) \in \mathbb{R}$, is defined as follows:

$$\tilde{z}_1^* \triangleq z_1^* - \hat{z}_1^* . \qquad (13.42)$$

13.3.4   Stability Analysis

Theorem 13.1
The control input designed in (13.39) along with the adaptive update law defined in (13.40) ensure asymptotic WMR tracking in the sense that $\lim_{t \to \infty} \|e(t)\| = 0$ provided the time derivative of the desired trajectory satisfies the following condition:

$$\lim_{t \to \infty} \dot{\eta}_{d1} \neq 0. \qquad (13.43)$$

Proof 13.1   To prove Theorem 13.3, the nonnegative function $V(t) \in \mathbb{R}$ is defined as follows:

$$V \triangleq \frac{1}{2} z_1^* e_1^2 + \frac{1}{2} \bar{e}_2^2 + \frac{1}{2} e_3^2 + \frac{1}{2\gamma_1} \tilde{z}_1^{*2} . \qquad (13.44)$$

The following simplified expression can be obtained by taking the time derivative of (13.44), substituting the closed-loop dynamics from (13.41) into the resulting expression, and then cancelling common terms

$$\dot{V} = -k_v e_1^2 + e_1 \tilde{z}_1^* (\eta_2 \omega_c - \dot{\eta}_{d1}) - k_\omega e_3^2 - \frac{1}{\gamma_1} \tilde{z}_1^* \dot{\hat{z}}_1^*. \qquad (13.45)$$

After substituting (13.40) into (15.23), the following expression can be obtained:

$$\dot{V} = -k_v e_1^2 - k_\omega e_3^2 . \qquad (13.46)$$

From (13.44) and (13.46), it is clear that $e_1(t)$, $\bar{e}_2(t)$, $e_3(t)$, $\tilde{z}_1^*(t) \in \mathcal{L}_\infty$ and that $e_1(t)$, $e_3(t) \in \mathcal{L}_2$. Since $\tilde{z}_1^*(t) \in \mathcal{L}_\infty$ and z_1^* is a constant, the expression in (13.42) can be used to determine that $\hat{z}_1^*(t) \in \mathcal{L}_\infty$. From the assumption that $\eta_{d1}(t)$, $\dot{\eta}_{d1}(t)$, $\eta_{d2}(t)$, $\theta_d(t)$, and $\dot{\theta}_d(t)$ are constructed as bounded functions, and the fact that $\bar{e}_2(t)$, $e_3(t) \in \mathcal{L}_\infty$, the expressions in (13.28), (13.37), and (13.39) can be used to prove that $e_2(t)$, $\eta_1(t)$, $\eta_2(t)$, $\theta(t)$, $\omega_c(t) \in \mathcal{L}_\infty$. Based on the previous development, the expressions in (13.39), (13.40), and (13.41) can be used to conclude that $v_c(t)$, $\dot{\hat{z}}_1^*(t)$,

$\dot{e}_1(t)$, $\ddot{e}_2(t)$, $\dot{e}_3(t) \in \mathcal{L}_\infty$. Based on the fact that $e_1(t)$, $e_3(t)$, $\dot{e}_1(t)$, $\dot{e}_3(t) \in \mathcal{L}_\infty$ and that $e_1(t)$, $e_3(t) \in \mathcal{L}_2$, Barbalat's lemma [192] can be employed to prove that

$$\lim_{t\to\infty} e_1(t), e_3(t) = 0. \tag{13.47}$$

From (13.47) and the fact that the signal $\dot{\eta}_{d1}(t)\bar{e}_2(t)$ is uniformly continuous (i.e., $\dot{\eta}_{d1}(t)$, $\ddot{\eta}_{d1}(t)$, $\bar{e}_2(t)$, $\dot{\bar{e}}_2(t) \in \mathcal{L}_\infty$), Extended Barbalat's Lemma (see Appendix E.1) can be applied to the last equation in (13.41) to prove that

$$\lim_{t\to\infty} \dot{e}_3(t) = 0 \qquad \lim_{t\to\infty} \dot{\eta}_{d1}(t)\bar{e}_2(t) = 0. \tag{13.48}$$

If the desired trajectory satisfies (13.43), then (13.48) can be used to prove that

$$\lim_{t\to\infty} \bar{e}_2(t) = 0. \tag{13.49}$$

Based on the definition of $\bar{e}_2(t)$ given in (13.37), the results in (13.47) and (13.49) can be used to conclude that $\lim_{t\to\infty} e_2(t) = 0$ provided the condition in (13.43) is satisfied.

Remark 13.5 Based on (13.44), (13.46), (13.47), and (13.49), the estimation error $\tilde{z}_1^*$ will converge to a constant. It does not necessarily converge to zero and the residual estimation error $\tilde{z}_1^*$ will not affect the control objectives.

Remark 13.6 For some cases, the signals $v_{cd}(t)$ and $\omega_{cd}(t)$ may be *a priori* known. For this case (13.14) and (13.35) can be substituted into (13.36) as follows:

$$z_1^*\dot{e}_1 = v_c - v_{cd} + z_1^*(\eta_2\omega_c - \eta_{d2}\omega_{cd}) \quad \dot{e}_2 = -\eta_1\omega_c + \eta_{d1}\omega_{cd} \quad \dot{e}_3 = \omega_c - \omega_{cd}. \tag{13.50}$$

Based on (13.50), the linear and angular velocity kinematic control inputs for the WMR are redesigned as follows:

$$v_c \triangleq -k_v e_1 + v_{cd} + \hat{z}_1^*(\eta_{d2}\omega_{cd} + \bar{e}_2\omega_c - \eta_2\omega_c)$$

$$\omega_c \triangleq -k_\omega e_3 + \omega_{cd} - \bar{e}_2 v_{cd} - \hat{z}_1^* \eta_{d2}\bar{e}_2\omega_{cd}$$

along with the following adaptation law:

$$\dot{\hat{z}}_1^* \triangleq -\gamma_1(e_1(\eta_{d2}\omega_{cd} + \bar{e}_2\omega_c - \eta_2\omega_c) - e_3\bar{e}_2\eta_{d2}\omega_{cd}).$$

13.3.5 Regulation Extension

Based on the assumption given in (13.43), it is clear that the controller developed in the previous section cannot be applied to solve the regulation problem. In this section, an extension is presented to illustrate how a visual servo controller can be developed to solve the regulation problem for the fixed camera configuration. To this end, the following kinematic model is obtained by examining Figure 9.3a:

$$\begin{bmatrix} \dot{x}_c \\ \dot{y}_c \\ \dot{\theta} \end{bmatrix} = \begin{bmatrix} \cos\theta & 0 \\ \sin\theta & 0 \\ 0 & 1 \end{bmatrix} \begin{bmatrix} v_c \\ \omega_c \end{bmatrix}, \tag{13.51}$$

where $\dot{x}_c(t)$, $\dot{y}_c(t)$, and $\dot{\theta}(t)$ denote the time derivative of $x_c(t)$, $y_c(t)$, and $\theta(t) \in \mathbb{R}$, respectively, where $x_c(t)$ and $y_c(t)$ denote the planar position of $\mathcal{F}$ expressed in $\mathcal{F}^*$, $\theta(t)$ was introduced in (13.8), and $v_c(t)$ and $\omega_c(t)$ denote the linear and angular velocity of $\mathcal{F}$. In addition to the kinematic model in (13.51), the geometric relationships between the coordinate frames (see Figure 9.1), can be used to develop the following expressions[4]

$$\bar{m}_1^* = d^* x_h^* + R^* s_1 \qquad \bar{m}_1 = d^* x_h + R s_1 \qquad \begin{bmatrix} x_c & y_c & 0 \end{bmatrix}^T = d^* R^{*T} (x_h - x_h^*). \tag{13.52}$$

After utilizing (13.4), (13.52), and the assumption that $s_1 = [0, 0, 0]^T$, the following expression can be obtained:

$$\frac{1}{z_1^*} \begin{bmatrix} x_c & y_c & 0 \end{bmatrix}^T = R^{*T} \begin{bmatrix} \frac{1}{\alpha_1} m_1 - m_1^* \end{bmatrix}, \tag{13.53}$$

where $\alpha_1(t)$ was defined in (13.5). After utilizing (13.6), the expression in (13.53) can be rewritten as follows:

$$\frac{1}{z_1^*} \begin{bmatrix} x_c & y_c & 0 \end{bmatrix}^T = R^{*T} A^{-1} \begin{bmatrix} \frac{1}{\alpha_1} p_1 - p_1^* \end{bmatrix}. \tag{13.54}$$

To facilitate the subsequent control objective, a global invertible transformation is defined as follows [55]:

$$\begin{bmatrix} e_1 \\ e_2 \\ e_3 \end{bmatrix} \triangleq \begin{bmatrix} \cos\theta & \sin\theta & 0 \\ -\sin\theta & \cos\theta & 0 \\ 0 & 0 & 1 \end{bmatrix} \begin{bmatrix} \frac{x_c}{z_1^*} \\ \frac{y_c}{z_1^*} \\ \theta \end{bmatrix}, \tag{13.55}$$

where $e(t) = [\ e_1(t) \quad e_2(t) \quad e_3(t)\ ]^T \in \mathbb{R}^3$ denotes the regulation error signal. From (13.8), (13.54), and (13.55), it is clear that $e(t)$ is a measurable signal, and that if $\|e(t)\| \to 0$, then $x_c(t) \to 0$, $y_c(t) \to 0$, and $\theta(t) \to 0$. After taking the time derivative of (13.55), the following open-loop error system can be obtained:

$$\begin{bmatrix} \dot{e}_1 \\ \dot{e}_2 \\ \dot{e}_3 \end{bmatrix} = \begin{bmatrix} \frac{v_c}{z_1^*} + \omega_c e_2 \\ -\omega_c e_1 \\ \omega_c \end{bmatrix}, \tag{13.56}$$

---

[4]See [28] for details regarding the development of the following relationships.

where (13.51) was utilized. Based on (13.56), the linear and angular velocity inputs are designed as follows:

$$v_c \triangleq -k_1 e_1 \qquad \omega_c \triangleq -k_2 e_3 + e_2^2 \sin t \qquad (13.57)$$

to yield the following closed-loop error system:

$$z_1^* \dot{e}_1 = -k_1 e_1 + z_1^* \omega_c e_2 \qquad \dot{e}_2 = -\omega_c e_1 \qquad \dot{e}_3 = -k_2 e_3 + e_2^2 \sin t.$$

Theorem 13.2

The control input designed in (13.57) ensures asymptotic WMR regulation in the sense that

$$\lim_{t \to \infty} \|e(t)\| = 0.$$

Similar to the proof for Theorem 13.3 with a new nonnegative function $V \triangleq \frac{1}{2} z_1^* \left(e_1^2 + e_2^2 \right)$, it is not difficult to show that the control input designed in (13.57) ensures asymptotic WMR regulation in the sense that $\lim_{t \to \infty} \|e(t)\| = 0$.

13.4 Control Development for Eye-in-Hand Configuration

The control objective is to ensure that the coordinate frame $\mathcal{F}$ tracks the time-varying trajectory of $\mathcal{F}_d$ (i.e., $\bar{m}_i(t)$ tracks $\bar{m}_{di}(t)$). This objective is naturally defined in terms of the Euclidean position/orientation of the WMR. Specifically, based on the previous development, the translation and rotation tracking error, denoted by $e(t) \triangleq \begin{bmatrix} e_1 & e_2 & e_3 \end{bmatrix}^T \in \mathbb{R}^3$, is defined as follows:

$$\begin{aligned} e_1 &\triangleq x_{h1} - x_{hd1} \\ e_2 &\triangleq x_{h2} - x_{hd2} \\ e_3 &\triangleq \theta - \theta_d, \end{aligned} \qquad (13.58)$$

where $x_{h1}(t)$, $x_{h2}(t)$, $x_{hd1}(t)$, and $x_{hd2}(t)$ are introduced in (13.23), and $\theta(t)$ and $\theta_d(t)$ are introduced in (13.13). Based on the definition in (13.58), it can be shown that the control objective is achieved if the tracking error $e(t) \to 0$. Specifically, it is clear from (13.23) that if $e_1(t) \to 0$ and $e_2(t) \to 0$, then $x_f(t) \to x_{fd}(t)$. If $e_3 \to 0$, then it is clear from (13.13) and (13.58) that $R(t) \to R_d(t)$. If $x_f(t) \to x_{fd}(t)$ and $R(t) \to R_d(t)$, then (13.12) can be used to prove that $\bar{m}_i(t) \to \bar{m}_{di}(t)$.

13.4.1 Open-Loop Error System

As a means to develop the open-loop tracking error system, the time derivative of the Euclidean position $x_f(t)$ is determined as follows [152]:

$$\dot{x}_f = -v + [x_f]_\times \omega, \qquad (13.59)$$

where $v(t)$, $\omega(t) \in \mathbb{R}^3$ denote the respective linear and angular velocity of the WMR expressed in $\mathcal{F}$ as:

$$v \triangleq [\; v_c \quad 0 \quad 0 \;]^T \qquad \omega \triangleq [\; 0 \quad 0 \quad \omega_c \;]^T , \qquad (13.60)$$

and $[x_f]_\times$ denotes the 3×3 skew-symmetric form of $x_f(t)$. After substituting (13.23) into (13.59), the time derivative of the translation vector $x_h(t)$ can be written in terms of the linear and angular velocity of the WMR as follows:

$$\dot{x}_h = -\frac{v}{d^*} + [x_h] \times \omega . \qquad (13.61)$$

After incorporating (13.60) into (13.61), the following expression can be obtained:

$$\dot{x}_{h1} = -\frac{v_c}{d^*} + x_{h2}\omega_c$$
$$\dot{x}_{h2} = -x_{h1}\omega_c, \qquad (13.62)$$

where (13.23) was utilized. Given that the desired trajectory is generated from a pre-recorded set of images taken by the on-board camera as the WMR was moving, a similar expression as (13.59) can be developed as follows:

$$\dot{x}_{fd} = -[\; v_{cd} \quad 0 \quad 0 \;]^T + [x_{fd}]_\times [\; 0 \quad 0 \quad \omega_{cd} \;]^T , \qquad (13.63)$$

where $v_{cd}(t) \in \mathbb{R}$ denotes the desired linear[5] velocity of the WMR expressed in $\mathcal{F}_d$. After substituting (13.23) into (13.63), the time derivative of the translation vector $x_{hd}(t)$ can be written as follows:

$$\dot{x}_{hd1} = -\frac{v_{cd}}{d^*} + x_{hd2}\omega_{cd}$$
$$\dot{x}_{hd2} = -x_{hd1}\omega_{cd}. \qquad (13.64)$$

After taking the time derivative of (13.58) and utilizing (13.14) and (13.62), the following open-loop error system can be obtained:

$$d^* \dot{e}_1 = -v_c + d^* (x_{h2}\omega_c - \dot{x}_{hd1})$$
$$\dot{e}_2 = -(x_{h1}\omega_c + x_{hd1}\dot{\theta}_d)$$
$$\dot{e}_3 = -(\omega_c + \dot{\theta}_d) , \qquad (13.65)$$

where the definition of $e_2(t)$ given in (13.58), and the second equation of (13.64) was utilized. To facilitate the subsequent development, the auxiliary variable $\bar{e}_2(t) \in \mathbb{R}$ is defined as:

$$\bar{e}_2 \triangleq e_2 - x_{hd1}e_3. \qquad (13.66)$$

After taking the time derivative of (13.66) and utilizing (13.65), the following expression is obtained:

$$\dot{\bar{e}}_2 = -(e_1\omega_c + \dot{x}_{hd1}e_3). \qquad (13.67)$$

---

[5]Note that $v_{cd}(t)$ is not measurable.

Based on (13.66), it is clear that if $\bar{e}_2(t)$, $e_3(t) \to 0$, then $e_2(t) \to 0$. Based on this observation and the open-loop dynamics given in (13.67), the following control development is based on the desire to prove that $e_1(t)$, $\bar{e}_2(t)$, $e_3(t)$ are asymptotically driven to zero.

13.4.2 Closed-Loop Error System

Based on the open-loop error systems in (13.65) and (13.67), the linear and angular velocity kinematic control inputs for the WMR are designed as follows:

$$v_c \triangleq k_v e_1 - \bar{e}_2 \omega_c + \hat{d}^* (x_{h2} \omega_c - \dot{x}_{hd1}) \tag{13.68}$$

$$\omega_c \triangleq k_\omega e_3 - \dot{\theta}_d - \dot{x}_{hd1} \bar{e}_{,2} \tag{13.69}$$

where k_v, $k_\omega \in \mathbb{R}$ denote positive, constant control gains. In (13.68), the parameter update law $\hat{d}^*(t) \in \mathbb{R}$ is generated by the following differential equation:

$$\dot{\hat{d}}^* = \gamma_1 e_1 (x_{h2} \omega_c - \dot{x}_{hd1}), \tag{13.70}$$

where $\gamma_1 \in \mathbb{R}$ is a positive, constant adaptation gain. After substituting the kinematic control signals designed in (13.68) and (13.69) into (13.65), the following closed-loop error systems are Obtained:

$$\begin{aligned} d^* \dot{e}_1 &= -k_v e_1 + \bar{e}_2 \omega_c + \tilde{d}^* (x_{h2} \omega_c - \dot{x}_{hd1}) \\ \dot{\bar{e}}_2 &= -(e_1 \omega_c + \dot{x}_{hd1} e_3) \\ \dot{e}_3 &= -k_\omega e_3 + \dot{x}_{hd1} \bar{e}_2, \end{aligned} \tag{13.71}$$

where (13.67) was utilized, and the depth-related parameter estimation error $\tilde{d}^*(t) \in \mathbb{R}$ is defined as follows:

$$\tilde{d}^* \triangleq d^* - \hat{d}^* . \tag{13.72}$$

13.4.3 Stability Analysis

Theorem 13.3
The adaptive update law defined in (13.70) along with the control input designed in (13.68) and (13.69) ensure that the WMR tracking error $e(t)$ is asymptotically driven to zero in the sense that

$$\lim_{t \to \infty} e(t) = 0 \tag{13.73}$$

provided the time derivative of the desired trajectory satisfies the following condition:

$$\lim_{t \to \infty} \dot{x}_{hd1} \neq 0. \tag{13.74}$$

Proof 13.2   To prove Theorem 13.3, the nonnegative function $V(t) \in \mathbb{R}$ is defined as follows:

$$V \triangleq \frac{1}{2}d^* e_1^2 + \frac{1}{2}\tilde{e}_2^2 + \frac{1}{2}e_3^2 + \frac{1}{2\gamma_1}\tilde{d}^{*2} . \tag{13.75}$$

The following simplified expression can be obtained by taking the time derivative of (13.75), substituting the closed-loop dynamics in (13.71) into the resulting expression, and then cancelling common terms

$$\dot{V} = -k_v e_1^2 + e_1 \tilde{d}^* (x_{h2}\omega_c - \dot{x}_{hd1}) - k_\omega e_3^2 - \frac{1}{\gamma_1}\tilde{d}^* \dot{\hat{d}}^* . \tag{13.76}$$

After substituting (13.70) into (15.23), the following expression can be obtained:

$$\dot{V} = -k_v e_1^2 - k_\omega e_3^2 . \tag{13.77}$$

From (13.75) and (13.77), it is clear that $e_1(t)$, $\tilde{e}_2(t)$, $e_3(t)$, $\tilde{d}^*(t) \in \mathcal{L}_\infty$ and that $e_1(t)$, $e_3(t) \in \mathcal{L}_2$. Since $\tilde{d}^*(t) \in \mathcal{L}_\infty$ and d^* is a constant, the expression in (13.72) can be used to determine that $\hat{d}^*(t) \in \mathcal{L}_\infty$. From the assumption that $x_{hd1}(t)$, $\dot{x}_{hd1}(t)$, $x_{hd2}(t)$, $\theta_d(t)$, and $\dot{\theta}_d(t)$ are constructed as bounded functions, and the fact that $\tilde{e}_2(t)$, $e_3(t) \in \mathcal{L}_\infty$, the expressions in (13.58), (13.66), and (13.69) can be used to prove that $e_2(t)$, $x_{h1}(t)$, $x_{h2}(t)$, $\theta(t)$, $\omega_c(t) \in \mathcal{L}_\infty$. Based on the previous development, the expressions in (13.68), (13.70), and (13.71) can be used to conclude that $v_c(t)$, $\dot{\hat{d}}^*(t)$, $\dot{e}_1(t)$, $\dot{\tilde{e}}_2(t)$, $\dot{e}_3(t) \in \mathcal{L}_\infty$. Based on the fact that $e_1(t)$, $e_3(t)$, $\dot{e}_1(t)$, $\dot{e}_3(t) \in \mathcal{L}_\infty$ and that $e_1(t)$, $e_3(t) \in \mathcal{L}_2$, Barbalat's lemma [192] can be employed to prove that

$$\lim_{t \to \infty} e_1(t), e_3(t) = 0. \tag{13.78}$$

From (13.78) and the fact that the signal $\dot{x}_{hd1}(t)\tilde{e}_2(t)$ is uniformly continuous (i.e., $\dot{x}_{hd1}(t)$, $\ddot{x}_{hd1}(t)$, $\tilde{e}_2(t)$, $\dot{\tilde{e}}_2(t) \in \mathcal{L}_\infty$), the Extended Barbalat's Lemma can be applied to the last equation in (13.71) to prove that

$$\lim_{t \to \infty} \dot{e}_3(t) = 0 \tag{13.79}$$

and that

$$\lim_{t \to \infty} \dot{x}_{hd1}(t)\tilde{e}_2(t) = 0. \tag{13.80}$$

If the desired trajectory satisfies (13.74), then (13.80) can be used to prove that

$$\lim_{t \to \infty} \tilde{e}_2(t) = 0. \tag{13.81}$$

Based on the definition of $\tilde{e}_2(t)$ given in (13.66), the results in (13.78) and (13.81) can be used to conclude that

$$\lim_{t \to \infty} e_2(t) = 0 \tag{13.82}$$

provided the condition in (13.74) is satisfied.

Remark 13.7 The condition given in (13.74) is in terms of the time derivative of the desired translation vector. Typically, for WMR tracking problems, this assumption is expressed in terms of the desired linear and angular velocity of the WMR. To this end, (13.64) can be substituted into (13.74) to obtain the following condition:

$$\lim_{t\to\infty} \frac{v_{cd}(t)}{d^*} \neq x_{hd2}(t)\omega_{cd}(t). \tag{13.83}$$

The condition in (13.83) is comparable to typical WMR tracking results that restrict the desired linear and angular velocity. For an in-depth discussion of this type of restriction including related previous results see [55].

13.5 Simulational and Experimental Verifications

13.5.1 Eye-to-Hand Case

A numerical simulation performed to illustrate the performance of the controller given in (13.39)–(13.70). The desired trajectory was generated by numerically integrating (13.32), where the unknown constant scaling term d^* and the desired linear and angular velocity were selected as follows

$$d^* = 7.86(\text{m}) \qquad v_{cd}(t) = 0.2\sin(t)\ (\text{m/s}) \qquad \omega_{cd}(t) = 0.1\sin(t)\ (\text{rad/sec}).$$
$$\tag{13.84}$$

For the simulation, desired pixel information from a pre-recorded sequence of images (and the corresponding spline function) were not utilized to generate the feedforward signals in (13.39)–(13.70); rather, (13.14) and (13.32) were utilized to calculate $\eta_d(t)$, $\dot{\eta}_{d1}(t)$, $\theta_d(t)$, and $\dot{\theta}_d(t)$. Specifically, by numerically integrating (13.32) with the values given in (13.84), the desired signals, denoted by $\eta_{d1}(t)$ and $\eta_{d2}(t)$, can be determined. By numerically integrating $\omega_{cd}(t)$, the value for $\theta_d(t)$ can also be determined. For the simulation, the reference pixels, initial pixels, and the initial parameter of $\hat{z}_1^*$ were selected as follows:

$$p_1^* = \begin{bmatrix} 146 & 145 & 1 \end{bmatrix} \qquad p_1(0) = \begin{bmatrix} 67 & 181 & 1 \end{bmatrix} \qquad \hat{z}_1^*(0) = 12.8.$$

Based on the initial condition and reference coordinates, the initial translation and rotation errors were determined as follows:

$$e_1(0) = 0.042 \quad e_2(0) = -0.194 \quad e_3(0) = 1.40\ (\text{rad}),$$

where

$$\eta(0) = \begin{bmatrix} 0.140 & -0.038 & -0.983 \end{bmatrix} \qquad \eta_d(0) = \begin{bmatrix} 0.098 & 0.156 & -0.983 \end{bmatrix}.$$

The control gains and the adaptation gain were adjusted to $k_v = 10$, $k_\omega = 5$, and $\gamma_1 = 745$ to yield the best performance. The resulting translational and rotational errors for the WMR are depicted in Figure 13.1, and the parameter estimate

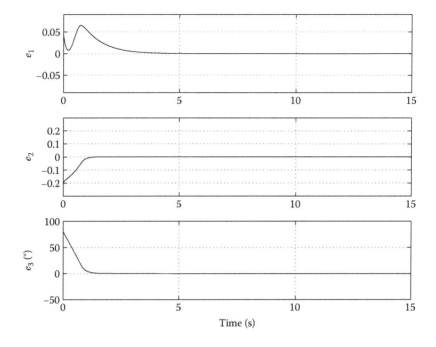

Figure 13.1: Translational and rotational tracking errors.

signal is depicted in Figure 13.3. The control input velocities $v_c(t)$ and $\omega_c(t)$ defined in (13.39) are depicted in Figure 13.2. Note that the angular velocity input was artificially saturated between $\pm 80.25 (^\circ \cdot s^{-1})$.

13.5.2 Eye-in-Hand Case

13.5.2.1 Experimental Configuration

To implement the adaptive tracking controller given by (13.68)–(13.70) an experimental testbed (see Figure 13.4) was constructed. The WMR testbed consists of the following components: a modified K2A WMR (with an inclusive Pentium 133 MHz personal computer (PC)) manufactured by Cybermotion Inc., a Dalsa CAD-6 camera that captures 955 frames per second with 8-bit gray scale at a 260×260 resolution, a Road Runner Model 24 video capture board, and two Pentium-based PCs. In addition to the WMR modifications described in detail in [55], additional modifications particular to this experiment included mounting a camera and the associated image processing Pentium IV 800 MHz PC (operating under QNX, a real-time micro-kernel based operating system) on the top of the WMR as depicted in Figure 13.4. The internal WMR computer (also operating under QNX) hosts the control algorithm that was written in "C/C++", and

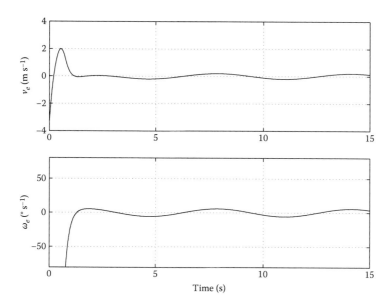

Figure 13.2: Linear and angular velocity control inputs.

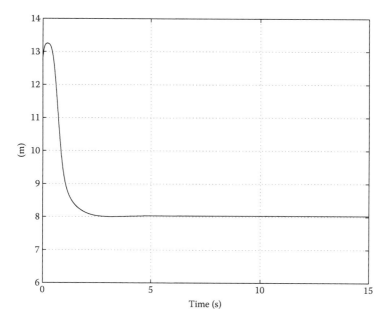

Figure 13.3: Parameter estimation of the depth z_1^*.

Figure 13.4: WMR testbed.

implemented using Qmotor 3.0 [139]. In addition to the image processing PC, a second PC (operating under the MS Windows 2000 operating system) was used to remotely login to the internal WMR PC via the QNX Phindows application. The remote PC was used to access the graphical user interface of Qmotor for execution of the control program, gain adjustment, and data management, plotting, and storage. Three light-emitting diodes (LEDs) were rigidly attached to a rigid structure that was used as the target, where the intensity of the LEDs contrasted sharply with the background. Due to the contrast in intensity, a simple thresholding algorithm was used to determine the coordinates of the centroid of the region of brightness values associated with each LED. The centroid was selected as the coordinates of the feature point.

The WMR is controlled by a torque input applied to the drive and steer motors. As subsequently described, to facilitate a torque controller the actual linear and angular velocity of the WMR is required. To acquire these signals, a backwards difference algorithm was applied to the drive and steering motor encoders. Encoder data acquisition and the control implementation were performed at a frequency of 1.0 kHz using the Quanser MultiQ I/O board. For simplicity, the electrical and mechanical dynamics of the system were not incorporated in the control design (i.e., the emphasis of this experiment is to illustrate the visual servo controller). However, since the developed kinematic controller is differentiable, standard backstepping techniques could be used to incorporate the mechanical and electrical dynamics. See [56] and [55] for several examples that incorporate the mechanical dynamics. Permanent magnet DC motors provide steering and drive actuation through a 106:1 and a 96:1 gear coupling, respectively. The dynamics for the modified K2A WMR are given as follows:

$$\frac{1}{r_o}\begin{bmatrix} 1 & 0 \\ 0 & \frac{L_o}{2} \end{bmatrix}\begin{bmatrix} \tau_1 \\ \tau_2 \end{bmatrix} = \begin{bmatrix} m_o & 0 \\ 0 & I_o \end{bmatrix}\begin{bmatrix} \dot{v}_1 \\ \dot{v}_2, \end{bmatrix} \tag{13.85}$$

where $\tau_1(t)$, $\tau_2(t) \in \mathbb{R}$ denote the drive and steering motor torques, respectively, $m_o = 165$ (kg) denotes the mass of the robot, $I_o = 4.643$ (kg m^2) denotes the inertia of the robot, $r_o = 0.010$ (m) denotes the radius of the wheels, and $L_o = 0.667$ (m) denotes the length of the axis between the wheels.

13.5.2.2 Experimental Results

To acquire the desired image trajectory, the WMR was driven by a joystick while the image processing PC acquired the camera images at 955 frames/s, determined the pixel coordinates of the feature points, and saved the pixel data to a file. The last image was also saved as the reference image. The desired image file and the reference image were read into a stand-alone program that computed $x_{hd}(t)$ and $\theta_d(t)$ offline. To determine the unique solution for $x_{hd}(t)$ and $\theta_d(t)$ (and likewise for $x_h(t)$ and $\theta(t)$) from the set of possible solutions generated by the homography decomposition using the Faugeras Decomposition Algorithm, a best-guess estimate of the constant normal n^* was selected as $n^* = \begin{bmatrix} 1 & 0 & 0 \end{bmatrix}^T$ (i.e., from the physical relationship between the camera and the plane defined by the object feature points, the focal axis of the camera mounted on the WMR was assumed to be roughly perpendicular to π). Of the possible solutions generated for n^* by the decomposition algorithm, the solution that yielded the minimum norm difference with the initial best-guess was determined as the correct solution. The solution that most closely matched the best-guess estimate was then used to determine the correct solutions for $x_{hd}(t)$ and $\theta_d(t)$ (or $x_h(t)$ and $\theta(t)$). The robustness of the system is not affected by the *a priori* estimate of n^* since the estimate is only used to resolve the ambiguity in the solutions generated by the decomposition algorithm, and the n^* generated by the decomposition algorithm is used to further decompose the homography. A Butterworth filter was applied to $x_{hd}(t)$ and $\theta_d(t)$ to reduce noise effects. A filtered backwards difference algorithm was used to compute $\dot{x}_{hd}(t)$ and $\dot{\theta}_d(t)$. Figures 13.5 and 13.6 depict the desired translation and rotation signals, respectively.

The desired trajectory signals $x_{hd}(t)$, $\dot{x}_{hd1}(t)$, $\theta_d(t)$, and $\dot{\theta}_d(t)$ were stored in a file that was opened by the control algorithm and loaded into memory when the control algorithm was loaded in Qmotor. Before the control program was executed, the image processing PC was set to acquire the live camera images at 955 frames/s, determine the pixel coordinates of the feature points, and transmit the coordinates via a server program over a dedicated 100 Mb/s network connection to the internal WMR computer. A client program was executed on the internal WMR computer to receive the pixel coordinates from the server program and write the current pixel information into a shared memory location. When the control program was executed, the current image information was acquired from the shared memory location (rather than directly from the network connection to maintain deterministic response and for program stability) and was compared to the reference image for online computation of the Euclidean homography.

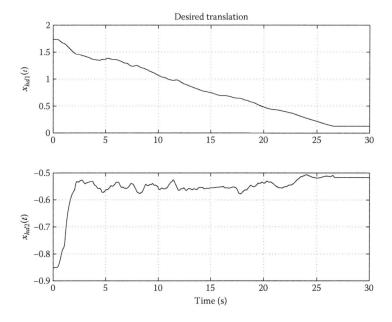

Figure 13.5: Desired translation of the mobile robot.

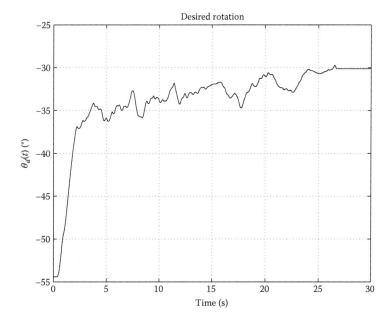

Figure 13.6: Desired rotation of the mobile robot.

The homography was decomposed using the Faugeras decomposition algorithm [67] to determine $x_h(t)$ and $\theta(t)$. After determining $x_h(t)$ and $\theta(t)$, comparisons with $x_{hd}(t)$ and $\theta_d(t)$ were made at each time instant to compute the error signal $e_1(t)$, $\bar{e}_2(t)$, and $e_3(t)$, which were subsequently used to compute $v_c(t)$, $\omega_c(t)$, and $\hat{d}^*(t)$ given in (13.68)–(13.70). To execute a torque level controller, a high-gain feedback loop was implemented as follows:

$$\tau = K_h \eta \tag{13.86}$$

where $K_h \in \mathbb{R}^{2\times2}$ is a diagonal high-gain feedback term, and $\eta(t) \in \mathbb{R}^2$ is a velocity mismatch signal defined as:

$$\eta = \begin{bmatrix} v_c & \omega_c \end{bmatrix}^T - \begin{bmatrix} v_a & \omega_a, \end{bmatrix}^T \tag{13.87}$$

where $v_c(t)$ and $\omega_c(t)$ denote the linear and angular velocity inputs computed in (13.68) and (13.69) and $v_a(t)$ and $\omega_a(t)$ denote actual linear and angular velocity of the WMR computed from the time derivative of the wheel encoders.

The control gains were adjusted to reduce the position/orientation tracking error with the adaptation gains set to zero and the initial adaptive estimate set to zero. After some tuning, we noted that the position/orientation tracking error response could not be significantly improved by further adjustments of the feedback gains. We then adjusted the adaptation gains to allow the parameter estimation to reduce the position/orientation tracking error. After the tuning process was completed, the final adaptation and feedback gain values were recorded as shown below:

$$k_v = 4.15, \quad k_\omega = 0.68, \quad \gamma = 40.1,$$
$$K_h = diag\{99.7, 23.27\}. \tag{13.88}$$

The unitless position/orientation tracking errors $e_1(t)$ and $e_2(t)$, are depicted in Figures 13.7 and 13.8, respectively. Figure 13.9 illustrates that the adaptive estimate for the depth parameter d^* approaches a constant. Figure 13.10 illustrates the linear and angular velocity of the WMR. The control torque inputs are presented in Figure 13.11 and represent the torques applied after the gearing mechanism.

13.5.2.3 Results Discussion

From Figures 13.7 and 13.8, it is clear that $e_2(t)$ is relatively unchanging in the first 8 s, whereas $e_1(t)$ and $e_3(t)$ are changing significantly. This phenomena is due to the nonholonomic nature of the vehicle. Specifically, since there is an initial position and orientation error, the controller moves the vehicle to minimize the error and align the WMR with the desired image trajectory. Since the WMR cannot move along both axes of the Cartesian plane simultaneously while also rotating (i.e., due to the nonholonomic motion constraints), the WMR initially moves to minimize $e_1(t)$ and $e_3(t)$. Likewise, when $e_2(t)$ undergoes change

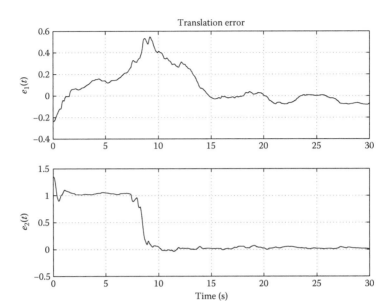

Figure 13.7: Translation error of the mobile robot.

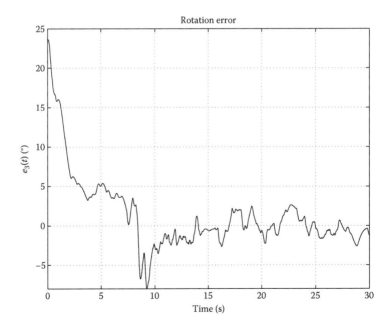

Figure 13.8: Rotation error of the mobile robot.

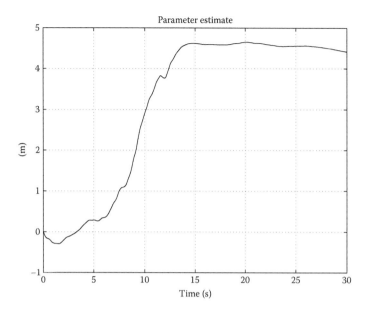

Figure 13.9: Parameter estimate.

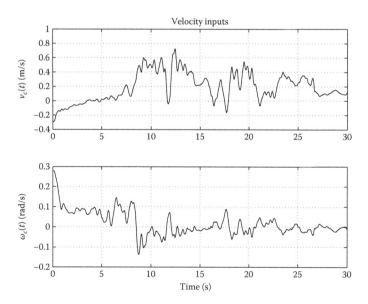

Figure 13.10: Linear and angular velocity control inputs.

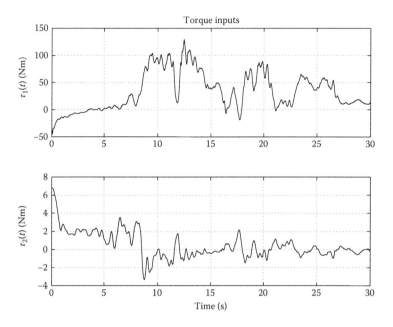

Figure 13.11: Drive and steer motor torque inputs.

between 8 and 10 s, $e_1(t)$ remains relatively unchanging. While performing the experiment, slightly different responses were obtained each run due to variations in the initial position and orientation of the WMR and variations in the control parameters as the gains were adjusted. With a constant set of control gains, the transient response still exhibited some variations due to differences in the initial conditions; however, the steady state response remained constant for each trial.

Note that $e_1(t)$ and $e_2(t)$ depicted in Figure 13.7 are unitless. From (13.23) and (13.58), it is clear that $e_1(t)$ and $e_2(t)$ are unitless because the translation $x_f(t)$ has units of meters, and the depth related constant d^* has units of meters. That is, $x_h(t)$ and $x_{hd}(t)$ are unitless translation terms computed from the homography decomposition (note that no units are provided in Figure 13.5). In practice, the WMR traversed an arc than approximately spanned a 6×1 m space, with an approximate speed of 0.22 m/s.

Based on the outcome of this experiment, several issues for future research and technology integration are evident. For example, the problem formulation in this chapter has a number of practical applications in environments where the reference object may not be stationary between each task execution. However, the result in this chapter does not address cases where an obstacle enters the task-space and inhibits the WMR from tracking the pre-recorded trajectory. To address this issue, there is a clear need for continued research that targets incorporating image-space path planning with the control design as in [46] and [158].

Additionally, the result in this chapter does not address a method to automatically reselect feature points. For example, methods to automatically determine new feature points if they become nearly aligned, or if a feature point leaves the field-of-view (e.g., becomes occluded) could add robustness to the implemented control system. Of course, an ad hoc approach of simply continuously tracking multiple redundant feature points could be utilized, but this approach may excessively restrict the image processing bandwidth.

13.6 Conclusion

In this chapter, vision-based control methods for wheeled mobile robots are proposed for camera-to-hand and camera-in-hand configurations.

- For the camera-to-hand case, with the visual feedback from a fixed monocular camera, the position/orientation of a WMR is controlled to follow a desired time-varying trajectory which is defined by a pre-recorded image sequence. The multiple views of four feature points were utilized for homography development and decomposition. The decomposed Euclidean information was utilized for the control development. An adaptive update law was designed to compensate for an unknown depth information. The homography-based analytical approach enables the position/orientation tracking control of a WMR subject to nonholonomic constraints without depth measurements. The optical axis of the camera is not required to be perpendicular to the WMR plane of motion. An extension is provided to illustrate the development of a visual servo regulation controller. Simulation results are provided to demonstrate the performance of the tracking control design.

- For the camera-in-hand case, the position/orientation of a WMR is forced to track a desired time-varying trajectory defined by a pre-recorded sequence of images. To achieve the result, multiple views of three target points were used to develop Euclidean homographies. By decomposing the Euclidean homographies into separate translation and rotation components, reconstructed Euclidean information was obtained for the control development. A Lyapunov-based stability argument was used to design an adaptive update law to compensate for the fact that the reconstructed translation signal was scaled by an unknown depth parameter. The impact that the development in this chapter makes is that a new analytical approach has been developed using homography-based concepts to enable the position/orientation of a WMR subject to nonholonomic constraints to track a desired trajectory generated from a sequence of images, despite the lack of depth measurements. Experimental results are provided to illustrate the performance of the controller.

Chapter 14

Trifocal Tensor Based Visual Control of Mobile Robots

14.1 Introduction

In this chapter, the trajectory tracking task is considered, in which the tracking of both pose and velocities are required with time constraints. Thus, the subgoal strategy in path following tasks is not acceptable. Besides, the workspace of trajectory tracking tasks is larger than pose regulation tasks, i.e., the start and final images don't need to have enough correspondences.

A trifocal tensor based visual control strategy is proposed to regulate a mobile robot to track the desired trajectory expressed by a set of pre-recorded images in large workspace. Trifocal tensor among three views is exploited to obtain the orientation and scaled position information, which is used for geometric reconstruction. For tasks in large workspace, key frame strategy, developed in Chapter 4, is used to perform global and continuous pose estimation, without the need of existing corresponding feature points in the start, current, and final frames. A series of key frames are selected along the desired trajectory; then, pose information is estimated with respect to two most similar key frames and then transformed to the final frame. Besides, unlike most researches that install the camera at the center of the robot wheel axis, the camera can be installed at an arbitrary position in this chapter, improving system flexibility. To compensate for the unknown depth and extrinsic parameters, an adaptive controller is developed using Lyapunov-based method, achieving asymptotic tracking with respect to the desired trajectory.

The proposed control strategy works for both classical trajectory and pose regulation tasks. Simulation results are provided for performance evaluation.

14.2 Geometric Modeling

To construct the feedback of the visual control system, the trifocal tensor among three views is exploited to extract geometric information (Figure 14.1). In this section, the model of trifocal tensor is introduced first, based on which scaled pose information can be calculated (Euclidean reconstruction). Without loss of generality, the geometric relationship among views C_0, C, and C^* is analyzed in this section. In the following, the geometric model can be generalized to other views.

Using the trifocal tensor among the three views, the translation vector can be estimated up to a scale d as:

$$x_h \triangleq \begin{bmatrix} x_{hx} & 0 & x_{hz} \end{bmatrix}^T = \frac{x_f}{d}, \tag{14.1}$$

where x_{hx}, x_{hz} are given by:

$$x_{hx} = -\bar{T}_{212}, \ x_{hz} = -\bar{T}_{232}. \tag{14.2}$$

The orientation angle θ can be determined by the following expressions:

$$\begin{aligned} \sin\theta &= \bar{T}_{221}(\bar{T}_{333} - \bar{T}_{131}) - \bar{T}_{223}(\bar{T}_{331} + \bar{T}_{133}), \\ \cos\theta &= \bar{T}_{221}(\bar{T}_{111} - \bar{T}_{313}) + \bar{T}_{223}(\bar{T}_{113} + \bar{T}_{311}). \end{aligned} \tag{14.3}$$

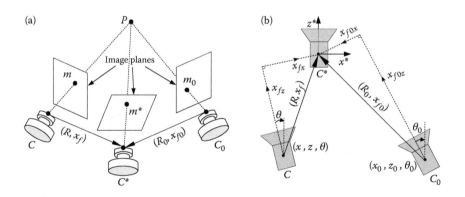

Figure 14.1: Geometric modeling of the trifocal tensor. (a) 3-D view of trifocal tensor and (b) planar view of trifocal tensor.

Then, the scaled position coordinates $x_m \triangleq \frac{x}{d}$, $z_m \triangleq \frac{z}{d}$ can be calculated as follows:

$$\begin{bmatrix} x_m \\ z_m \end{bmatrix} = - \begin{bmatrix} \cos\theta & -\sin\theta \\ \sin\theta & \cos\theta \end{bmatrix} \cdot \begin{bmatrix} x_{hx} \\ x_{hz} \end{bmatrix}. \tag{14.4}$$

In the subsequent development, $(x_{hx}(t), x_{hz}(t), \theta(t))$ is used to express the pose of the robot. Similarly, denote $x_{hd} \triangleq \begin{bmatrix} x_{hdx} & 0 & x_{hdz} \end{bmatrix}^T$ as the scaled estimation of the desired translation vector, and θ_d as the desired orientation. $(x_{hdx}(t), x_{hdz}(t), \theta_d(t))$ can be used to express the desired pose of the robot, which can be estimated in the same manner with the current pose $(x_{hx}(t), x_{hz}(t), \theta(t))$.

Assumption 14.1 *The desired trajectory $C_d(t)$ is feasible and sufficiently smooth, i.e., its first and second time derivatives $\dot{x}_{hd}(t), \dot{\theta}_d(t), \ddot{x}_{hd}(t), \ddot{\theta}_d(t)$ exist and are bounded.*

Assumption 14.2 *In this chapter, both tracking and regulation tasks are considered, and they are generally under one of the following two conditions:*

1. $\lim_{t\to\infty} \dot{x}_{hdz}(t) \neq 0$;

2. $\lim_{t\to\infty} \dot{x}_{hdz}(t) = 0$ *and there exists $\chi > 0$ such that $\int_0^\infty |\dot{x}_{hdz}(t)| dt \leq \chi$, i.e., $\dot{x}_{hdz} \in \mathcal{L}_1$.*

Remark 14.1 The first and second conditions mean that the reference linear velocity either doesn't vanish or converges to zero with a finite excitation (absolute value integrable). Pose regulation tasks can be formulated into trajectory tracking tasks with time-invariant desired trajectories, which are under the second condition. Practically, almost all circumstances are covered in these two conditions, which are comparable with the conditions used in previous works [58,59].

Remark 14.2 The exceptional case of Assumption 14.2 is that $\lim_{t\to\infty} \dot{x}_{hdz}(t) = 0$ and $\int_0^\infty |\dot{x}_{hdz}(t)| dt = \infty$, i.e., the reference forward velocity converges to zero with an infinite excitation. A typical circumstance is that the robot moves to infinity with a desired forward velocity converging to zero (e.g. $\dot{x}_{hdz}(t) = 1/(t+1)$), which is almost not practical and can be hardly implemented for real robotic systems.

14.3 Control Development

As illustrated in Figure 14.2, using the key frame strategy presented in Chapter 4, the current pose information and the desired pose information with respect to the final frame C^* can be computed efficiently. In this section, an adaptive controller is developed using the measurable pose information.

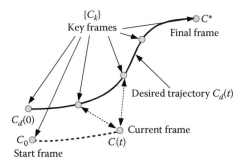

Figure 14.2: Key frame strategy.

As shown in Figure 14.3, the control objective is to regulate the robot so that $C(t)$ tracks $C_d(t)$, i.e., $(x_{hx}(t), x_{hz}(t), \theta(t))$ tracks $(x_{hdx}(t), x_{hdz}(t), \theta_d(t))$. Then, it is straightforward to define the error vector $e(t)$ as follows:

$$
e(t) \triangleq \begin{bmatrix} e_1(t) \\ e_2(t) \\ e_3(t) \end{bmatrix} = \begin{bmatrix} x_{hx}(t) - x_{hdx}(t) \\ x_{hz}(t) - x_{hdz}(t) \\ \theta(t) - \theta_d(t). \end{bmatrix}
\tag{14.5}
$$

14.3.1 Error System Development

As shown in Figure 9.3, the camera is mounted at an arbitrary position $(x = D, z = L)$ with respect to $\mathcal{F}$, where $\circ$ could be an arbitrary constant denoting the installing height of the camera and doesn't affect the kinematic model.

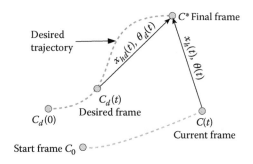

Figure 14.3: Definition of unified control objective.

Denoting the current coordinate of the robot as $(x_r(t), z_r(t), \theta(t))$, the coordinate of the camera $(x(t), z(t), \theta(t))$ can be obtained as:

$$\begin{bmatrix} x \\ z \end{bmatrix} = \begin{bmatrix} x_r \\ z_r \end{bmatrix} + \begin{bmatrix} \cos\theta & -\sin\theta \\ \sin\theta & \cos\theta \end{bmatrix} \begin{bmatrix} D \\ L \end{bmatrix}. \tag{14.6}$$

The classical kinematic model of a nonholonomic robot is given by:

$$\begin{aligned} \dot{x}_r &= -v\sin\theta \\ \dot{z}_r &= v\cos\theta \\ \dot{\theta} &= -\omega. \end{aligned} \tag{14.7}$$

After taking the time derivative of (14.6) and utilize (14.7), the kinematic model of the camera is given by:

$$\begin{aligned} \dot{x} &= -v\sin\theta + (D\sin\theta + L\cos\theta)\omega \\ \dot{z} &= v\cos\theta - (D\cos\theta - L\sin\theta)\omega \\ \dot{\theta} &= -\omega. \end{aligned} \tag{14.8}$$

The measurable pose information $(x_{hx}(t), x_{hz}(t), \theta(t))$ can be expressed by the coordinates $(x(t), z(t), \theta(t))$ as follows:

$$\begin{aligned} x_{hx} &= \frac{-x\cos\theta - z\sin\theta}{d} \\ x_{hz} &= \frac{x\sin\theta - z\cos\theta}{d}, \end{aligned} \tag{14.9}$$

where d is the distance between the last two key frames, as indicated in the pose estimation process in Algorithm 4.2 After taking the derivative of (14.9) and utilizing (14.8), the kinematic model of $(x_{hx}(t), x_{hz}(t), \theta(t))$ is given by:

$$\begin{aligned} \dot{x}_{hx} &= -(x_{hz} + L_m)\omega \\ \dot{x}_{hz} &= -\frac{v}{d} + (x_{hx} + \frac{D}{d})\omega \\ \dot{\theta} &= -\omega, \end{aligned} \tag{14.10}$$

where $L_m \triangleq L/d$. Similarly, for the desired trajectory, the following kinematic model can be obtained:

$$\begin{aligned} \dot{x}_{hdx} &= -(x_{hdz} + L_m)\omega_d \\ \dot{x}_{hdz} &= -\frac{v_d}{d} + (x_{hdx} + \frac{D}{d})\omega_d \\ \dot{\theta}_d &= -\omega_d. \end{aligned} \tag{14.11}$$

Since d, D are unknown, $v_d(t)$ is not measurable and will not be used in the controller, while $\omega_d(t)$ can be directly computed from the desired trajectory $\dot{\theta}_d(t)$ and will be used in the controller.

After taking the derivative of (14.5) and utilizing (14.10), (14.11), the open-loop error system is obtained as follows:

$$\dot{e}_1 = -e_2\omega + (x_{hdz} + L_m)(\omega_d - \omega)$$
$$d\dot{e}_2 = -v + d(x_{hx}\omega - \dot{x}_{hdz}) + D\omega \tag{14.12}$$
$$\dot{e}_3 = -\omega + \omega_d.$$

To facilitate the controller design, an auxiliary error signal $\bar{e}_1(t) \in \mathbb{R}$ is defined as:

$$\bar{e}_1 = e_1 - (x_{hdz} + \hat{L}_m)e_3, \tag{14.13}$$

where $\hat{L}_m(t)$ is the estimate of L_m. It is clear that if $\bar{e}_1(t), e_3(t) \to 0$, then $e_1(t) \to 0$. Taking the time derivative of (14.13) and utilizing (14.12), it can be obtained that

$$\dot{\bar{e}}_1 = -e_2\omega + \tilde{L}_m(\omega_d - \omega) - (\dot{x}_{hdz} + \dot{\hat{L}}_m + \rho)e_3 + \rho e_3, \tag{14.14}$$

where $\tilde{L}_m(t) \triangleq L_m - \hat{L}_m(t)$ is the estimate error of L_m, and $\rho(t)$ is an auxiliary signal defined as:

$$\rho(t) \triangleq \beta \cdot e^{-\int_0^t \left(\frac{|\dot{x}_{hdz}(\tau)|}{\sigma_1} + \frac{|e_3(\tau)|}{\sigma_2} \right) d\tau} \tag{14.15}$$

in which $\beta \neq 0$ and $\sigma_1, \sigma_2 > 0$ are constant scale factors.

14.3.2 Controller Design

The above open-loop system can be stabilized by the following control law:

$$v = k_2 e_2 + \hat{d}(x_{hx}\omega - \dot{x}_{hdz}) + \hat{D}\omega - \eta_1\omega$$
$$\omega = \omega_d + k_3\eta_2 e_3 - k_1\eta_1\eta_2(\dot{x}_{hdz} + \rho), \tag{14.16}$$

where $k_1, k_2, k_3 \in \mathbb{R}^+$ are constant control gains, $\hat{d}(t)$ and $\hat{D}(t)$ are the estimate of d and D, respectively. The update rules for $\hat{d}(t)$, $\hat{D}(t)$ and $\hat{L}_m(t)$ are given by:

$$\dot{\hat{d}} = \gamma_1 e_2(x_{hx}\omega - \dot{x}_{hdz})$$
$$\dot{\hat{D}} = \gamma_2 e_2\omega \tag{14.17}$$
$$\dot{\hat{L}}_m = \gamma_3\eta_1(\omega_d - \omega),$$

where $\gamma_1, \gamma_2, \gamma_3 \in \mathbb{R}^+$ are constant update gains. In (14.16), $\eta_1(t), \eta_2(t) \in \mathbb{R}$ are two auxiliary signals defined as:

$$\eta_1(t) = \frac{\bar{e}_1}{\bar{e}_1^2 + \mu}$$
$$\eta_2(t) = \frac{(\bar{e}_1^2 + \mu)^2}{(\bar{e}_1^2 + \mu)^2 - k_1\gamma_3\bar{e}_1^2}, \tag{14.18}$$

where $\mu \in \mathbb{R}^+$ is a constant satisfying $\mu > \max(1.5, \frac{k_1 \gamma_3}{4})$, which is a sufficient condition to ensure that $(\bar{e}_1^2 + \mu)^2 - k_1 \gamma_3 \bar{e}_1^2 > 0$ and $\eta_2(t) > 0$.

Remark 14.3 The auxiliary signal $\rho(t)$ is utilized to deal with the second condition in *Assumption 14.2* to remove the requirements on the desired velocity in [31] and [104], as will be analyzed in the subsequent stability analysis.

After substituting (14.16), (14.17) into (14.12) and (14.14), the closed-loop system is given by:

$$
\begin{aligned}
\dot{e}_1 =& -e_2 \omega + (\tilde{L}_m - \gamma_3 \eta_1 e_3)(\omega_d - \omega) \\
& - (\dot{x}_{hdz} + \rho)e_3 + \rho e_3 \\
d\dot{e}_2 =& -k_2 e_2 + \eta_1 \omega + \tilde{d}(x_{hx}\omega - \dot{x}_{hdz}) + \tilde{D}\omega \\
\dot{e}_3 =& -k_3 \eta_2 e_3 + k_1 \eta_1 \eta_2 (\dot{x}_{hdz} + \rho),
\end{aligned}
\tag{14.19}
$$

where $\tilde{d}(t) = d - \hat{d}(t)$ and $\tilde{D}(t) = D - \hat{D}(t)$ are the estimate errors of d and D, respectively.

14.3.3 Stability Analysis

Theorem 1: The control input in (14.16) along with the update law in (14.17) ensure that the tracking error $e(t)$ asymptotically converges to zero in the sense that

$$
\lim_{t \to \infty} e(t) = 0
\tag{14.20}
$$

if one of the conditions in *Assumption 14.2* is satisfied.

Proof 14.1 To prove *Theorem 1*, a nonnegative function $V(t) \in \mathbb{R}$ is defined as:

$$
V = \frac{1}{2}\left(\ln(\bar{e}_1^2 + \mu) + de_2^2 + \frac{e_3^2}{k_1} + \frac{\tilde{d}^2}{\gamma_1} + \frac{\tilde{D}^2}{\gamma_2} + \frac{\tilde{L}_m^2}{\gamma_3}\right),
\tag{14.21}
$$

with its time derivative given by:

$$
\dot{V} = \frac{\bar{e}_1}{\bar{e}_1^2 + \mu}\dot{e}_1 + de_2\dot{e}_2 + \frac{1}{k_1}e_3\dot{e}_3 - \frac{\tilde{d}}{\gamma_1}\dot{\hat{d}} - \frac{\tilde{D}}{\gamma_2}\dot{\hat{D}} - \frac{\tilde{L}_m}{\gamma_3}\dot{\hat{L}}_m.
\tag{14.22}
$$

Substituting (14.17), (14.19) into (14.22) and after some simplifications, it can be obtained that

$$
\begin{aligned}
\dot{V} &= -k_2 e_2^2 - \frac{k_3}{k_1}e_3^2 + \rho e_3 \eta_1 \\
&\leq -k_2 e_2^2 - \frac{k_3}{k_1}e_3^2 + |\rho e_3 \eta_1|.
\end{aligned}
\tag{14.23}
$$

According to the proof in Appendix F.2, $\frac{1}{2}\ln(\bar{e}_1^2 + \mu) > \frac{\bar{e}_1}{\bar{e}_1^2 + \mu}$ for $\mu > 1.5$, then (14.23) can be rewritten as:

$$\dot{V} \leq -k_2 e_2^2 - \frac{k_3}{k_1} e_3^2 + |\rho e_3| V. \tag{14.24}$$

Then, it is obvious that $\dot{V} \leq |\rho e_3| V$, which implies that

$$V(t) \leq V(0) e^{\int_0^t |\rho e_3| d\tau}. \tag{14.25}$$

According to the development in Appendix F.3, $\rho e_3 \in \mathcal{L}_1$, then $\int_0^t |\rho e_3| d\tau$ is bounded, and it is straightforward to conclude from (14.25) that $V(t) \in \mathcal{L}_\infty$. Then, it can be concluded from (14.21) that $\bar{e}_1(t), e_2(t), e_3(t), \tilde{d}(t), \tilde{D}(t), \tilde{L}_m(t) \in \mathcal{L}_\infty$, and hence $\eta_1(t), \eta_2(t) \in \mathcal{L}_\infty$. Based on *Assumption 14.1*, it is clear that $x_{hdx}(t), x_{hdz}(t), \dot{x}_{hdz}(t), \dot{\theta}_d(t) \in \mathcal{L}_\infty$. From (14.16), it can be inferred that $v(t), \omega(t) \in \mathcal{L}_\infty$. Then it can inferred from (14.19) that $\dot{\bar{e}}_1(t), \dot{e}_2(t), \dot{e}_3(t) \in \mathcal{L}_\infty$. Since $V(t) \in \mathcal{L}_\infty$, there exists a constant $\delta > 0$ for each initial condition, such that $V(t) \leq \delta$. Integrating both sides of (14.24) it can be obtained that

$$\int_0^t \left(k_2 e_2^2(\tau) + \frac{k_3}{k_1} e_3^2(\tau) \right) d\tau$$
$$\leq V(0) - V(t) + \int_0^t |\rho(\tau) e_3(\tau)| V(\tau) d\tau \tag{14.26}$$
$$\leq V(0) - V(t) + \delta \int_0^t |\rho(\tau) e_3(\tau)| d\tau.$$

Since $\rho(t) e_3(t) \in \mathcal{L}_1$ and $V(t) \in \mathcal{L}_\infty$; then, it can be inferred from (14.26) that $e_2(t), e_3(t) \in \mathcal{L}_2$. Barbalat's lemma [192] can be exploited to conclude that

$$\lim_{t \to \infty} e_2(t), e_3(t) = 0. \tag{14.27}$$

Taking the time derivative of $\eta_1(t), \eta_2(t)$, it can be obtained that

$$\dot{\eta}_1(t) = \frac{(\mu - \bar{e}_1^2)\dot{\bar{e}}_1}{(\bar{e}_1^2 + \mu)^2},$$
$$\dot{\eta}_2(t) = \frac{2k_1 \gamma_3 \bar{e}_1 (\mu^2 - \bar{e}_1^4)\dot{\bar{e}}_1}{\left[(\bar{e}_1^2 + \mu)^2 - k_1 \gamma_3 \bar{e}_1^2 \right]^2}. \tag{14.28}$$

Since $\bar{e}_1(t), \dot{\bar{e}}_1(t) \in \mathcal{L}_\infty$ and $(\bar{e}_1^2 + \mu)^2 - k_1 \gamma_3 \bar{e}_1^2 \neq 0$ from previous facts, it can be inferred that $\dot{\eta}_1(t), \dot{\eta}_2(t) \in \mathcal{L}_\infty$. After taking the time derivative of $\dot{e}_3(t)$, it can be obtained that

$$\ddot{e}_3(t) = -k_3 \dot{\eta}_2 e_3 - k_3 \eta_2 \dot{e}_3 + k_1 \dot{\eta}_1 \eta_2 (\dot{x}_{hdz} + \rho)$$
$$+ k_1 \eta_1 \dot{\eta}_2 (\dot{x}_{hdz} + \rho) + k_1 \eta_1 \eta_2 (\ddot{x}_{hdz} + \dot{\rho}). \tag{14.29}$$

Based on previous facts, $e_3(t), \dot{e}_3(t) \in \mathcal{L}_\infty$ and $\eta_1(t), \dot{\eta}_1(t), \eta_2(t), \dot{\eta}_2(t) \in \mathcal{L}_\infty$. From *Assumption 14.1*, $\dot{x}_{hdz}(t), \ddot{x}_{hdz}(t) \in \mathcal{L}_\infty$. From (14.15), $\rho(t), \dot{\rho}(t) \in \mathcal{L}_\infty$. Then, it can be inferred that $\ddot{e}_3(t) \in \mathcal{L}_\infty$, Barbalat's lemma [192] can be exploited to conclude that

$$\lim_{t \to \infty} \dot{e}_3(t) = 0. \tag{14.30}$$

Since $e_3(t) \to 0$ as $t \to \infty$, it can be inferred from (14.19) that $k_1 \eta_1 \eta_2 (\dot{x}_{hdz} + \rho) \to 0$ as $t \to \infty$, which can be rewritten as:

$$\lim_{t \to \infty} k_1 \frac{\bar{e}_1^2 + \mu}{(\bar{e}_1^2 + \mu)^2 - k_1 \gamma_3 \bar{e}_1^2} (\dot{x}_{hdz} + \rho)\bar{e}_1 = 0. \tag{14.31}$$

In *Assumption 14.2*, if the first condition is satisfied, i.e., $\lim_{t \to \infty} \dot{x}_{hdz}(t) \neq 0$; then, $\lim_{t \to \infty} \rho(t) = 0$ according to (14.15); if the second condition is satisfied, i.e., $\lim_{t \to \infty} \dot{x}_{hdz}(t) = 0$ and $\dot{x}_{hdz}(t) \in \mathcal{L}_1$, since $e_3(t) \in \mathcal{L}_1$ from Appendix F.4; then, $\lim_{t \to \infty} \rho(t) \neq 0$ according to (14.15). Hence, under *Assumption 3*, $\lim_{t \to \infty} (\dot{x}_{hdz} + \rho) \neq 0$ is guaranteed; then, it can be concluded that $\bar{e}_1(t) \to 0$ as $t \to \infty$.

From the above, it can be concluded that $e(t) \to 0$ as $t \to \infty$ if one of the conditions in *Assumption 14.2* is satisfied.

14.4 Simulation Verification

In this section, the performance of the proposed scheme is evaluated via simulations based on the dynamic robot simulator V-REP [175]. As shown in Figure 14.4a, the simulation platform is a Pioneer P3-DX robot, which has two differential drive wheels and one free wheel. The simulated perspective camera is mounted at an arbitrary position ($D = 0.1$ m, $L = 0.1$ m) of the robot, which is shown as the blue block in Figure 14.4b. The field of view of the camera is $90°$,

Figure 14.4: Simulation configuration for mobile robot unified control. (a) Simulation environment and (b) robot setup.

and gray scale images are captured by the camera, with their resolutions being 1024×1024. The environment is made of textured walls, and feature points can be extracted using speeded-up robust feature (SURF) [11]. Corresponding feature points in three relevant images are matched and trifocal tensor is estimated using RANSAC based method [86]. The control system is implemented in MATLAB and communicates with the simulator via the remote API of V-REP.

14.4.1 Pose Estimation

To show the necessity and effectiveness of key frame strategy, simulations are made for a comparison among the proposed key frame strategy and standard multiple-view geometry-based methods. In the following, pose estimation methods based on trifocal tensor without key frame strategy, homography, and epipolar geometry are used for comparison, with the same image feature extraction algorithm (i.e., SURF) and the same parameters. The top 300 strongest feature points according to the magnitude of the SURF descriptor [11] are selected to reduce the effect of image noise.

Since standard multiple-view geometry methods require the corresponding feature points among the relevant views, the comparison is made with a task that the start and final views have only a few correspondences. As shown in Figure 14.5, the demonstrated trajectory is performed for 25 s and frames are captured

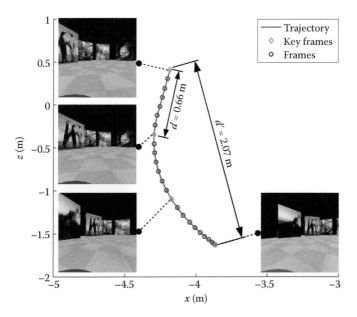

Figure 14.5: Demonstrated trajectory of the mobile robot.

every 1 s. Key frames are selected using a threshold $\tau = 0.55$. In Figure 14.5, d is denoted as the distance between the last two frames and is the scale factor in the pose information estimated by key frame strategy; d' is denoted as the distance between the start and final frames which is the scale factor in the pose information estimated by classical trifocal tensor based methods. Their actual values are obtained from the simulator, as shown in Figure 14.5. They are only used for the evaluation of measuring precision, and won't be used in the pose estimation methods.

As shown in Figure 14.6a, homography-based methods rely on the decomposition of homography matrix between the current and final frames, and d'' is denoted as the distance from the final view to the captured plane, which is the scaled factor in the estimated pose information. Readers can refer to [31] for a detailed introduction on the scaled pose estimation based on homography. For simplicity, only planar object is considered for comparison in this chapter, readers can refer to [150] for the eight-points algorithm for nonplanar scenes. In the demonstration process for the homography-based method, the trajectory is the same as before, but only the wall which is visible in both start and final images is retained in the scene, the distance d'' can be obtained from the simulator and will only be used for the evaluation of measuring precision.

In the literature, epipolar geometry has not been used for scaled pose estimation. While the strategy based on three views proposed in [15] can be extended for scaled pose estimation, as shown in Figure 14.6b. To avoid the short baseline problem, a simple strategy similar to the "virtual target" [144] is used, the virtual views C_0' and $C^{*\prime}$ are obtained by performing a translation along the z-axis of the robot from C_0 and C^*, respectively. In Figure 14.6b, $\varepsilon_1, \varepsilon_2, \ldots, \varepsilon_6$ are denoted as

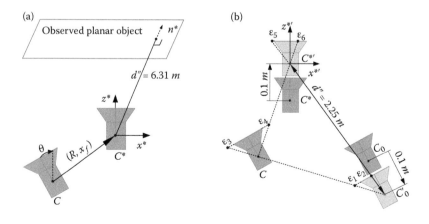

Figure 14.6: Pose estimation based on two-view geometry. (a) Homography-based method and (b) epipolar geometry-based method.

the horizontal coordinates of the epipoles, d''' is denoted as the distance from C_0' to $C^{*'}$ which is the scale factor in the estimated pose information. Assuming the images are normalized to unit focal length, the scaled pose information (x_m', z_m', θ) of C can be calculated with respect to $C^{*'}$ after some geometry operations as follows:

$$x_m' = -\frac{\varepsilon_6(\varepsilon_2 - \varepsilon_1)\sqrt{(\varepsilon_3^2 + 1)(\varepsilon_4^2 + 1)}}{(\varepsilon_4 - \varepsilon_3)\sqrt{(\varepsilon_1^2 + 1)(\varepsilon_2^2 + 1)(\varepsilon_6^2 + 1)}}$$

$$z_m' = -\frac{(\varepsilon_2 - \varepsilon_1)\sqrt{(\varepsilon_3^2 + 1)(\varepsilon_4^2 + 1)}}{(\varepsilon_4 - \varepsilon_3)\sqrt{(\varepsilon_1^2 + 1)(\varepsilon_2^2 + 1)(\varepsilon_6^2 + 1)}} \qquad (14.32)$$

$$\theta = \arctan \varepsilon_4 - \arctan \varepsilon_6$$

From (14.32) it is obvious that singularity exists when $\varepsilon_3 = \varepsilon_4$, which means that C_0', C, and $C^{*'}$ are collinear. This is another limitation of epipolar geometry-based methods but doesn't exist in the demonstrated trajectory as shown in Figure 14.5.

Conventional multiple-view geometry based methods directly use corresponding points in relevant views to calculate the geometric constraints, based on which scaled pose information is estimated. In Figure 14.7, the matched pairs are validated by the actual pose information from the simulator. It is obvious that most of the matched pairs are invalid. In fact, the reliability of image features degrades significantly with large displacements and viewpoint changes, which will greatly affect the precision of pose estimation.

To evaluate the measuring precision of the above methods, pose information of the frames captured along the demonstrated trajectory is estimated up to scale only using image measurements, and the actual pose information is calculated by the data obtained from the simulator. The measuring errors are also expressed up to the scale, including the translational error $e_t \triangleq \left\| \begin{bmatrix} \bar{x}_m - x_m & \bar{z}_m - z_m \end{bmatrix} \right\|$ and rotational error $e_r = |\bar{\theta} - \theta|$, with (x_m, z_m, θ_m) being the estimated scaled pose information and $(\bar{x}_m, \bar{z}_m, \bar{\theta}_m)$ being the actual scaled pose information. Pose estimation

Figure 14.7: Matched features between start and final images.

errors of the above methods are presented in Figure 14.8. From the results, it is obvious that pose estimation based on key frame strategy outperforms the other methods even with a smaller scale factor. Considering the scale factor, the precision of pose estimation based on trifocal tensor is improved by about 12 times owing to the key frame strategy. The result based on epipolar geometry in Figure 14.8d is comparable to that based on trifocal tensor without key frame strategy in Figure 14.8b, which is due to the fact that they both rely on the matching of the three views. As shown in Figure 14.8c, an advantage of the homography-based method is that the precision increases as closing the final frame. From Figure 14.8a it can be seen that the key frame strategy also has a similar characteristic as mentioned in Remark 4.3, which is suitable for such a trajectory tracking task.

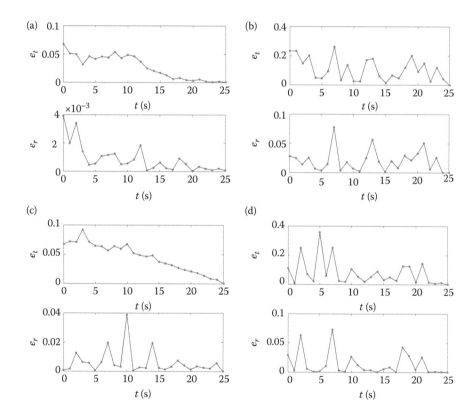

Figure 14.8: Scaled pose estimation errors based on multiple-view geometry. (a) Pose estimation error based on trifocal tensor with key frame strategy (scaled by $d = 0.66$ m). (b) Pose estimation error based on trifocal tensor without key frame strategy (scaled by $d' = 2.07$ m). (c) Pose estimation error based on homography (scaled by $d'' = 6.31$ m). Pose estimation error based on epipolar geometry (scaled by $d''' = 2.25$ m)

From the above results, the effectiveness of key frame strategy has been evaluated, which performs much better than other conventional multiple-view geometry-based methods when the start and final views only have a few correspondences. In addition, for tasks in larger workspace, i.e., the start and final views have no correspondences, the conventional multiple-view geometry methods are not applicable any more. This is the specific condition in which key frame strategy is necessary, and the workspace can be significantly enlarged.

14.4.2 Visual Trajectory Tracking and Pose Regulation

As analyzed above, global scaled pose information is estimated by the key frame strategy, based on which the robot is regulated to track the desired trajectory using only visual feedback. The parameters involved in the controller are the control gains (k_1, k_2, and k_3), update gains (γ_1, γ_2, and γ_3), and scale factors in auxiliary signals (μ, β, σ_1, and σ_2). Obviously, the magnitude of the control gains and update gains regulate the convergence speed of the corresponding signals. Due to the nonholonomic constraints, the robot has three degrees of freedom to control but only has two control inputs, there're trade-offs among the convergence speed of each error signal. The conflict mainly exists between the convergence of e_1 and e_3, which mean the lateral position error and orientation error, respectively. For example, with a larger k_1, the convergence speed of e_1 will be faster and the convergence speed of e_3 will be slower, and vice versa. The control paramters are set after tuning as follows:

$$k_1 = 2.7, k_2 = 1.2, k_3 = 1, \gamma_1 = 1, \gamma_2 = 1, \gamma_3 = 5,$$
$$\mu = 10, \beta = 1.5, \sigma_1 = 10, \sigma_2 = 10$$

(14.33)

The initial values of the unknown parameters ($\hat{d}, \hat{D}$, and $\hat{L}_m$) are set as zeros.

Simulation results are presented in Figure 14.9. To evaluate system performance, the real pose of the robot is obtained from the simulator, but is not used in the controller. From Figure 14.9a it can be seen that the tracking error is eliminated during the control process. Finally, the robot tracks the desired trajectory accurately. The evolution of the error vector $e(t)$ can be clearly seen in Figure 14.9b, $\bar{e}_1, e_2, e_3$ all converge to small values that are very close to zero in the presence of image noise. The evolution of unknown parameters is shown in Figure 14.9c, and they all converge to certain values. The control signals applied to the robot are shown in Figure 14.9d. The convergence time is about 5 s, the final position and rotation error is about 1.5 cm and 0.01 rad.

Another tracking result from a different initial pose is shown in Figure 14.10, which has a larger initial error. From the tracking result it can be seen that the tracking controller is aggressive, so that the robot converges to the desired path quickly. However, due to the nonholonomic constraints, a fast convergence of the lateral error $\bar{e}_1$ causes a large overshoot of orientation error e_3. According to the evolution of e_2 and e_3, the real trajectory is a little ahead of the desired

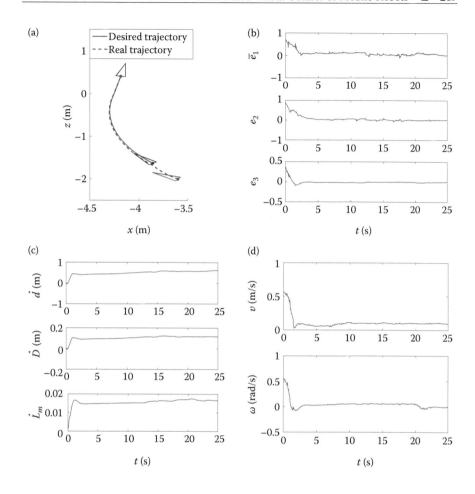

Figure 14.9: Visual trajectory tracking results. (a) Tracking process, (b) system error vector, (c) parameter estimation, and (d) control signals.

trajectory before 10 s; then, the error vector converges after 10 s. The position error and rotation error with respect to the final view are about 2.5 cm and 0.01 rad, respectively.

A visual regulation task is defined by the final view C^* as the target frame, and the robot is placed at C_0 in the beginning. Generally, the pose regulation task can be formulated into a special case of the trajectory tracking task, i.e., the desired trajectory consists of only one view C^*. In this case, the key frame strategy works as the conventional pose estimation method since there are only two key frames (start view C_0 and the final view C^*). In this case, the noise is larger as analyzed in Section 14.4.1. The proposed control strategy is also applicable. Regulation results from the initial pose are shown in Figure 14.11. The regulation

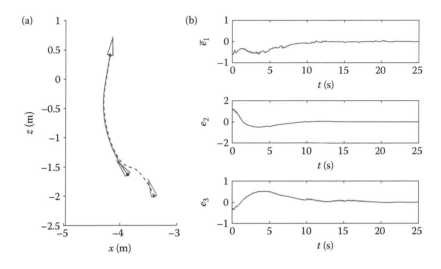

Figure 14.10: Visual trajectory tracking results. (a) Tracking process and (b) error convergence.

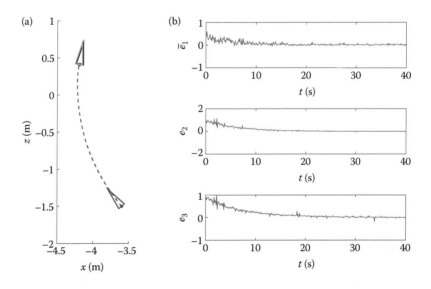

Figure 14.11: Pose regulation task. (a) Regulation process and (b) error convergence.

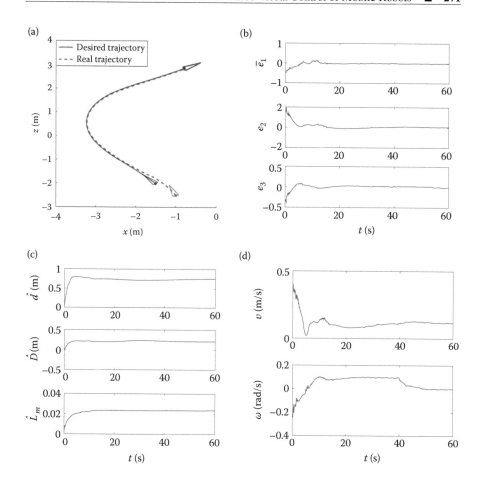

Figure 14.12: Visual trajectory tracking results for longer range. (a) Tracking process, (b) system error vector, (c) parameter estimation, and (d) control signals.

process takes about 30 s to converge. The position error and rotation error after convergence are about 2 cm and 0.01 rad, respectively.

14.4.3 Trajectory Tracking with Longer Range

To evaluate the performance of trajectory tracking with longer range, the demonstration is run for 1 minute as shown in Figure 14.13. On the desired trajectory, the robot firstly goes straightly, makes a turn, and then goes straightly toward the target. It can be clearly seen that the start and final views have no correspondence. It should be noted that previous works are not applicable to this scenario for visual trajectory tracking, as mentioned in Section 14.1. Owing to the key frame

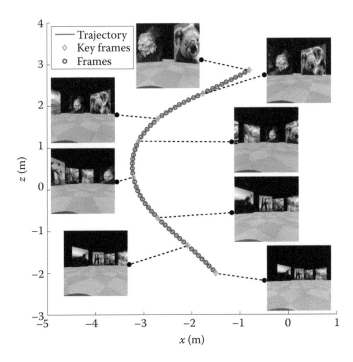

Figure 14.13: Visual trajectory tracking with longer range.

strategy, the scaled pose information can be estimated efficiently with respect to the final frame and can be used as the feedback globally. Several key frames are selected from the demonstrated frames on the desired trajectory, so that neighbouring key frames have enough correspondences for geometric reconstruction reliably. The tracking performance from an arbitrary pose is shown in Figure 14.12, it can be seen that the trajectory converges fast to the desired trajectory using only visual feedback. The position and orientation errors at last are 5 cm and 0.01 rad, respectively.

From the above simulations based on the V-REP platform, the improvements of key frame strategy with respect to conventional approaches are shown and the effectiveness of the control strategy is evaluated. To the best of our knowledge, the unknown extrinsic parameters (the installing position of the camera) have not been considered in the control strategy in previous works on the visual trajectory tracking of mobile robots, thus comparisons are not made with previous methods in terms of control strategy. The video of the simulations can be seen at https://youtu.be/gjV0gDlU6PY, in which the relevant views (current view, desired view, final view, and key frame view) and the current and desired trajectories in workspace are shown.

14.5   Conclusion

This chapter presents a visual trajectory tracking control strategy for nonholo-nomic mobile robots. Trifocal tensor is exploited to obtain the orientation and scaled position information used in the control system. To overcome the limita-tion of sharing visual information among the start, current, and final images, key frame strategy is proposed, extending the workspace of the system. To improve implementing flexibility, the camera can be mounted at an arbitrary position of the robot. An adaptive controller is developed considering the unknown depth and extrinsic parameters, and asymptotical convergence is guaranteed for almost all practical circumstances including both tracking and regulation tasks based on Lyapunov analysis. Simulation results show that the proposed key frame strat-egy improves the precision and extends the workspace of scaled pose estimation. Besides, the performance of trajectory tracking and pose regulation is evaluated from the results. The proposed control strategy can be applied for visual naviga-tion tasks conveniently, with large workspace and high tracking accuracy.

Chapter 15

Unified Visual Control of Mobile Robots with Euclidean Reconstruction

15.1 Introduction

In this chapter, a vision-based approach is proposed to address the unified tracking and regulation problem of the wheeled mobile robot. Specifically, a monocular camera is installed at the center of the wheeled mobile robot (i.e., camera-in-hand configuration). Exploiting the homography-based techniques, the orientation and scaled position information used in the error system is extracted from the current, the reference, and the desired images. An adaptive continuous controller is developed by fully taking the nonholonomic constraint and the unknown depth constant into consideration. The asymptotic stability of the error system is proven via the Lyapunov-based analysis. Differing from [148], no *a priori* geometric knowledge is required for the target object (i.e., model-free), and the proposed controller can achieve asymptotic tracking and regulation. Meanwhile, compared with [127], the control gains can be selected flexibly without any additional constraint. Furthermore, to estimate the unknown depth constant, an update law is designed in this chapter. If a persistent excitation condition is satisfied, the estimated value of the depth constant will converge to its actual one. Once the unknown depth constant is identified, the Euclidean space can be reconstructed. Simulation results are provided to demonstrate the performance of the proposed approach.

15.2   Control Development

Based on the homography techniques, the current pose information $\bar{x}_f(t)$, $\theta(t)$ and the desired pose information $\bar{x}_{fd}(t)$, $\theta_d(t)$ with respect to the reference frame $\mathcal{F}^*$ can be computed efficiently. In this section, an adaptive unified tracking and regulation controller is developed using the measurable pose information.

15.2.1   Kinematic Model

Based on the classical kinematic model of a nonholonomic wheeled mobile robot, the time derivative of $x_{fx}(t)$, $x_{fz}(t)$, and $\theta(t)$ is given by:

$$\dot{x}_{fx} = -x_{fz}\omega$$
$$\dot{x}_{fz} = -v + x_{fx}\omega \tag{15.1}$$
$$\dot{\theta} = -\omega,$$

where $v(t) \in \mathbb{R}$ is the linear velocity of the wheeled mobile robot along the z-axis, and $\omega(t) \in \mathbb{R}$ is the angular velocity along the y-axis. After taking the time derivative of the measurable signals $\bar{x}_{fx}(t)$, $\bar{x}_{fz}(t)$, $\theta(t)$ and utilizing (15.1), the following kinematic model can be obtained:

$$\dot{\bar{x}}_{fx} = -\bar{x}_{fz}\omega$$
$$\dot{\bar{x}}_{fz} = -\frac{v}{d} + \bar{x}_{fx}\omega \tag{15.2}$$
$$\dot{\theta} = -\omega.$$

Similarly, the kinematic model of the desired trajectory $\bar{x}_{fdx}(t)$, $\bar{x}_{fdz}(t)$, $\theta_d(t)$ is given by:

$$\dot{\bar{x}}_{fdx} = -\bar{x}_{fdz}\omega_d$$
$$\dot{\bar{x}}_{fdz} = -\frac{v_d}{d} + \bar{x}_{fdx}\omega_d \tag{15.3}$$
$$\dot{\theta}_d = -\omega_d,$$

where $v_d(t)$, $\omega_d(t) \in \mathbb{R}$ are the desired linear and angular velocities of the wheeled mobile robot, respectively. Based on (15.3), it is clear that the desired angular velocity $\omega_d(t)$ and the scaled desired linear velocity $\bar{v}_d(t) \triangleq \frac{v_d}{d} \in \mathbb{R}$ can be calculated by:

$$\omega_d = -\dot{\theta}_d$$
$$\bar{v}_d = \bar{x}_{fdx}\omega_d - \dot{\bar{x}}_{fdz}. \tag{15.4}$$

It should be pointed out that the desired linear velocity $v_d(t)$ is unmeasurable since the depth constant d is unknown. Besides, it is assumed that the desired signals $\bar{x}_{fd}(t)$, $\theta_d(t)$ are continuously differentiable and bounded up to their second-order time derivative. Then it can be concluded from (15.4) that $\bar{v}_d(t)$ and $\omega_d(t)$ are bounded. In the following, the measurable and bounded signals $\bar{v}_d(t)$ and $\omega_d(t)$ will be utilized to facilitate the controller design.

15.2.2   Open-Loop Error System

Let $e_\theta(t) \triangleq \theta - \theta_d$, then the system errors can be defined as follows:

$$
\begin{bmatrix} e_x \\ e_z \\ e_\theta \end{bmatrix} \triangleq \begin{bmatrix} \bar{x}_{fx} \\ \bar{x}_{fz} \\ \theta \end{bmatrix} - \begin{bmatrix} \cos(e_\theta) & \sin(e_\theta) & 0 \\ -\sin(e_\theta) & \cos(e_\theta) & 0 \\ 0 & 0 & 1 \end{bmatrix} \begin{bmatrix} \bar{x}_{fdx} \\ \bar{x}_{fdz} \\ \theta_d \end{bmatrix}. \tag{15.5}
$$

Based on (15.5), it is clear that $\theta(t) \to \theta_d(t)$, $\bar{x}_f(t) \to \bar{x}_{fd}(t)$ as $e_x(t), e_z(t)$, $e_\theta(t) \to 0$. Taking the time derivative of (15.5) and substituting from (15.2), (15.3), and $\bar{v}_d(t) = \frac{v_d}{d}$, it can be determined that

$$
\begin{aligned}
\dot{e}_x &= -e_z \omega + \bar{v}_d \sin(e_\theta) \\
\dot{e}_z &= -\frac{v}{d} + e_x \omega + \bar{v}_d \cos(e_\theta) \\
\dot{e}_\theta &= -\omega + \omega_d.
\end{aligned} \tag{15.6}
$$

To facilitate the following development, two auxiliary time-varying signals $\alpha(t)$, $\rho(t) \in \mathbb{R}$ are introduced.

$$
\alpha = \exp\left(-\int_0^t (\gamma |\bar{v}_d(\tau)|) d\tau \right) \tag{15.7}
$$

$$
\rho = \alpha h_0 \tanh\left(\lambda_1 (e_x^2 + e_z^2)^{\frac{\lambda_2}{2}} \right) \sin(\lambda_3 t), \tag{15.8}
$$

where $\gamma, h_0, \lambda_1, \lambda_2, \lambda_3 \in \mathbb{R}$ are positive constants. Furthermore, inspired by [110, 214], an auxiliary error signal $\bar{e}_\theta(t) \in \mathbb{R}$ is defined as follows:

$$
\bar{e}_\theta \triangleq e_\theta - \rho. \tag{15.9}
$$

From (15.7)–(15.9), it is not difficult to conclude that $e_x(t)$, $e_z(t)$, $\bar{e}_\theta(t) \to 0$ indicates $e_x(t)$, $e_z(t)$, $e_\theta(t) \to 0$. So $e_x(t)$, $e_z(t)$, and $\bar{e}_\theta(t)$ can be regarded as the new system errors. Based on (15.6)–(15.9), the open-loop error system can be obtained.

$$
\begin{aligned}
\dot{e}_x &= -e_z \omega + \bar{v}_d \sin(e_\theta) \\
\dot{e}_z &= -\frac{v}{d} + e_x \omega + \bar{v}_d \cos(e_\theta) \\
\dot{\bar{e}}_\theta &= -\omega + \omega_d - f_1 - f_2 \left(\bar{v}_d \sin(e_\theta) e_x \right. \\
&\quad \left. -\frac{v}{d} e_z + \bar{v}_d \cos(e_\theta) e_z \right),
\end{aligned} \tag{15.10}
$$

where $f_1(t)$, $f_2(t) \in \mathbb{R}$ are two auxiliary functions given by:

$$
\begin{aligned}
f_1 &\triangleq -(\gamma |\bar{v}_d|)\rho + \alpha h_0 \lambda_3 \tanh\left(\lambda_1 (e_x^2 + e_z^2)^{\frac{\lambda_2}{2}} \right) \cos(\lambda_3 t) \\
f_2 &\triangleq \alpha h_0 \lambda_1 \lambda_2 \sin(\lambda_3 t) \cosh^{-2}\left(\lambda_1 (e_x^2 + e_z^2)^{\frac{\lambda_2}{2}} \right) \\
&\quad \times (e_x^2 + e_z^2)^{\frac{\lambda_2}{2} - 1}.
\end{aligned} \tag{15.11}
$$

15.2.3 Controller Design

According to the structure of the open-loop error system in (15.10) and the Lyapunov stability analysis, the control inputs $v(t)$ and $\omega(t)$ are designed as follows:

$$v = k_1 e_z + \hat{d}\bar{v}_d \cos(e_\theta)$$
$$\omega = k_2\bar{e}_\theta + (\omega_d - f_1 - f_2(\bar{v}_d \sin(e_\theta)e_x + \bar{v}_d \cos(e_\theta)e_z)) \qquad (15.12)$$
$$+ \hat{D}f_2 v e_z + k_3 f_3 \hat{d}\bar{v}_d e_x,$$

where $k_1, k_2, k_3 \in \mathbb{R}$ are positive control gains, $f_3(t) \in \mathbb{R}$ is an auxiliary function defined as:

$$f_3 \triangleq \frac{\sin(e_\theta) - \sin(\rho)}{\bar{e}_\theta} = \frac{\sin(e_\theta) - \sin(\rho)}{e_\theta - \rho}. \qquad (15.13)$$

Since $\sin(e_\theta) - \sin(\rho) = \sin(\bar{e}_\theta)\cos\rho + (\cos(\bar{e}_\theta) - 1)\sin(\rho)$, it can be concluded from (15.13) that $f_3(t)$ is a continuous and bounded function in $\bar{e}_\theta(t)$ [214]. Besides, in (15.12), $\hat{d}(t)$, $\hat{D}(t) \in \mathbb{R}$ are the estimate of d and $\frac{1}{d}$, respectively. To compensate for the unknown depth information, the update laws for $\hat{d}(t)$ and $\hat{D}(t)$ are given by:

$$\dot{\hat{d}} = \Gamma_1 k_3 \text{Proj}_1(e_z\bar{v}_d \cos(e_\theta) + e_x\bar{v}_d(\sin(e_\theta) - \sin(\rho)))$$
$$\dot{\hat{D}} = \Gamma_2 \text{Proj}_2(\bar{e}_\theta f_2 v e_z), \qquad (15.14)$$

where $\Gamma_1, \Gamma_2 \in \mathbb{R}$ are positive update gains and $\text{Proj}_1(\star)$, $\text{Proj}_2(\star)$ denote the projection functions defined as follows:

$$\text{Proj}_1(\star) = \begin{cases} 0, & \text{for } \hat{d} \leq \sigma_1 \text{ and } \star < 0 \\ \star, & \text{else} \end{cases}$$
$$\qquad (15.15)$$
$$\text{Proj}_2(\star) = \begin{cases} 0, & \text{for } \hat{D} \leq \sigma_2 \text{ and } \star < 0 \\ \star, & \text{else} \end{cases}.$$

In (15.15), $\sigma_1, \sigma_2 \in \mathbb{R}$ are arbitrarily small positive constants satisfying $\sigma_1 < d$ and $\sigma_2 < \frac{1}{d}$. It can be concluded from (15.14), (15.15) that if the initial values $\hat{d}(0)$ and $\hat{D}(0)$ are selected to satisfy $\hat{d}(0) \geq \sigma_1$ and $\hat{D}(0) \geq \sigma_2$, then for $t \geq 0$, $\hat{d}(t) \geq \sigma_1$ and $\hat{D}(t) \geq \sigma_2$ always hold.

After substituting (15.12) into (15.10), the closed-loop error system can be obtained.

$$\dot{e}_x = -e_z\omega + \bar{v}_d \sin(e_\theta)$$
$$\dot{e}_z = -\frac{1}{d}k_1 e_z + e_x\omega + \frac{\tilde{d}}{d}\bar{v}_d \cos(e_\theta) \qquad (15.16)$$
$$\dot{\bar{e}}_\theta = -k_2\bar{e}_\theta + \tilde{D}f_2 v e_z - k_3 f_3 \hat{d}\bar{v}_d e_x,$$

where $\tilde{d}(t) \triangleq d - \hat{d}$, $\tilde{D}(t) \triangleq \frac{1}{d} - \hat{D} \in \mathbb{R}$ are the estimate errors of d and $\frac{1}{d}$, respectively.

Remark 15.1   The control gains in (15.12) and the update gains in (15.14) are required to be positive without any additional constraint. Thanks to the flexible gain selection, the proposed controller is more practicable than [59,98,127], in which some control gains should be adjusted discreetly to satisfy certain inequalities.

Remark 15.2   The update laws for $\hat{d}(t)$ and $\hat{D}(t)$ designed in (15.14) not only can compensate for the unknown depth information, but also ensure $\hat{d}(t)$ converges to the actual value d on condition that a persistent excitation is satisfied. Nevertheless, in most existing works [31,64,104,127,223], the convergence of the depth estimate is not guaranteed.

15.2.4   Stability Analysis

Theorem 15.1
The control inputs in (15.12) and the update laws in (15.14), (15.15) ensure that the system errors $e_x(t)$, $e_z(t)$, and $\bar{e}_\theta(t)$ asymptotically converge to zero in the sense that

$$\lim_{t \to \infty} e_x(t), e_z(t), \bar{e}_\theta(t) = 0. \tag{15.17}$$

Proof 15.1   To proof Theorem 15.1, a nonnegative Lyapunov function $V(t) \in \mathbb{R}$ is defined as follows:

$$V \triangleq \frac{1}{2}k_3 d\left(e_x^2 + e_z^2\right) + \frac{1}{2}\bar{e}_\theta^2 + \frac{1}{2\Gamma_1}\tilde{d}^2 + \frac{1}{2\Gamma_2}\tilde{D}^2. \tag{15.18}$$

After taking the time derivative of (15.18) and substituting from (15.13), (15.14), and (15.16), it can be obtained that

$$
\begin{aligned}
\dot{V} &= k_3 d(e_x \dot{e}_x + e_z \dot{e}_z) + \bar{e}_\theta \dot{\bar{e}}_\theta + \tilde{d}\frac{\dot{\tilde{d}}}{\Gamma_1} + \tilde{D}\frac{\dot{\tilde{D}}}{\Gamma_2} \\
&= -k_1 k_3 e_z^2 - k_2 \bar{e}_\theta^2 + k_3 d e_x \bar{v}_d \sin(\rho) + \tilde{D}\left(\bar{e}_\theta f_2 v e_z - \frac{\dot{\hat{D}}}{\Gamma_2}\right) \\
&\quad + \tilde{d}\left(k_3\left(e_z \bar{v}_d \cos(e_\theta) + e_x \bar{v}_d\left(\sin(e_\theta) - \sin(\rho)\right)\right) - \frac{\dot{\hat{d}}}{\Gamma_1}\right).
\end{aligned}
\tag{15.19}
$$

Based on (15.14) and (15.15), it is not difficult to determine that $\tilde{D}(\bar{e}_\theta f_2 v e_z - \frac{\dot{\hat{D}}}{\Gamma_2}) \leq 0$ and $\tilde{d}(k_3(e_z \bar{v}_d \cos(e_\theta) + e_x \bar{v}_d(\sin(e_\theta) - \sin(\rho))) - \frac{\dot{\hat{d}}}{\Gamma_1}) \leq 0$. Besides, by utilizing

(15.18), $|e_x(t)| \leq \sqrt{\frac{2V}{k_3 d}}$ can be obtained. Hence, $\dot{V}(t)$ can be upper bounded as follows:

$$
\begin{aligned}
\dot{V} &\leq -k_1 k_3 e_z^2 - k_2 \bar{e}_\theta^2 + k_3 d e_x \bar{v}_d \sin(\rho) \\
&\leq -k_1 k_3 e_z^2 - k_2 \bar{e}_\theta^2 + k_3 d |\bar{v}_d \sin(\rho)||e_x| \\
&\leq -k_1 k_3 e_z^2 - k_2 \bar{e}_\theta^2 + k_3 d |\bar{v}_d \sin(\rho)| \sqrt{\frac{2V}{k_3 d}}.
\end{aligned}
\tag{15.20}
$$

According to (15.7) and (15.8), it can be found that $0 \leq \alpha(t) \leq 1$, $\dot{\alpha} = -\gamma |\bar{v}_d| \alpha$, and $|\rho| \leq \alpha h_0$. After integrating $|\bar{v}_d \sin(\rho)|$, it can be concluded that

$$
\begin{aligned}
\int_0^t |\bar{v}_d(\tau) \sin(\rho(\tau))| d\tau &\leq \int_0^t |\bar{v}_d(\tau)||\rho(\tau)| d\tau \\
&\leq h_0 \int_0^t |\bar{v}_d(\tau)| \alpha(\tau) d\tau \\
&\leq h_0 \int_0^t -\frac{\dot{\alpha}(\tau)}{\gamma} d\tau \\
&\leq h_0 \frac{\alpha(0) - \alpha(t)}{\gamma} \leq \frac{h_0}{\gamma} < \infty,
\end{aligned}
\tag{15.21}
$$

which implies that $|\bar{v}_d \sin(\rho)| \in \mathcal{L}_1$. Based on (15.20), we have $\dot{V} \leq k_3 d |\bar{v}_d \sin(\rho)| \sqrt{\frac{2V}{k_3 d}}$ indicating that

$$
\frac{d(\sqrt{V})}{dt} \leq \sqrt{\frac{k_3 d}{2}} |\bar{v}_d \sin(\rho)|.
\tag{15.22}
$$

Since $|\bar{v}_d \sin(\rho)| \in \mathcal{L}_1$, it can be concluded from (15.22) that $V(t) \in \mathcal{L}_\infty$. Then, it can be obtained from (15.18) that $e_x(t)$, $e_z(t)$, $\bar{e}_\theta(t)$, $\tilde{d}(t)$, $\tilde{D}(t) \in \mathcal{L}_\infty$. Based on $\tilde{d}(t) = d - \hat{d}$, $\tilde{D}(t) = \frac{1}{d} - \hat{D}$, and d are positive constant, $\hat{d}(t)$, $\hat{D}(t) \in \mathcal{L}_\infty$ can be obtained. Furthermore, it can be inferred from (15.12) and (15.16) that $v(t)$, $\omega(t)$, $\dot{e}_x(t)$, $\dot{e}_z(t)$, $\dot{\bar{e}}_\theta(t) \in \mathcal{L}_\infty$. Since $V(t) \in \mathcal{L}_\infty$, there exists a positive constant $\delta \in \mathbb{R}$ satisfying $V(t) \leq \delta$. Using (15.20), (15.21), and $V(t) \leq \delta$, the following inequality can be determined:

$$
\begin{aligned}
\int_0^t \left(k_1 k_3 e_z^2(\tau) + k_2 \bar{e}_\theta^2(\tau) \right) d\tau \\
\leq -\int_0^t \dot{V}(\tau) d\tau + \sqrt{2k_3 d \delta} \int_0^t |\bar{v}_d(\tau) \sin(\rho(\tau))| d\tau \\
\leq V(0) - V(t) + \frac{h_0 \sqrt{2k_3 d \delta}}{\gamma} \leq \delta + \frac{h_0 \sqrt{2k_3 d \delta}}{\gamma} < \infty.
\end{aligned}
\tag{15.23}
$$

Obviously, it can be concluded from (15.23) that $e_z(t)$, $\bar{e}_\theta(t) \in \mathcal{L}_2$. Since $e_z(t)$, $\bar{e}_\theta(t) \in \mathcal{L}_2$ and $\dot{e}_z(t)$, $\dot{\bar{e}}_\theta(t) \in \mathcal{L}_\infty$, Barbalat's lemma [192] can be exploited to infer that

$$
\lim_{t \to \infty} e_z(t), \bar{e}_\theta(t) = 0.
\tag{15.24}
$$

In the following, the proof is divided into two cases to show that $\lim_{t\to\infty} e_x(t) = 0$ with the aid of the extended Barbalat's lemma (see Appendix A).

Case 15.1

The desired trajectory keeps moving, and $v_d(t)$ satisfies the persistent excitation condition, i.e., there exist T, $\mu_1 > 0$ such that $\forall t > 0$

$$\int_t^{t+T} (|v_d(\tau)|)\, d\tau \geq \mu_1. \tag{15.25}$$

Since $\bar{v}_d(t) = \frac{v_d}{d}$, $\bar{v}_d(t)$ is persistently exciting. Then, based on (15.7), it can be concluded that $\alpha(t)$ converges to zero asymptotically, i.e., $\lim_{t\to\infty} \alpha(t) = 0$. According to the expression of $\rho(t)$ given in (15.8), $\lim_{t\to\infty} \rho(t) = 0$ can be obtained. Since $\bar{e}_\theta = e_\theta - \rho$ and $\lim_{t\to\infty} \bar{e}_\theta(t), \rho(t) = 0$, it is clear that $\lim_{t\to\infty} e_\theta(t) = 0$. Moreover, $\lim_{t\to\infty} f_3(t) = \lim_{t\to\infty} \frac{\sin(e_\theta) - \sin(\rho)}{e_\theta - \rho} = 1$ can be inferred from $\lim_{t\to\infty} e_\theta(t)$, $\rho(t) = 0$ and (15.13). Since $\lim_{t\to\infty} e_z(t), \bar{e}_\theta(t) = 0$ and $\lim_{t\to\infty} f_3(t) = 1$, applying the extended Barbalat's lemma to the last equation of (15.16) yields

$$\lim_{t\to\infty} k_3 f_3 \hat{d} \bar{v}_d e_x = \lim_{t\to\infty} k_3 \hat{d} \bar{v}_d e_x = 0. \tag{15.26}$$

Because $k_3 > 0$ and the update laws designed in (15.14), (15.15) can guarantee that $\hat{d}(t) \geq \sigma_1 > 0$, it can be obtained from (15.26) that $\lim_{t\to\infty} \bar{v}_d e_x = 0$. Furthermore, since $\bar{v}_d(t)$ is persistently exciting, it can be concluded that $\lim_{t\to\infty} e_x = 0$ [180].

Case 15.2

The desired trajectory stops finally, i.e., $\lim_{t\to\infty} v_d(t), \omega_d(t) = 0$, and there exists $\mu_2 \geq 0$ such that

$$\int_0^\infty (|v_d(\tau)|)\, d\tau \leq \mu_2. \tag{15.27}$$

Note that the regulation task belongs to this case.

Based on (15.7), (15.27), and $\bar{v}_d(t) = \frac{v_d}{d}$ it can be concluded that $0 < \exp(-\frac{\gamma\mu_2}{d}) \leq \alpha(t) \leq 1$ and $\lim_{t\to\infty} \bar{v}_d(t) = 0$. Since $\lim_{t\to\infty} \bar{v}_d(t), \omega_d(t), e_z(t)$, $\bar{e}_\theta(t) = 0$, it can be inferred from (15.11) and (15.12) that $\lim_{t\to\infty} f_1(t) = \lim_{t\to\infty} \alpha h_0 \lambda_3 \tanh\left(\lambda_1 e_x^{\lambda_2}\right) \cos(\lambda_3 t)$ and $\lim_{t\to\infty} \omega(t) = \lim_{t\to\infty} -f_1(t)$. Utilizing $\lim_{t\to\infty} \bar{v}_d(t), e_z(t) = 0$ and exploiting the extended Barbalat's lemma to the second equation of (15.16) yields

$$\begin{aligned}
\lim_{t\to\infty} e_x \omega &= \lim_{t\to\infty} -e_x f_1 \\
&= \lim_{t\to\infty} -\alpha h_0 \lambda_3 \cos(\lambda_3 t) \tanh\left(\lambda_1 e_x^{\lambda_2}\right) e_x = 0.
\end{aligned} \tag{15.28}$$

Since $\alpha(t), h_0, \lambda_1, \lambda_2, \lambda_3 > 0$ and $\cos(\lambda_3 t)$ satisfies the persistent excitation condition, it can be concluded from (15.28) that $\lim_{t\to\infty} \tanh\left(\lambda_1 e_x^{\lambda_2}\right) e_x = 0$, which implies $\lim_{t\to\infty} e_x = 0$.

15.3 Euclidean Reconstruction

Proposition 15.1
*The update laws designed in (15.14) and (15.15) ensure that the depth estimate $\hat{d}(t)$
converge to its actual value d in the sense that*

$$\lim_{t \to \infty} \tilde{d}(t) = 0 \tag{15.29}$$

*on condition that the desired linear velocity $v_d(t)$ satisfies the persistent excitation
condition given in (15.25).*

Proof 15.2   In Theorem 15.1, it has been proved that if (15.25) holds, then
$\lim_{t \to \infty} e_x(t), e_z(t), \bar{e}_\theta(t) = 0$. Furthermore, it is obvious that $\lim_{t \to \infty} e_\theta(t) = 0$.
Since $\lim_{t \to \infty} e_x(t), e_z(t), e_\theta(t) = 0$, applying the extended Barbalat's lemma to the
second equation of (15.16) yields that

$$\lim_{t \to \infty} \frac{\tilde{d}}{d} \bar{v}_d \cos(e_\theta) = \lim_{t \to \infty} \frac{\tilde{d}}{d} \bar{v}_d = 0. \tag{15.30}$$

Because $\bar{v}_d(t) = \frac{v_d}{d}$ and (15.25) holds, $\bar{v}_d(t)$ is persistently exciting. Since $\bar{v}_d(t)$
satisfies the persistent excitation condition, it can be inferred from (15.30) that
$\lim_{t \to \infty} \tilde{d} = 0$, i.e., $\hat{d}(t) \to d$ as $t \to \infty$ [180].

Once $\hat{d}(t)$ converges to d, based on the facts that $\bar{x}_f(t)$, $\bar{x}_{fd}(t)$ are measurable
signals, $x_f(t)$ and $x_{fd}(t)$ can be reconstructed as follows:

$$\hat{x}_f = \hat{d}\bar{x}_f \qquad \hat{x}_{fd} = \hat{d}\bar{x}_{fd}, \tag{15.31}$$

where $\hat{x}_f(t)$, $\hat{x}_{fd}(t) \in \mathbb{R}^3$ denote the estimate of $x_f(t)$ and $x_{fd}(t)$, respectively.
According to Proposition 15.1 and (15.31), it is clear that $\hat{x}_f(t) \to x_f(t)$, $\hat{x}_{fd}(t) \to
x_{fd}(t)$ as $t \to \infty$. Now, since the rotation and translation information between
$\mathcal{F}$ and $\mathcal{F}^*$ (i.e., $\theta(t)$, $x_f(t)$), and the rotation and translation information between
$\mathcal{F}_d$ and $\mathcal{F}^*$ (i.e., $\theta_d(t)$, $x_{fd}(t)$) can be obtained, $\bar{m}_i(t)$, $\bar{m}_{di}(t)$, and $\bar{m}_i^*$ can be
calculated based on triangulation [86].

Remark 15.3   The persistent excitation condition given in (15.25) implies that the
wheeled mobile robot's desired linear velocity should not be identically zero. Similar
PE conditions exist extensively in the vision-based estimation methods [33,48,49].

15.4 Simulation Results

Simulation studies are performed to illustrate the performance of the controller given in (15.12) and the update laws given in (15.14), (15.15). For the simulation, the intrinsic camera calibration matrix is shown as follows:

$$A = \begin{bmatrix} 500 & 0 & 640 \\ 0 & 400 & 480 \\ 0 & 0 & 1 \end{bmatrix}.$$

To incorporate homography-based techniques, four coplanar but non-collinear feature points are selected to calculate $H(t)$ and $H_d(t)$. Utilizing classical decomposition algorithm [86], the signals $\theta(t)$, $\theta_d(t)$, $\bar{x}_f(t)$, and $\bar{x}_{fd}(t)$ can be obtained. Without loss of generality, the position of $\mathcal{F}^*$ expressed in terms of $\mathcal{I}$ is selected as: $\begin{bmatrix} 1 & 0 & 0 \end{bmatrix}$, and the reference image is captured at the origin of the coordinate system $\mathcal{F}^*$. Besides, the initial position of $\mathcal{F}$ is chosen as: $\begin{bmatrix} -0.5 & 0 & -5 \end{bmatrix}$ under the coordinate system of $\mathcal{I}$, and the initial rotation angle between $\mathcal{F}$ and $\mathcal{F}^*$ is selected as $\theta(0) = -\frac{\pi}{6}$. In the following, to test the proposed approach, both the trajectory tracking and the regulation are studied.

∎ *Trajectory Tracking Task:* For the trajectory tracking task, the wheeled mobile robot equipped with a monocular camera is firstly driven by the desired velocities $v_d(t)$ and $\omega_d(t)$ to capture a set of images. Specifically, the desired velocities are chosen as follows:

$$\begin{cases} v_d = 0.01t + 0.05, \omega_d = 0 & 0 \le t < 10 \\ v_d = \frac{3}{20}, \omega_d = 0 & t \ge 10 \end{cases}.$$

Then, according to Remark 15.1, the prerecorded images are utilized to calculate $\bar{x}_{fd}(t)$, $\theta_d(t)$, $\dot{\bar{x}}_{fd}(t)$, and $\dot{\theta}_d(t)$. Finally, the scaled desired linear velocity $\bar{v}_d(t)$ and the desired angular velocity $\omega_d(t)$ can be obtained based on (15.4). The initial position of $\mathcal{F}_d$ expressed in terms of $\mathcal{I}$ is selected as: $\begin{bmatrix} 0 & 0 & -3 \end{bmatrix}$, and the initial rotation angle between $\mathcal{F}_d$ and $\mathcal{F}^*$ is chosen as $\theta_d(0) = 0$. The initial values of $\hat{d}(t)$ and $\hat{D}(t)$ are set as $\hat{d}(0) = 0.5$, $\hat{D}(0) = 0.75$ while the actual value of the depth constant d is 1 m. The gains are adjusted to the following values:

$$\gamma = 1, h_0 = 1.5, \lambda_1 = 0.1, \lambda_2 = 1, \lambda_3 = \frac{\pi}{36}$$
$$k_1 = 0.5, k_2 = 0.7, k_3 = 5, \Gamma_1 = 5, \Gamma_2 = 1.$$

The corresponding simulation results of the tracking task are shown in Figure 15.1. From Figure 15.1a, it can be seen that the tracking errors $e_x(t)$, $e_z(t)$, and $\bar{e}_\theta(t)$ all asymptotically converge to zero. The control inputs are depicted in Figure 15.1b, and the updating process of the unknown parameters is presented in Figure 15.1c. It should be noted that

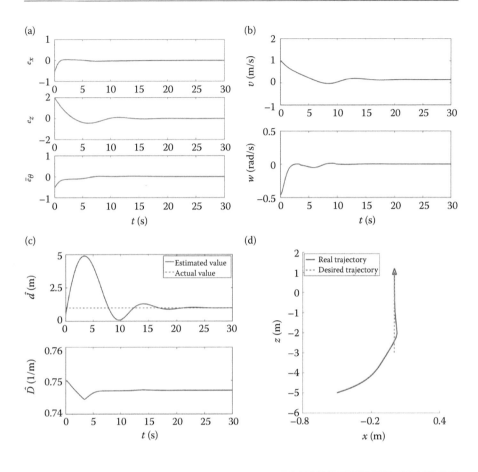

Figure 15.1: Simulation results of the trajectory tracking task. (a) Error convergence, (b) control inputs, (c) parameter estimation, and (d) tracking process.

since the desired velocity $v_d(t)$ satisfies the persistent excitation condition given in (15.25), the estimate of the unknown depth constant $\hat{d}(t)$ converges to its actual value d, which is shown in Figure 15.1c. What is more, the desired trajectory and the tracking trajectory are both illustrated in Figure 15.1d. According to the simulation results, it is clear that the proposed approach can achieve the tacking task effectively.

■ *Regulation Task:* For the regulation task, the desired stationary pose is described by only one image. To test the robustness of the controller, a random noise, obeying Gaussian distribution with a mean of 0 pixel and a standard deviation of 1 pixels, is injected into the image coordinates of feature points. The position of $\mathcal{F}_d$ expressed in terms of $\mathcal{I}$ is selected as: $\begin{bmatrix} 0 & 0 & -1 \end{bmatrix}$, and the rotation angle between $\mathcal{F}_d$ and $\mathcal{F}^*$ is chosen as

$\theta_d(t) = 0$. For the regulation task, the desired pose $\mathcal{F}_d$ does not change during the control process, i.e., for $t \geq 0$, $\bar{v}_d(t) = 0$ and $\omega_d(t) = 0$ always hold. The initial values of $\hat{a}(t)$ and $\hat{D}(t)$ are set as $\hat{a}(0) = 0.75$ and $\hat{D}(0) = 0.75$. Furthermore, the gains are tuned to the following values:

$$\gamma = 1, h_0 = 1.5, \lambda_1 = 0.8, \lambda_2 = 1, \lambda_3 = \frac{\pi}{600}$$
$$k_1 = 1, k_2 = 3, k_3 = 3, \Gamma_1 = 2, \Gamma_2 = 20.$$

The simulation results of the regulation task are depicted in Figure 15.2. It can be concluded from (15.14) that for the regulation task (i.e., $\bar{v}_d(t), \omega_d(t) = 0$), $\hat{a}(t)$ does not update during control process. So the updating process of the unknown parameters are not shown in Figure 15.2. Based on Figure 15.2a, it

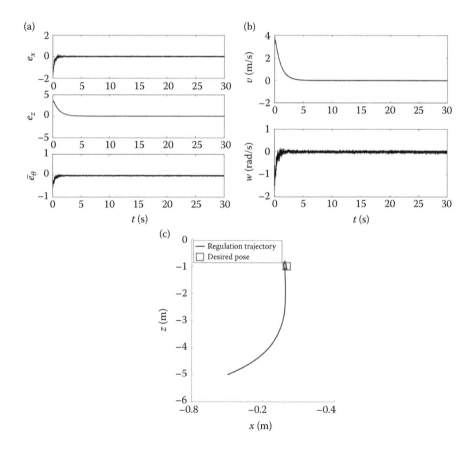

Figure 15.2: Simulation results of the regulation task. (a) Error convergence, (b) control inputs, and (c) regulation process.

can be obtained that in the presence of noise, the regulation errors $e_x(t)$, $e_z(t)$, and $\bar{e}_\theta(t)$ converge to small values that are very close to zero. Besides, the control inputs and the regulation process are shown in Figure 15.2b and c, respectively. From the simulation results, it can be concluded that the regulation task can be accomplished by the proposed controller.

15.5 Conclusion

This chapter presents an approach for the unified tracking and regulation visual servoing of the wheeled mobile robot. A monocular camera is equipped at the center of the wheeled mobile robot. Based on the current, the reference, and the desired images, the orientation and scaled position can be obtained by using the homography techniques. To overcome the nonhonolomic constraint and compensate for the unknown depth constant, an adaptive unified tracking and regulation controller is developed. The Lyapunov-based method is utilized to prove the asymptotic stability of the control system. Moreover, it is shown that the estimate of the unknown depth constant will converge to its real value if a persistent excitation condition is satisfied. The performance of the proposed vision-based unified tracking and regulation approach is evaluated from simulation results.

APPENDICES IV

Appendix A: Chapter 7

A.1 Proof of Theorem 7.1

Proof: Let $V_1(t) \in \mathbb{R}$ denote a nonnegative scalar function defined as follows:

$$V_1 \triangleq \frac{1}{2} \tilde{\theta}_i^T L_i^{-1} \tilde{\theta}_i + \frac{1}{2} \eta_i^T \eta_i. \tag{A.1}$$

After taking the time derivative of (A.1) and substituting (7.20), (7.24), (7.25), and (7.26), the following expression can be obtained

$$
\begin{aligned}
\dot{V}_1 \leq & -\frac{1}{4} \left\| W_{fi} \tilde{\theta}_i \right\|^2 - \beta_i \left\| \eta_i \right\|^2 + k_{2i} \left\| W_i \right\|_\infty^2 \left\| \eta_i \right\|^2 \\
& + \left[|\mu| e^{-\beta_i t} \left\| W_{fi} \tilde{\theta}_i \right\| - \frac{1}{4} \left\| W_{fi} \tilde{\theta}_i \right\|^2 \right] \\
& + \left[\left\| \theta_i \right\| \left\| W_i \right\|_\infty \left\| \tilde{V}_{vw} \right\|_\infty \left\| \eta_i \right\| - k_{2i} \left\| W_i \right\|_\infty^2 \left\| \eta_i \right\|^2 \right] \\
& + \left[\left\| W_{fi} \tilde{\theta}_i \right\| \left\| \eta_i \right\| - k_{1i} \left\| \eta_i \right\|^2 \right] + k_{1i} \left\| \eta_i \right\|^2,
\end{aligned}
\tag{A.2}
$$

where $k_{1i}, k_{2i} \in \mathbb{R}$ are positive constants as previously mentioned in (7.28). After utilizing the nonlinear damping argument [120], we can simplify (A.2) further as follows:

$$
\begin{aligned}
\dot{V}_1 \leq & -\left(\frac{1}{4} - \frac{1}{k_{1i}} \right) \left\| W_{fi} \tilde{\theta}_i \right\|^2 + \frac{1}{k_{2i}} \left\| \theta_i \right\|^2 \left\| \tilde{V}_{vw} \right\|_\infty^2 \\
& -\left(\beta_i - k_{1i} - k_{2i} \left\| W_i \right\|_\infty^2 \right) \left\| \eta_i \right\|^2 + 4\mu^2 e^{-2\beta_i t}.
\end{aligned}
\tag{A.3}
$$

The gains k_{1i}, k_{2i}, and β_i must be selected to ensure that

$$\frac{1}{4} - \frac{1}{k_{1i}} \geq \mu_{1i} > 0 \tag{A.4}$$

$$\beta_i - k_{1i} - k_{2i} \|W_i\|_\infty^2 \geq \mu_{2i} > 0, \tag{A.5}$$

where $\mu_{1i}, \mu_{2i} \in \mathbb{R}$ are positive constants. The gain conditions given by (A.4) and (A.5) allow us to formulate the conditions given by (7.28) with ignoring μ_{1i} and μ_{2i} for simplicity, as well as allowing us to further upper bound the time derivative of (A.1) as follows:

$$\dot{V}_1 \leq -\mu_{1i} \left\|W_{fi}\tilde{\theta}_i\right\|^2 - \mu_{2i} \|\eta_i\|^2$$
$$+ 4\mu^2 e^{-2\beta_i t} + \frac{1}{k_{2i}} \|\theta_i\|^2 \left\|\tilde{V}_{vw}\right\|_\infty^2. \tag{A.6}$$

From the discussion given in Remark 7.1, we can see that the last term in (A.6) is $\mathcal{L}_1$; hence,

$$\int_0^\infty \frac{1}{k_{2i}} \|\theta_i(\tau)\|^2 \left\|\tilde{V}_{vw}(\tau)\right\|_\infty^2 d\tau \leq \varepsilon, \tag{A.7}$$

where $\varepsilon \in \mathbb{R}$ is a positive constant. From (A.1), (A.6) and (A.7), we can conclude that

$$\int_0^\infty \left(\mu_{1i} \left\|W_{fi}(\tau)\tilde{\theta}_i(\tau)\right\|^2 + \mu_{2i} \|\eta_i(\tau)\|^2\right) d\tau$$
$$\leq V_1(0) - V_1(\infty) + \frac{2\mu^2}{\beta_i} + \varepsilon. \tag{A.8}$$

It can be concluded from (A.8) that $W_{fi}(t)\tilde{\theta}_i(t), \eta_i(t) \in \mathcal{L}_2$. From (A.8) and the fact that $V_1(t)$ is non-negative, it can be concluded that $V_1(t) \leq V_1(0) + \varepsilon + \frac{2\mu^2}{\beta_i}$ for any t, and hence $V_1(t) \in \mathcal{L}_\infty$. Signal chasing can be utilized to prove that $W_{fi}(t)\tilde{\theta}_i(t)$ is uniformly continuous [192] (see [26] for the details). Since we also have that $W_{fi}(t)\tilde{\theta}_i(t) \in \mathcal{L}_2$, we can conclude that [192]

$$W_{fi}(t)\tilde{\theta}_i(t) \to 0 \text{ as } t \to \infty. \tag{A.9}$$

As shown in Appendix B in [26], if the signal $W_{fi}(t)$ satisfies the persistent excitation condition [192] given in (7.27), then it can be concluded from (A.9) that

$$\tilde{\theta}_i(t) \to 0 \text{ as } t \to \infty. \tag{A.10}$$

Appendix B: Chapter 10

B.1

To obtain geometric insight into the structure of $\bar{R}(t)$ and $\bar{x}_f(t)$ defined in (13.2) can be obtained from Figure 9.1 by placing a fictitious camera that has a frame $\mathcal{I}^*$ attached to its center such that $\mathcal{I}^*$ initially coincides with $\mathcal{I}$. Since $\mathcal{I}$ and $\mathcal{I}^*$ coincide, the relationship between $\mathcal{I}^*$ and $\mathcal{F}^*$ can be denoted by rotational and translational parameters (x_f^*, R^*) as is evident from Figure 9.1. Without relative translational or rotational motion between $\mathcal{I}^*$ and $\mathcal{F}^*$, the two coordinate frames are moved until $\mathcal{F}^*$ aligns with $\mathcal{F}$, resulting in Figure B.1. It is now evident that the fixed camera problem reduces to a stereo vision problem with the parameters $\left(x_f - R(R^*)^T x_f^*, R(R^*)^T\right)$ denoting the translation and rotation between $\mathcal{I}$ and $\mathcal{I}^*$.

B.2

To develop the open-loop error system for $e_v(t)$, the time derivative of (10.13) is obtained as follows:

$$\dot{e}_v = \dot{p}_e - \dot{p}_{ed} = \frac{1}{z_1} A_e L_v \dot{\bar{m}}_1 - \dot{p}_{ed}, \tag{B.1}$$

where (10.1), (13.4), (13.6), (10.14), and the definition of $\alpha_i(t)$ in (13.5) were utilized. From the geometry between the coordinate frames depicted in Figure 9.1, the following relationships can be developed:

$$\begin{aligned}
\bar{m}_i &= x_f + R s_i \\
\bar{m}_{di} &= x_{fd} + R_d s_i \\
\bar{m}_i^* &= x_f^* + R^* s_i.
\end{aligned} \tag{B.2}$$

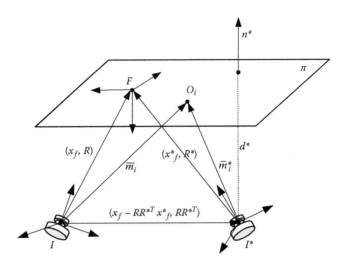

Figure B.1: Geometric relationships for $\bar{R}(t)$ and $\bar{x}_f(t)$.

After taking the time derivative of the first equation in (B.2), $\dot{\bar{m}}_1(t)$ can be determined as follows:

$$\dot{\bar{m}}_1 = Rv_e + R[\omega_e]_\times s_1, \tag{B.3}$$

where the following property have been utilized [68]:

$$[R\omega_e]_\times = R[\omega_e]_\times R^T. \tag{B.4}$$

After substituting (B.3) into (B.1), multiplying the resulting expression by z_1^*, and utilizing the definition of $\alpha_i(t)$ in (13.5), the open-loop error system given in (A.3) is obtained.

B.3

As stated in Section 10.2.2, to ensure that $\bar{m}_i(t)$ tracks $\bar{m}_{di}(t)$ from the Euclidean reconstruction given in (10.1), the tracking control objective can be stated as follows: $\bar{R}(t) \to \bar{R}_d(t)$, $m_1(t) \to m_{d1}(t)$, and $z_1(t) \to z_{d1}(t)$. This appendix describes how this objective can be achieved provided the result given in (13.73) is obtained. To this end, the expressions in (13.19) and (10.14) can be used to conclude that if $\|e_v(t)\| \to 0$ then $p_1(t) \to p_{d1}(t)$ and the ratio $\alpha_1(t)/\alpha_{d1}(t) \to 1$; hence, from (13.6) and the definition of the depth ratios in (13.5) and (12.6), it can be shown that $m_1(t) \to m_{d1}(t)$ and $z_1(t) \to z_{d1}(t)$. Given that $m_1(t) \to m_{d1}(t)$ and $z_1(t) \to z_{d1}(t)$, then (13.4) and (10.6) can be used to prove that $\bar{m}_1(t) \to \bar{m}_{d1}(t)$.

To examine if $\bar{R}(t) \to \bar{R}_d(t)$, we first take the difference between them based on the definition in (1.25).

$$\bar{R} - \bar{R}_d = \sin\theta \, [u]_\times - \sin\theta_d \, [u_d]_\times + 2\sin^2 \frac{\theta}{2} [u]_\times^2 - 2\sin^2 \frac{\theta_d}{2} [u_d]_\times^2. \tag{B.5}$$

To continue the analysis, we can see from the result in (13.73) that $\|e_\omega(t)\| \to 0$, and hence, we can use (11.6) and (10.16) to show that

$$u(t)\theta(t) \to u_d(t)\theta_d(t) \text{ as } t \to \infty. \tag{B.6}$$

To see if (B.6) implies that $\bar{R}(t) \to \bar{R}_d(t)$, we note that (13.73) and (B.6) imply that

$$\|u(t)\theta(t)\|^2 \to \|u_d(t)\theta_d(t)\|^2 \text{ as } t \to \infty, \tag{B.7}$$

which implies that

$$\theta^2(t) \|u(t)\|^2 \to \theta_d^2(t) \|u_d(t)\|^2 \text{ as } t \to \infty. \tag{B.8}$$

Since $\|u(t)\| = \|u_d(t)\| = 1$, we can see from (B.8) that

$$\theta(t) \to \pm\theta_d(t) \text{ as } t \to \infty. \tag{B.9}$$

We can now see from (B.6) that

$$\text{Case 1)} \quad u(t) \to u_d(t) \text{ when } \theta(t) \to \theta_d(t) \tag{B.10}$$
$$\text{Case 2)} \quad u(t) \to -u_d(t) \text{ when } \theta(t) \to -\theta_d(t).$$

After substituting each case given in (B.10) into (B.5) and then passing the limit, it is clear that $\bar{R}(t) \to \bar{R}_d(t)$. Based on the results that $\bar{m}_1(t) \to \bar{m}_{d1}(t)$ and that $\bar{R}(t) \to \bar{R}_d(t)$, it is clear that $\bar{m}_i(t) \to \bar{m}_{di}(t)$.

Appendix C: Chapter 11

C.1

Definition C.1 (Filippov Set-Valued Map [165]) For a vector field $f(x,t)$: $\mathbb{R}^n \times \mathbb{R} \to \mathbb{R}^n$, the Filippov set-valued map is defined as:

$$K[f](x,t) \triangleq \bigcap_{\delta > 0} \bigcap_{\mu(S)=0} \overline{co} f(B(x,\delta) - S, t)$$

where μ denotes Lebesgue measure, $\bigcap_{\mu(S)=0}$ denotes the intersection over sets S of Lebesgue measurable zero, $\overline{co}$ denotes convex closure, and $B(x,\delta) \triangleq \{v \in \mathbb{R}^n \mid \|x - v\| < \delta\}$.

Definition C.2 (Filippov Solution [69]) A function $x(t)$: $[0,\infty) \to \mathbb{R}^n$ is called a Filippov solution of $\dot{x} = f(x,t)$ on the interval of $[0,\infty)$ if $x(t)$ is absolutely continuous and for almost all $t \in [0,\infty)$

$$\dot{x} \in K[f](x,t).$$

Definition C.3 (Clarke's Generalized Gradient [41,71]) For a function $V(x,t)$: $\mathbb{R}^n \times [0,\infty) \to \mathbb{R}$ that is locally Lipschitz in (x,t), define the generalized gradient of V at (x,t) by

$$\partial V(x,t) \triangleq \overline{co} \{\lim \nabla V(x,t) \mid (x_i,t_i) \to (x,t), (x_i,t_i) \notin \Omega_V\}$$

where Ω_V is the set of measure zero where the gradient of V is not defined.

Lemma C.1 (Chain Rule [71,165])

Let $x(t)$ be a Filippov solution of $\dot{x} = f(x,t)$ and $V(x,t)$: $\mathbb{R}^n \times [0,\infty) \to \mathbb{R}$ be a locally Lipschitz regular function. Then, V is absolutely continuous, $\frac{d}{dt}V(x,t)$ exists almost everywhere (a.e.), i.e., for almost all $t \in [0,\infty)$, and $\dot{V}(x,t) \stackrel{a.e.}{\in} \dot{\tilde{V}}(x,t)$, where

$$\dot{V} \triangleq \bigcap_{\xi \in \partial V} \xi^T \begin{bmatrix} K[f](x,t) \\ 1 \end{bmatrix}.$$

C.2 Proof of Property 11.1

To facilitate the subsequent development, $f_{s1}(t), f_{s2}(t) \in \mathbb{R}$ are defined as follows:

$$f_{s1} \triangleq \left(1 - \frac{\text{sinc}(\phi)}{\text{sinc}^2\left(\frac{\phi}{2}\right)}\right)\frac{1}{\phi} \qquad f_{s2} \triangleq \frac{\phi}{4\sin^2\left(\frac{\phi}{2}\right)} - \frac{1}{2}\cot\left(\frac{\phi}{2}\right). \tag{C.1}$$

For any $\theta \in \mathbb{R}$, $\text{sinc}(\theta) \triangleq \frac{\sin(\theta)}{\theta}$. In (11.8), the Jacobian-like matrix $L_\omega(t)$ is defined as [152]

$$L_\omega \triangleq I_3 - \frac{\phi}{2}[\mu]_\times + f_{s1}\phi[\mu]_\times^2, \tag{C.2}$$

where the notation I_i denotes an $i \times i$ identity matrix. Based on the axis-angle representation of rotation matrix, $\bar{R}(t)$ can be written as [86]:

$$\bar{R} = \exp([e_r]_\times) = I_3 + \sin(\phi)[\mu]_\times + 2\sin^2\left(\frac{\phi}{2}\right)[\mu]_\times^2. \tag{C.3}$$

From (11.6), (C.2), and (C.3), the following expression can be obtained:

$$L_\omega \bar{R} = I_3 + \frac{1}{2}[e_r]_\times + f_{s1}[\mu]_\times[e_r]_\times = L_\omega + [e_r]_\times, \tag{C.4}$$

where $[\mu]_\times^2 = \mu\mu^T - I_3$ and $[\mu]_\times \mu\mu^T = 0$ are utilized. Taking the time derivative of (C.4), then substituting from $\dot{\phi}\mu = (I_3 + [\mu]_\times^2)\dot{e}_r$ and $\phi\dot{\mu} = -[\mu]_\times^2\dot{e}_r$, we obtain that

$$\dot{L}_\omega\bar{R} + L_\omega\dot{\bar{R}} = \dot{L}_\omega + [\dot{e}_r]_\times = \frac{1}{2}[\dot{e}_r]_\times + f_{s2}[\mu]_\times\left[\left(I_3 + [\mu]_\times^2\right)\dot{e}_r\right]_\times$$
$$+ f_{s1}\left([\mu]_\times\left[-[\mu]_\times^2\dot{e}_r\right]_\times + \left[-[\mu]_\times^2\dot{e}_r\right]_\times[\mu]_\times\right). \tag{C.5}$$

Based on $\bar{R} = R(R^*)^T$, the first equation of (11.13) can be rewritten as:

$$\delta_r = L_\omega\bar{R}R^*\omega_l. \tag{C.6}$$

After taking the time derivative of (C.6) and using the fact that $\dot{R}^* = 0$, it can be obtained that

$$\dot{\delta}_r = \left(\dot{L}_\omega\bar{R} + L_\omega\dot{\bar{R}}\right)R^*\omega_l + L_\omega\bar{R}R^*\dot{\omega}_l. \tag{C.7}$$

From (C.4) and (C.5), it is not difficult to see that $L_\omega \bar{R} = I_3$ and $\dot{L}_\omega \bar{R} + L_\omega \dot{\bar{R}} = 0$ when e_r, $\dot{e}_r = 0$. Based on (11.16), (11.17), and (C.7), $N_{rd}(t)$ and $\tilde{N}_r(t)$ can be reviewed as:

$$N_{rd} = R^* \dot{\omega}_l \tag{C.8}$$

$$\tilde{N}_r = e_r + K_{Ir}\dot{e}_r + \left(\dot{L}_\omega \bar{R} + L_\omega \dot{\bar{R}}\right) R^* \omega_l + (L_\omega \bar{R} - I_3) R^* \dot{\omega}_l. \tag{C.9}$$

To facilitate further development, some properties are listed as follow

$$\left\| \left[-[\mu]_\times^2 \dot{e}_r \right]_\times \right\|_2 \le \|\dot{e}_r\| \qquad \left\| [\mu]_\times \right\|_2 = 1$$

$$\left\| \left[\left(I_3 + [\mu]_\times^2 \right) \dot{e}_r \right]_\times \right\|_2 \le \|\dot{e}_r\| \qquad \|R^*\|_2 = 1 \tag{C.10}$$

$$\left\| [e_r]_\times \right\|_2 = \|e_r\| \qquad \left\| [\dot{e}_r]_\times \right\|_2 = \|\dot{e}_r\|,$$

where $\| \cdot \|_2$ denotes the spectral norm for matrices [95]. Utilizing (C.4), (C.5), (C.9), and (C.10), $\tilde{N}_r(t)$ can be bounded as follows:

$$\|\tilde{N}_r\| \le \left(1 + \left(\frac{1}{2} + |f_{s1}| \right) \|\dot{\omega}_l\| \right) \|e_r\|$$

$$+ \left(\lambda_{\max}(K_{Ir}) + \left(\frac{1}{2} + 2|f_{s1}| + |f_{s2}| \right) \|\omega_l\| \right) \|\dot{e}_r\|. \tag{C.11}$$

According to the region of $\phi(t)$ defined in (11.7), it can be determined that

$$|f_{s1}| < \frac{1}{\pi} \qquad |f_{s2}| < \frac{\pi}{4}. \tag{C.12}$$

Based on Assumption 11.1, (11.14), (C.11), and (C.12), $\|\tilde{N}_r\| \le \rho_r \|z_r\|$ can be obtained, where

$$\rho_r = (1 + \lambda_{\max}(K_{Ir})) \left(\frac{1}{2} + \frac{2}{\pi} + \frac{\pi}{4} \right) \bar{\omega}_l + \left(\frac{1}{2} + \frac{1}{\pi} \right) \dot{\bar{\omega}}_l + 1$$

$$+ \lambda_{\max}(K_{Ir}) + \lambda_{\max}^2(K_{Ir}). \tag{C.13}$$

C.3 Proof of Lemma 11.1

Based on (C.8) and the fact that $\dot{R}^* = 0$, it is easy to get that

$$\dot{N}_{rd} = R^* \ddot{\omega}_l. \tag{C.14}$$

From Assumption 1, (C.8), and (C.14), the following relation can be obtained:

$$\|N_{rd}\|, \|\dot{N}_{rd}\| \in \mathcal{L}_\infty. \tag{C.15}$$

Then, it is straightforward that the right-hand side of (11.22) is bounded.

After substituting the first equation of (11.14) into (11.21) and integrating in time, it can be determined that

$$
\int_0^t \Lambda_r(\tau)d\tau = \int_0^t -\dot{e}_r^T(\tau)K_{tr}\mathrm{Tanh}(e_r(\tau))d\tau + \int_0^t -\dot{e}_r^T(\tau)K_{sr}\mathrm{Sgn}(e_r(\tau))d\tau
$$
$$
+ \int_0^t \dot{e}_r^T(\tau)N_{rd}(\tau)d\tau
$$
$$
+ \int_0^t e_r^T(\tau)K_{Ir}^T\left(N_{rd}(\tau) - K_{sr}\mathrm{Sgn}(e_r(\tau)) - K_{tr}\mathrm{Tanh}(e_r(\tau))\right)d\tau
$$
$$
= - a^T K_{tr}\mathrm{Lncosh}(e_r)\big|_0^t - e_r^T K_{sr}\mathrm{Sgn}(e_r)\big|_0^t + e_r^T N_{rd}\big|_0^t
$$
$$
+ \int_0^t e_r^T(\tau)\left(-\dot{N}_{rd}(\tau) + K_{Ir}^T\left(N_{rd}(\tau)\right.\right.
$$
$$
\left.\left. -K_{sr}\mathrm{Sgn}(e_r(\tau)) - K_{tr}\mathrm{Tanh}(e_r(\tau))\right)\right)d\tau.
$$

(C.16)

Since $a^T K_{tr}\mathrm{Lncosh}(e_r) \geq 0$, $e_r^T K_{Ir}^T K_{tr}\mathrm{Tanh}(e_r) \geq 0.5\lambda_{\min}(K_{Ir})\lambda_{\min}(K_{tr})\|e_r\|$ $\|\mathrm{Tanh}(e_r)\| \geq 0$, and $e_r^T K_{Ir}^T K_{sr}\,\mathrm{Sgn}(e_r) \geq \lambda_{\min}(K_{Ir})\lambda_{\min}(K_{sr})\|e_r\| \geq 0$, the right-hand side of (C.16) can be upper bounded as follows

$$
\int_0^t \Lambda_r(\tau)d\tau \leq (\|N_{rd}(t)\| - \lambda_{\min}(K_{sr}))\,\|e_r(t)\| + a^T K_{tr}\mathrm{Lncosh}(e_r(0))
$$
$$
- e_r^T(0)\left(N_{rd}(0) - K_{sr}\mathrm{Sgn}(e_r(0))\right)
$$
$$
+ \int_0^t \left(\lambda_{\max}(K_{Ir})\|N_{rd}\| + \|\dot{N}_{rd}\| - \lambda_{\min}(K_{Ir})\lambda_{\min}(K_{sr})\right.
$$
$$
\left. -0.5\lambda_{\min}(K_{Ir})\lambda_{\min}(K_{tr})\|\mathrm{Tanh}(e_r)\|\right)\|e_r\|d\tau.
$$

(C.17)

From (C.17), it is clear that if $\lambda_{\min}(K_{sr})$ is chosen to satisfy (11.22), then (11.23) holds. Furthermore, since the right-hand side of (11.22) is bounded and does not contain K_{sr}, the existence of K_{sr} satisfying the condition of (11.22) is guaranteed.

C.4   Proof of Property 11.2

Based on $\bar{R} = R(R^*)^T$ and $[R\omega_l]_\times = R[\omega_l]_\times(R)^T$, the second equation of (11.13) can be rewritten as

$$
\delta_t = \tilde{d}^*[e_t]_\times \omega + \bar{R}R^* v_l - \bar{R}R^*[\omega_l]_\times(R^*)^T x_f^*.
$$

(C.18)

Taking the time derivative of (C.18), the following expression can be obtained:

$$
\dot{\delta}_t = \tilde{d}^*\left([\dot{e}_t]_\times \omega + [e_t]_\times \dot{\omega}\right) + \dot{\bar{R}}R^* v_l + \bar{R}R^* \dot{v}_l - \dot{\bar{R}}R^*[\omega_l]_\times(R^*)^T x_f^*
$$
$$
- \bar{R}R^*[\dot{\omega}_l]_\times(R^*)^T x_f^*.
$$

(C.19)

From (11.16), (11.17), and (C.19), $N_{td}(t)$ and $\tilde{N}_{td}(t)$ can be rewritten as:

$$N_{td} = \dot{\bar{R}}R^* v_l + \bar{R}R^* \dot{v}_l - \dot{\bar{R}}R^* [\omega_l]_\times (R^*)^T x_f^* - \bar{R}R^* [\dot{\omega}_l]_\times (R^*)^T x_f^* \tag{C.20}$$

$$\tilde{N}_t = d^* e_t + d^* K_{It} \dot{e}_t + \tilde{d}^* \left([\dot{e}_t]_\times \omega + [e_t]_\times \dot{\omega}\right). \tag{C.21}$$

In Theorem 11.1, $\omega(t) \in \mathcal{L}_\infty$ is proved. Furthermore, from (11.8), it can be determined that $\dot{\omega} = -\dot{L}_\omega^{-1} \dot{e}_r - L_\omega^{-1} \ddot{e}_r + \dot{R}\omega_l + R\dot{\omega}_l$. Utilizing Assumption 1, (C.2), and $\omega(t)$, $e_r(t)$, $\dot{e}_r(t)$, $\dot{\eta}_r(t) \in \mathcal{L}_\infty$, it can be concluded that $\dot{\omega}(t) \in \mathcal{L}_\infty$. Since $\omega(t)$, $\dot{\omega}(t) \in \mathcal{L}_\infty$, there exist positive constants $\bar{\omega}$, $\bar{\dot{\omega}} \in \mathbb{R}$ satisfying $\|\omega\| \leq \bar{\omega}$, $\|\dot{\omega}\| \leq \bar{\dot{\omega}}$. According to (C.21), $\tilde{N}_t(t)$ can be bounded as follows:

$$\left\|\tilde{N}_t\right\| \leq \left(d^* + |\tilde{d}^*|\bar{\dot{\omega}}\right)\|e_t\| + \left(d^* \lambda_{\max}(K_{It}) + |\tilde{d}^*|\bar{\omega}\right)\|\dot{e}_t\|, \tag{C.22}$$

where $\left\|[e_t]_\times\right\|_2 = \|e_t\|$, $\left\|[\dot{e}_t]_\times\right\|_2 = \|\dot{e}_t\|$ are used to obtain the expression.

Based on (11.14) and (C.22), $\|\tilde{N}_t\| \leq \rho_t \|z_t\|$ can be obtained, where

$$\rho_t = (1 + \lambda_{\max}(K_{It})) \left(d^* \lambda_{\max}(K_{It}) + |\tilde{d}^*|\bar{\omega}\right) + \left(d^* + |\tilde{d}^*|\bar{\dot{\omega}}\right). \tag{C.23}$$

Appendix D: Chapter 12

D.1 Open-Loop Dynamics

The extended image coordinates $p_{e1}(t)$ of (12.12) can be written as follows:

$$p_{e1} = \begin{bmatrix} a_1 & a_2 & 0 \\ 0 & a_3 & 0 \\ 0 & 0 & 1 \end{bmatrix} \begin{bmatrix} \frac{x_1}{z_1} \\ \frac{y_1}{z_1} \\ \ln(z_1) \end{bmatrix} + \begin{bmatrix} a_4 \\ a_5 \\ -\ln(z_1^*) \end{bmatrix}, \qquad \text{(D.1)}$$

where (12.7), (13.6), and (12.9) were utilized. After taking the time derivative of (D.1), the following expression can be obtained:

$$\dot{p}_{e1} = \frac{1}{z_1} A_{e1} \dot{\bar{m}}_1 .$$

By exploiting the fact that $\dot{\bar{m}}_1(t)$ can be expressed as follows:

$$\dot{\bar{m}}_1 = -v_c + [\bar{m}_1]_\times \omega_c,$$

the open-loop dynamics for $p_{e1}(t)$ can be rewritten as follows:

$$\dot{p}_{e1} = -\frac{1}{z_1} A_{e1} v_c + A_{e1} [m_1]_\times \omega_c.$$

The open-loop dynamics for $\Theta(t)$ can be expressed as follows [29]

$$\dot{\Theta} = -L_\omega \omega_c.$$

D.2 Image Jacobian-Like Matrix

Similar to (12.14), the dynamics for $\Upsilon_d(t)$ can be expressed as:

$$\dot{\Upsilon}_d = \begin{bmatrix} \dot{p}_{ed1} \\ \dot{\Theta}_d \end{bmatrix} = \begin{bmatrix} -\frac{1}{z_{d1}} A_{ed1} & A_{ed1} [m_{d1}]_\times \\ 0_3 & -L_{\omega d} \end{bmatrix} \begin{bmatrix} v_{cd} \\ \omega_{cd} \end{bmatrix}, \qquad \text{(D.2)}$$

where $\Theta_d(t)$ is defined in (12.22), $z_{di}(t)$ is introduced in (13.4), $A_{edi}(u_{di}, v_{di})$ is defined in the same manner as in (12.15) with respect to the desired pixel coordinates $u_{di}(t)$, $v_{di}(t)$, $m_{di}(t)$ is given in (12.1), $L_{\omega d}(\theta_d, \mu_d)$ is defined in the same manner as in (C.2) with respect to $\theta_d(t)$ and $\mu_d(t)$, and $v_{cd}(t)$, $\omega_{cd}(t) \in \mathbb{R}^3$ denote the desired linear and angular velocity signals that ensure compatibility with (D.2). The signals $v_{cd}(t)$ and $\omega_{cd}(t)$ are not actually used in the trajectory generation scheme presented in this chapter as similarly done in [29]; rather, these signals are simply used to clearly illustrate how $\dot{\bar{p}}_d(t)$ can be expressed in terms of $\dot{\Upsilon}_d(t)$ as required in (12.23). Specifically, we first note that the top block row in (D.2) can be used to write the time derivative of $p_{ed2}(t)$ in terms of $v_{cd}(t)$ and $\omega_{cd}(t)$ with $i = 2$

$$\dot{p}_{ed2} = \begin{bmatrix} -\frac{1}{z_{d2}}A_{ed2} & A_{ed2}[m_{d2}]_\times \end{bmatrix} \begin{bmatrix} v_{cd} \\ \omega_{cd} \end{bmatrix}, \tag{D.3}$$

where $p_{edi}(t)$ is defined in the same manner as (12.21) $\forall i = 1, 2, 3, 4$. After inverting the relationship given by (D.2), we can also express $v_{cd}(t)$ and $\omega_{cd}(t)$ as a function of $\dot{\Upsilon}_d(t)$ as follows:

$$\begin{bmatrix} v_{cd} \\ \omega_{cd} \end{bmatrix} = \begin{bmatrix} -z_{d1}A_{ed1}^{-1} & -z_{d1}[m_{d1}]_\times L_{\omega d}^{-1} \\ 0 & -L_{\omega d}^{-1} \end{bmatrix} \dot{\Upsilon}_d. \tag{D.4}$$

After substituting (D.4) into (D.3), $\dot{p}_{ed2}(t)$ can be expressed in terms of $\dot{\Upsilon}_d(t)$ as follows

$$\dot{p}_{ed2} = \begin{bmatrix} \frac{z_{d1}}{z_{d2}}A_{ed2}A_{ed1}^{-1} & A_{ed2}\left[\frac{z_{d1}}{z_{d2}}m_{d1} - m_{d2}\right]_\times L_{\omega d}^{-1} \end{bmatrix} \dot{\Upsilon}_d. \tag{D.5}$$

After formulating similar expressions for $\dot{p}_{ed3}(t)$ and $\dot{p}_{ed4}(t)$ as the one given by (D.5) for $\dot{p}_{ed2}(t)$, we can compute the expression for $L_{\Upsilon_d}(\bar{p}_d)$ in (12.24) by utilizing the definitions of $p_{di}(t)$ and $p_{edi}(t)$ given in (12.7) and (12.21), respectively (i.e., we must eliminate the bottom row of the expression given by (D.5)).

D.3 Image Space Navigation Function

Inspired by the framework developed in [46], an image space NF is constructed by developing a diffeomorphism[1] between the image space and a model space, developing a model space NF, and transforming the model space NF into an image space NF through the diffeomorphism (since NFs are invariant under diffeomorphism [117]). To this end, a diffeomorphism is defined that maps the desired image feature vector $\bar{p}_d$ to the auxiliary model space signal $\zeta(\bar{p}_d) \triangleq [\zeta_1(\bar{p}_d)\ \zeta_2(\bar{p}_d) \dots \zeta_8(\bar{p}_d)]^T : [-1, 1]^8 \to \mathbb{R}^8$ as follows:

---

[1] A diffeomorphism is a map between manifolds which is differentiable and has a differentiable inverse.

$$\zeta = diag\left\{\frac{2}{u_{max}-u_{min}}, \frac{2}{v_{max}-v_{min}}, \ldots, \frac{2}{v_{max}-v_{min}}\right\}\bar{p}_d \qquad (D.6)$$

$$-\left[\frac{u_{max}+u_{min}}{u_{max}-u_{min}} \quad \frac{v_{max}+v_{min}}{v_{max}-v_{min}} \cdots \frac{v_{max}+v_{min}}{v_{max}-v_{min}}\right]^T.$$

In (D.6), u_{max}, u_{min}, v_{max}, and $v_{min} \in \mathbb{R}$ denote the maximum and minimum pixel values along the u and v axes, respectively. The model space NF, denoted by $\tilde{\varphi}(\zeta) \in \mathbb{R}^8 \to \mathbb{R}$, is defined as follows [46]:

$$\tilde{\varphi}(\zeta) \triangleq \frac{\bar{\varphi}}{1+\bar{\varphi}}. \qquad (D.7)$$

In (D.7), $\bar{\varphi}(\zeta) \in \mathbb{R}^8 \to \mathbb{R}$ is defined as:

$$\bar{\varphi}(\zeta) \triangleq \frac{1}{2}f(\zeta)^T K f(\zeta), \qquad (D.8)$$

where the auxiliary function $f(\zeta) : (-1,1)^8 \to \mathbb{R}^8$ is defined similar to [46] as Follows:

$$f(\zeta) = \left[\frac{\zeta_1 - \zeta_1^*}{\left(1-\zeta_1^{2\kappa}\right)^{1/2\kappa}} \cdots \frac{\zeta_8 - \zeta_8^*}{\left(1-\zeta_8^{2\kappa}\right)^{1/2\kappa}}\right]^T, \qquad (D.9)$$

where $K \in \mathbb{R}^{8\times 8}$ is a positive definite, symmetric matrix, and κ is a positive parameter. The reason we use κ instead of 1 as in [46] is to get an additional parameter to change the potential field formed by $f(\zeta)$. See [46] for a proof that (D.7) satisfies the properties of a NF as described in Definition 12.1. The image space NF, denoted by $\varphi(\bar{p}_d) \in \mathcal{D} \to \mathbb{R}$, can then be developed as follows:

$$\varphi(\bar{p}_d) \triangleq \tilde{\varphi} \circ \zeta(\bar{p}_d), \qquad (D.10)$$

where $\circ$ denotes the composition operator. The gradient vector $\nabla\varphi(p_d)$ can be expressed as follows:

$$\nabla\varphi \triangleq \left(\frac{\partial\varphi}{\partial\bar{p}_d}\right)^T = \left(\frac{\partial\tilde{\varphi}}{\partial\zeta}\frac{\partial\zeta}{\partial\bar{p}_d}\right)^T. \qquad (D.11)$$

In (D.11), the partial derivative expressions $\frac{\partial\zeta(\bar{p}_d)}{\partial\bar{p}_d}$, $\frac{\partial\tilde{\varphi}(\zeta)}{\partial\zeta}$, and $\frac{\partial f(\zeta)}{\partial\zeta}$ can be expressed as follows:

$$\frac{\partial\zeta}{\partial\bar{p}_d} = diag\left\{\frac{2}{u_{max}-u_{min}}, \frac{2}{v_{max}-v_{min}}, \ldots, \frac{2}{v_{max}-v_{min}}\right\} \qquad (D.12)$$

$$\frac{\partial\tilde{\varphi}}{\partial\zeta} = \frac{1}{(1+\bar{\varphi})^2}f^T K \frac{\partial f}{\partial\zeta} \qquad (D.13)$$

$$\frac{\partial f}{\partial\zeta} = diag\left\{\frac{1-\zeta_1^{2\kappa-1}\zeta_1^*}{\left(1-\zeta_1^{2\kappa}\right)^{(2\kappa+1)/2\kappa}}, \ldots, \frac{1-\zeta_8^{2\kappa-1}\zeta_8^*}{\left(1-\zeta_8^{2\kappa}\right)^{(2\kappa+1)/2\kappa}}\right\}. \qquad (D.14)$$

It is clear from (D.6) to (D.14) that $\bar{p}_d(t) \to \bar{p}^*$ when $\nabla\varphi(\bar{p}_d) \to 0$.

Appendix E: Chapter 13

The Extended Barbalat's Lemma was utilized in the stability analysis for Theorem 13.3. This lemma stated as follows, and a proof for the Lemma can be found in [55].

Lemma E.1

If a differentiable function $f(t) \in \mathbb{R}$ has a finite limit as $t \to \infty$, and its time derivative can be written as follows:

$$\dot{f}(t) = g_1(t) + g_2(t), \tag{E.1}$$

where $g_1(t)$ is a uniformly continuous function and

$$\lim_{t \to \infty} g_2(t) = 0, \tag{E.2}$$

then,

$$\lim_{t \to \infty} \dot{f}(t) = 0 \quad and \quad \lim_{t \to \infty} g_1(t) = 0. \tag{E.3}$$

Appendix F: Chapter 14

F.1

Theorem 13.2 [20]: Let f be a bounded function that is integrable on $[a,b]$, let further m_ϕ and M_ϕ be the minimum and maximum values of the function $\phi(t) = \int_t^b f(\tau)d\tau$ on $[a,b]$. If a function $g(t)$ is non-increasing with $g(t) \geq 0$ on $[a,b]$, then there is a number $\Lambda \in [m_\phi, M_\phi]$ such that

$$\int_a^b f(t)g(t)dt = g(b) \cdot \Lambda. \tag{F.1}$$

F.2 Proof of $\frac{1}{2}\ln(\bar{e}_1^2 + \mu) > \frac{\bar{e}_1}{\bar{e}_1^2 + \mu}$ for $\mu > 1.5$ and $\bar{e}_1 \in \mathbb{R}$

Define the function $f(\bar{e}_1, \mu)$ as:

$$f(\bar{e}_1, \mu) \triangleq \frac{1}{2}\ln(\bar{e}_1^2 + \mu) - \frac{\bar{e}_1}{\bar{e}_1^2 + \mu}. \tag{F.2}$$

Take the partial derivative of $f(\bar{e}_1, \mu)$ with respect to $\bar{e}_1$.

$$\frac{\partial f}{\partial \bar{e}_1} = \frac{\bar{e}_1^3 + \bar{e}_1^2 + \mu\bar{e}_1 - \mu}{(\bar{e}_1^2 + \mu)^2}. \tag{F.3}$$

If $\mu = 1.5$, the minimum value can be obtained as $f_{\min} \approx 0.3 > 0$. Take the partial derivative of $f(\bar{e}_1, \mu)$ with respect to μ.

$$\frac{\partial f}{\partial \mu} = \frac{\bar{e}_1^2 - 2\bar{e}_1 + \mu}{2(\bar{e}_1^2 + \mu)^2}. \tag{F.4}$$

Obviously, if $\mu \geq 1.5$, $\frac{\partial f}{\partial \mu} > 0$ for $\bar{e}_1 \in \mathbb{R}$. That means for $\mu \geq 1.5$, the minimum value for function $f(\bar{e}_1, \mu)$ is not less than $f_{\min}$. Thus $\mu > 1.5$ is a sufficient condition for $\frac{1}{2}\ln(\bar{e}_1^2 + \mu) > \frac{\bar{e}_1}{\bar{e}_1^2 + \mu}$.

F.3   Proof of $\rho(t)e_3(t) \in \mathcal{L}_1$

From (14.15) it can be obtained that

$$\int_0^t |\rho(s)e_3(s)|ds \leq \beta \int_0^t |e_3(s)| \cdot e^{-\int_0^s \frac{|e_3(\tau)|}{\sigma_2}d\tau} ds. \tag{F.5}$$

Since $|e_3(t)|$ is bounded from Appendix D, then according to *Theorem 14.2* in Appendix 14.A, (F.5) can be rewritten as:

$$\int_0^t |\rho(s)e_3(s)|ds \leq \beta e^{-\int_0^t \frac{|e_3(\tau)|}{\sigma_2}d\tau} \cdot \int_0^t |e_3(\tau)|d\tau \tag{F.6}$$

$$\leq \beta \frac{\sigma_2}{e},$$

which implies that $\int_0^\infty |\rho(s)e_3(s)|ds$ is bounded, and hence $\rho(t)e_3(t) \in \mathcal{L}_1$.

F.4   Proof of $e_3(t) \in \mathcal{L}_1$ in the Closed-Loop System (14.19)

Solving the differential equation of $e_3(t)$ in (14.19):

$$
\begin{aligned}
e_3(t) \\
= \left(\int_0^t k_1 \eta_1 \eta_2 (\dot{x}_{hdz} + \rho) e^{\int_0^s k_3 \eta_2 d\tau} ds + e_3(0) \right) e^{-\int_0^t k_3 \eta_2 ds} \\
= \int_0^t k_1 \eta_1 \eta_2 (\dot{x}_{hdz} + \rho) e^{-\int_s^t k_3 \eta_2 d\tau} ds + e_3(0) e^{-\int_0^t k_3 \eta_2 ds}.
\end{aligned}
\tag{F.7}
$$

Then, it can be obtained that

$$
\begin{aligned}
|e_3(t)| \leq \int_0^t \left| k_1 \eta_1 \eta_2 (\dot{x}_{hdz} + \rho) e^{-\int_s^t k_3 \eta_2 d\tau} \right| ds \\
+ \left| e_3(0) e^{-\int_0^t k_3 \eta_2 ds} \right|.
\end{aligned}
\tag{F.8}
$$

In (F.8), since $k_3 > 0$ and $\eta_2(t) > 1$ from (14.18), then it is obvious that $\left| e_3(0) e^{-\int_0^t k_3 \eta_2 ds} \right| \in \mathcal{L}_1$. Developing (F.8) it can be obtained that

$$
\begin{aligned}
\int_0^t \left| k_1 \eta_1 \eta_2 (\dot{x}_{hdz} + \rho) e^{-\int_s^t k_3 \eta_2 d\tau} \right| ds \\
= \int_0^t |k_1 \eta_1 (\dot{x}_{hdz} + \rho)| \eta_2 e^{-\int_s^t k_3 \eta_2 d\tau} ds \\
= \int_0^t f(t-s) \eta_2(t-s) e^{-\int_{t-s}^t k_3 \eta_2 d\tau} ds,
\end{aligned}
\tag{F.9}
$$

where $f(t) \triangleq |k_1 \eta_1(t)(\dot{x}_{hdz}(t) + \rho(t))|$. Since in the closed-loop system, $\eta_1(t), \rho(t), \dot{x}_{hdz}(t) \in \mathcal{L}_\infty$, then $f(t) \in \mathcal{L}_\infty$, which means that for each initial condition, there exists a constant δ such that $f(t) \leq \delta$. Then by using *Theorem 14.2* in Appendix 14.A, (F.9) can be rewritten as:

$$\int_0^t f(t-s)\eta_2(t-s)e^{-\int_{t-s}^t k_3\eta_2 d\tau} ds$$

$$\leq \delta \cdot \int_0^t \eta_2(t-s)e^{-\int_{t-s}^t k_3\eta_2 d\tau} ds$$

$$\leq \delta \cdot e^{-\int_0^t k_3\eta_2 d\tau} \cdot \int_0^t \eta_2(t-s) ds \qquad (F.10)$$

$$= \delta \cdot e^{-\frac{1}{2}\int_0^t k_3\eta_2 d\tau} \cdot e^{-\frac{1}{2}\int_0^t k_3\eta_2 d\tau} \cdot \int_0^t \eta_2(s) ds$$

$$\leq \frac{2\delta}{k_3 e} e^{-\frac{1}{2}\int_0^t k_3\eta_2 d\tau} \in \mathcal{L}_1.$$

Based on the above developments, $e_3 \in \mathcal{L}_1$.

References

[1] M. Ajmal, M. H. Ashraf, M. Shakir, Y. Abbas, and F. A. Shah. Video summarization: Techniques and classification. In: Bolc L., Tadeusiewicz R., Chmielewski L. J., Wojciechowski K. (eds) *Computer Vision and Graphics*, Lecture Notes in Computer Science, volume 7594, pp. 1–13. Berlin, Heidelberg: Springer, 2012.

[2] J. Álvarez and A. M. López. Road detection based on illuminant invariance. *IEEE Transactions on Intelligent Transportation Systems*, 12(1):184–193, 2011.

[3] J. M. Álvarez, T. Gevers, Y. LeCun, and A. M. Lopez. Road scene segmentation from a single image. In: Fitzgibbon A., Lazebnik S., Perona P., Sato Y., Schmid C. (eds) *Computer Vision ECCV 2012*, Lecture Notes in Computer Science, pp. 376–389. Berlin, Heidelberg: Springer, 2012.

[4] J. M. Álvarez, A. M. López, T. Gevers, and F. Lumbreras. Combining priors, appearance, and context for road detection. *IEEE Transactions on Intelligent Transportation Systems*, 15(3):1168–1178, 2014.

[5] J. Arróspide, L. Salgado, M. Nieto, and R. Mohedano. Homography-based ground plane detection using a single on-board camera. *IET Intelligent Transport Systems*, 4(2):149–160, 2010.

[6] A. Assa and F. Janabi-Sharifi. A robust vision-based sensor fusion approach for real-time pose estimation. *IEEE Transactions on Cybernetics*, 44(2):217–227, 2014.

[7] M. Azizian, M. Khoshnam, N. Najmaei, and R. V. Patel. Visual servoing in medical robotics: A survey. Part I: Endoscopic and direct vision imaging—techniques and applications. *International Journal of Medical Robotics and Computer Assisted Surgery*, 10(3):263–74, 2014.

[8] M. Azizian, N. Najmaei, M. Khoshnam, and R. Patel. Visual servoing in medical robotics: A survey. Part II: Tomographic imaging modalities— techniques and applications. *International Journal of Medical Robotics and Computer Assisted Surgery*, 11(1):67–79, 2015.

[9] B. Bandyopadhayay, S. Janardhanan, and S. K. Spurgeon. *Advances in Sliding Mode Control: Concept, Theory and Implementation*. Berlin: Springer, 2013.

[10] R. Basri, E. Rivlin, and I. Shimshoni. Visual homing: Surfing on the epipoles. *International Journal of Computer Vision*, 33(2):117–137, 1999.

[11] H. Bay, A. Ess, T. Tuytelaars, and L. Van Gool. Speeded-up robust features (SURF). *Computer Vision and Image Understanding*, 110(3):346–359, 2008.

[12] H. M. Becerra. Fuzzy visual control for memory-based navigation using the trifocal tensor. *Intelligent Automation & Soft Computing*, 20(2):245–262, 2014.

[13] H. M. Becerra, J. B. Hayet, and C. Sagüés. A single visual-servo controller of mobile robots with super-twisting control. *Robotics and Autonomous Systems*, 62(11):1623–1635.

[14] H. M. Becerra, G. López-Nicolás, and C. Sagüés. Omnidirectional visual control of mobile robots based on the 1D trifocal tensor. *Robotics and Autonomous Systems*, 58(6):796–808, 2010.

[15] H. M. Becerra, G. López-Nicolás, and C. Sagüés. A sliding-mode-control law for mobile robots based on epipolar visual servoing from three views. *IEEE Transactions on Robotics*, 27(1):175–183, 2011.

[16] H. M. Becerra and C. Sagüés. Visual control for memory-based navigation using the trifocal tensor. In *World Automation Congress*, pp. 1–6. IEEE, 2012.

[17] H. M. Becerra and C. Sagüés. Exploiting the trifocal tensor in dynamic pose estimation for visual control. *IEEE Transactions on Control Systems Technology*, 21(5):1931–1939, 2013.

[18] H. M. Becerra, C. Sagüés, Y. Mezouar, and J. B. Hayet. Visual navigation of wheeled mobile robots using direct feedback of a geometric constraint. *Autonomous Robots*, 37(2):137–156, 2014.

[19] S. Bhattacharya, R. Murrieta-Cid, and S. Hutchinson. Optimal paths for landmark-based navigation by differential-drive vehicles with field-of-view constraints. *IEEE Transactions on Robotics*, 23(1):47–59, 2007.

[20] M. Bohner and G. Guseinov. Improper integrals on time scales. *Dynamic Systems and Applications*, 12(1/2):45–66, 2003.

[21] R. Brockett. The early days of geometric nonlinear control. *Automatica*, 50(9):2203–2224, 2014.

[22] M. Bueno-López and M. A. Arteaga-Pérez. Fuzzy vs nonfuzzy in 2D visual servoing for robot manipulators. *International Journal of Advanced Robotic Systems*, 10(2):108, 2013.

[23] F. Chaumette. Potential problems of stability and convergence in image-based and position-based visual servoing. In: Kriegman D. J., Hager G. D., Morse A. S. (eds) *The Confluence of Vision and Control*, Lecture Notes in Control and Information Sciences, volume 237, pp. 66–78. London: Springer, 1998.

[24] F. Chaumette and S. Hutchinson. Visual servo control part I: Basic approaches. *IEEE Robotics and Automation Magazine*, 13(4):82–90, 2006.

[25] J. Chen. Visual servo control with a monocular camera. PhD thesis, Clemson University, 2005.

[26] J. Chen, V. Chitrakaran, and D. M. Dawson. Range identification of features on an object using a single camera. *Clemson University CRB Technical Report*, 2010.

[27] J. Chen and D. M. Dawson. UAV tracking with a monocular camera. In *Proceedings of IEEE Conference on Decision and Control*, pp. 3873–3878, San Diego, CA, 13–15 December 2006.

[28] J. Chen, D. M. Dawson, W. E. Dixon, and A. Behal. Adaptive homography-based visual servo tracking. In *Proceedings of IEEE/RSJ International Conference on Intelligent Robots and Systems*, pp. 230–235. IEEE, Las Vegas, NV, 27–31 October 2003.

[29] J. Chen, D. M. Dawson, W. E. Dixon, and A. Behal. Adaptive homography-based visual servo tracking for a fixed camera configuration with a camera-in-hand extension. *IEEE Transactions on Control Systems Technology*, 13(5):814–825, 2005.

[30] J. Chen, D. M. Dawson, W. E. Dixon, and V. K. Chitrakaran. Navigation function-based visual servo control. *Automatica*, 43(7):1165–1177, 2007.

[31] J. Chen, W. E. Dixon, D. M. Dawson, and M. McIntyre. Homography-based visual servo tracking control of a wheeled mobile robot. *IEEE Transactions on Robotics*, 22(2):406–415, 2006.

[32] J. S. Paris, and F. Durand. Real-time edge-aware image processing with the bilateral grid. *ACM Transactions on Graphics*, 26(3):103, 2007.

[33] J. Chen, V. K. Chitrakaran, and D. M. Dawson. Range identification of features on an object using a single camera. *Automatica*, 47(1):201–206, 2011.

[34] X. Chen and H. Kano. A new state observer for perspective systems. *IEEE Transactions on Automatic Control*, 47(4):658–663, 2002.

[35] X. Chen and H. Kano. State observer for a class of nonlinear systems and its application to machine vision. *IEEE Transactions on Automatic Control*, 49(11):2085–2091, 2004.

[36] G. Chesi. Optimal object configurations to minimize the positioning error in visual servoing. *IEEE Transactions on Robotics*, 26(3):584–589, 2010.

[37] G. Chesi and K. Hashimoto. *Visual Servoing via Advanced Numerical Methods*. London: Springer, 2010.

[38] V. K. Chitrakaran, D. M. Dawson, J. Chen, and H. Kannan. Velocity and structure estimation of a moving object using a moving monocular camera. In *Proceedings of American Control Conference*, pp. 5159–5164, Minneapolis, MN, 14–16 June 2006.

[39] V. K. Chitrakaran, D. M. Dawson, W. E. Dixon, and J. Chen. Identification of a moving object's velocity with a fixed camera. *Automatica*, 41(3): 553–562, 2005.

[40] D. Chwa, A. P. Dani, and W. E. Dixon. Range and motion estimation of a monocular camera using static and moving objects. *IEEE Transactions on Control Systems Technology*, 24(4):1174–1183, 2016.

[41] F. H. Clarke. *Optimization and Nonsmooth Analysis*. New York: Wiley, 1983.

[42] C. Collewet and E. Marchand. Photometric visual servoing. *IEEE Transactions on Robotics*, 27(4):828–834, 2011.

[43] A. I. Comport, E. Marchand, and F. Chaumette. Statistically robust 2-D visual servoing. *IEEE Transactions on Robotics*, 22(2):415–420, 2006.

[44] D. Conrad and G. N. DeSouza. Homography-based ground plane detection for mobile robot navigation using a modified EM algorithm. In *IEEE International Conference on Robotics and Automation*, pp. 910–915, 2010.

[45] P. Corke, R. Paul, W. Churchill, and P. Newman. Dealing with shadows: Capturing intrinsic scene appearance for image-based outdoor localisation. In *Proceedings of IEEE/RSJ International Conference on Intelligent Robots and Systems*, pp. 2085–2092, Tokyo, Japan, 3–7 November 2013.

[46] N. J. Cowan, J. D. Weingarten, and D. E. Koditschek. Visual servoing via navigation functions. *IEEE Transactions on Robotics and Automation*, 18(4):521–533, 2002.

[47] A. Dame and E. Marchand. Mutual information-based visual servoing. *IEEE Transactions on Robotics*, 27(5):958–969, 2011.

[48] A. P. Dani, N. R. Fischer, and W. E. Dixon. Single camera structure and motion. *IEEE Transactions on Automatic Control*, 57(1):241–246, 2012.

[49] A. P. Dani, Z. Kan, N. R. Fischer, and W. E. Dixon. Structure and motion estimation of a moving object using a moving camera. In *Proceedings of American Control Conference*, pp. 6962–6967, Baltimore, MD, 30 June–2 July 2010.

[50] A. P. Dani, Z. Kan, N. R. Fischer, and W. E. Dixon. Structure estimation of a moving object using a moving camera: An unknown input observer approach. In *Proceedings of the IEEE Conference on Decision and Control*, pp. 5005–5010, 2011.

[51] M. de Queiroz, D. M. Dawson, S. P. Nagarkatti, and F. Zhang. *Lyapunov-Based Control of Mechanical Systems*. Cambridge, MA: Birkhauser, 2000.

[52] R. DeCarlo, M. S. Branicky, S. Pettersson, and B. Lennartson. Perspectives and results on the stability and stabilizability of hybrid systems. *Proceedings of IEEE*, 88(7):1069–1082, 2000.

[53] C. A. Desoer and M. Vidyasagar. *Feedback Systems: Input–Output Properties*. New York: Academic Press, 1975.

[54] A. Diosi, S. Segvic, A. Remazeilles, and F. Chaumette. Experimental evaluation of autonomous driving based on visual memory and image-based visual servoing. *IEEE Transactions on Intelligent Transportation Systems*, 12(3):870–883, 2011.

[55] W. E. Dixon, D. M. Dawson, E. Zergeroglu, and A. Behal. *Nonlinear Control of Wheeled Mobile Robots*. New York: Springer, 2001.

[56] W. E. Dixon, M. S. de Queiroz, D. M. Dawson, and T. J. Flynn. Adaptive tracking and regulation of a wheeled mobile robot with controller/update law modularity. *IEEE Transactions on Control Systems Technology*, 12(1):138–147, 2004.

[57] W. E. Dixon, Y. Fang, D. M. Dawson, and T. J. Flynn. Range identification for perspective vision systems. *IEEE Transactions on Automatic Control*, 48(12):2232–2238, 2003.

[58] K. D. Do, Z. P. Jiang, and J. Pan. A global output-feedback controller for simultaneous tracking and stabilization of unicycle-type mobile robots. *IEEE Transactions on Robotics and Automation*, 20(3):589–594, 2004.

[59] K. D. Do, Z. P. Jiang, and J. Pan. Simultaneous tracking and stabilization of mobile robots: An adaptive approach. *IEEE Transactions on Automatic Control*, 49(7):1147–1151, 2004.

[60] J. Douret and R. Benosman. A multi-cameras 3D volumetric method for outdoor scenes: A road traffic monitoring application. In *Proceedings of IEEE International Conference on Pattern Recognition*, pp. 334–337, Cambridge, UK, 26 August 2004.

[61] X. Du and K. K. Tan. Vision-based approach towards lane line detection and vehicle localization. *Machine Vision and Applications*, 27(2):175–191, 2015.

[62] H. Durrant-Whyte and T. Bailey. Simultaneous localization and mapping: Part I. *IEEE Robotics and Automation Magazine*, 13(2):99–110, 2006.

[63] Y. Fang, A. Behal, W. E. Dixon, and D. M. Dawson. Adaptive 2.5d visual servoing of kinematically redundant robot manipulators. In *Proceedings of the 41st IEEE Conference on Decision and Control*, pp. 2860–2865, Las Vegas, NV, 10–13 December 2002.

[64] Y. Fang, W. E. Dixon, D. M. Dawson, and P. Chawda. Homography-based visual servo regulation of mobile robots. *IEEE Transactions on System, Man, and Cybernetics: Part B (Cybernetics)*, 35(5):1041–1050, 2005.

[65] Y. Fang, X. Liu, and X. Zhang. Adaptive active visual servoing of non-holonomic mobile robots. *IEEE Transactions on Industrial Electronics*, 59(1):486–497, 2012.

[66] O. Faugeras. *Three-Dimensional Computer Vision*. Cambridge, MA: MIT Press, 2001.

[67] O. D. Faugeras and F. Lustman. Motion and structure from motion in a piecewise planar environment. *International Journal of Pattern Recognition and Artificial Intelligence*, 2(3):485–508, 1988.

[68] C. A. Felippa. A systematic approach to the element-independent corotational dynamics of finite elements. Technical report, Technical Report CU-CAS-00-03, Center for Aerospace Structures, 2000.

[69] A. F. Filippov. *Differential Equations with Discontinuous Righthand Sides*. Norwell, MA: Kluwer, 1988.

[70] N. Fischer, D. Hughes, P. Walters, E. M. Schwartz, and W. E. Dixon. Non-linear RISE-based control of an autonomous underwater vehicle. *IEEE Transactions on Robotics*, 30(4):845–852, 2014.

[71] N. Fischer, R. Kamalapurkar, and W. E. Dixon. Lasalle-Yoshizawa corollaries for nonsmooth systems. *IEEE Transactions on Automatic Control*, 58(9):2333–2338, 2013.

[72] M. A. Fischler and R. C. Bolles. Random sample consensus: A paradigm for model fitting with applications to image analysis and automated cartography. *Communications of the ACM*, 24(6):381–395, 1981.

[73] S. A. Flórez, V. Frémont, P. Bonnifait, and V. Cherfaoui. Multi-modal object detection and localization for high integrity driving assistance. *Machine Vision and Applications*, 25(3):583–598, 2011.

[74] C. S. Fraser. Photogrammetric camera component calibration: A review of analytical techniques. In: Gruen A., Huang T. S. (eds) *Calibration and Orientation of Cameras in Computer Vision*, pp. 95–121. Berlin, Heidelberg: Springer, 2001.

[75] J. Fritsch, T. Kuhnl, and A. Geiger. A new performance measure and evaluation benchmark for road detection algorithms. In *Proceedings of IEEE International Conference on Intelligent Transportation Systems*, pp. 1693–1700. Hague, the Netherlands, 6–9 October 2013.

[76] N. R. Gans, G. Hu, and J. Shen. Adaptive visual servo control to simultaneously stabilize image and pose error. *Mechatronics*, 22(4):410–422, 2012.

[77] N. R. Gans, S. Hutchinson, and P. I. Corke. Performance tests for visual servo control systems, with application to partitioned approaches to visual servo control. *International Journal of Robotics Research*, 22(10–11):955–981, 2003.

[78] N. Garcia-Aracil, E. Malis, R. Aracil-Santonja, and C. Perez-Vidal. Continuous visual servoing despite the changes of visibility in image features. *IEEE Transactions on Robotics*, 21(6):1214–1220, 2005.

[79] A. Geiger, P. Lenz, C. Stiller, and R. Urtasun. Vision meets robotics: The KITTI dataset. *International Journal of Robotics Research*, 32(11):1231–1237, 2013.

[80] A. Geiger, P. Lenz, and R. Urtasun. Are we ready for autonomous driving? The KITTI vision benchmark suite. In *Proceedings of IEEE Conference on Computer Vision and Pattern Recognition*, pp. 3354–3361. IEEE, Providence, RI, 16–21 June 2012.

[81] A. Gil, O. M. Mozos, M. Ballesta, and O. Reinoso. A comparative evaluation of interest point detectors and local descriptors for visual SLAM. *Machine Vision and Applications*, 21(6):905–920, 2010.

[82] I. Grave and Y. Tang. A new observer for perspective vision systems under noisy measurements. *IEEE Transactions on Automatic Control*, 60(2):503–508, 2015.

[83] M. D. Grossberg and S. K. Nayar. Determining the camera response from images: What is knowable? *IEEE Transactions on Pattern Analysis and Machine Intelligence*, 25(11):1455–1467, 2003.

[84] R. Guo, Q. Dai, and D. Hoiem. Paired regions for shadow detection and removal. *IEEE Transactions on Pattern Analysis and Machine Intelligence*, 35(12):2956–2967, 2013.

[85] M. Han and T. Kanade. Reconstruction of a scene with multiple linearly moving objects. *International Journal of Computer Vision*, 59(3):285–300, 2004.

[86] R. Hartley and A. Zisserman. *Multiple View Geometry in Computer Vision*. Cambridge University Press, 2003.

[87] Y. He, H. Wang, and B. Zhang. Color-based road detection in urban traffic scenes. *IEEE Transactions on Intelligent Transportation Systems*, 5(4):309–318, 2004.

[88] J. Heikkila and O. Silven. A four-step camera calibration procedure with implicit image correction. In *Proceedings of IEEE Computer Society Conference on Computer Vision and Pattern Recognition*, pp. 1106–1112, San Juan, PR, 17–19 June 1997.

[89] E. E. Hemayed. A survey of camera self-calibration. In *Proceedings of the IEEE Conference on Advanced Video and Signal Based Surveillance, 2003*, pp. 351–357, Miami, FL, 22–22 July 2003.

[90] Y. S. Heo, K. M. Lee, and S. U. Lee. Robust stereo matching using adaptive normalized cross-correlation. *IEEE Transactions on Pattern Analysis and Machine Intelligence*, 33(4):807–822, 2011.

[91] Y. S. Heo, K. M. Lee, and S. U. Lee. Joint depth map and color consistency estimation for stereo images with different illuminations and cameras. *IEEE Transactions on Pattern Analysis and Machine Intelligence*, 35(5):1094–1106, 2013.

[92] A. B. Hillel, R. Lerner, D. Levi, and G. Raz. Recent progress in road and lane detection: A survey. *Machine Vision and Applications*, 25(3):727–745, 2012.

[93] H. Hirschmüller. Stereo processing by semiglobal matching and mutual information. *IEEE Transactions on Pattern Analysis and Machine Intelligence*, 30(2):328–341, 2008.

[94] R. Horaud. New methods for matching 3-D objects with single perspective views. *IEEE Transactions on Pattern Analysis and Machine Intelligence (PAMI)* 9(3):401–412, 1987.

[95] R. A. Horn and C. R. Johnson. *Matrix Analysis*. Cambridge University Press, 2012.

[96] G. Hu, N. Gans, and W. Dixon. Quaternion-based visual servo control in the presence of camera calibration error. *International Journal of Robust and Nonlinear Control*, 20(5):489–503, 2010.

[97] C. Hua, Y. Makihara, Y. Yagi, S. Iwasaki, K. Miyagawa, and B. Li. Onboard monocular pedestrian detection by combining spatio-temporal hog with structure from motion algorithm. *Machine Vision and Applications*, 26(2):161–183, 2015.

[98] J. Huang, C. Wen, W. Wang, and Z. P. Jiang. Adaptive stabilization and tracking control of a nonholonomic mobile robot with input saturation and disturbance. *Systems & Control Letters*, 62(3):234–241, 2013.

[99] P. J. Huber. Robust statistics. In: Lovric M. (ed) *International Encyclopedia of Statistical Science*, pp. 1248–1251. Berlin, Heidelberg: Springer, 2011.

[100] A. Ibarguren, J. M. Martínez-Otzeta, and I. Maurtua. Particle filtering for industrial 6DOF visual servoing. *Journal of Intelligent & Robotic Systems*, 74(3–4):689, 2014.

[101] Y. Iwatani. Task selection for control of active-vision systems. *IEEE Transactions on Robotics*, 26(4):720–725, 2010.

[102] M. Jagersand, O. Fuentes, and R. Nelson. Experimental evaluation of uncalibrated visual servoing for precision manipulation. In *Proceedings of IEEE International Conference on Robotics and Automation*, pp. 2874–2880. Albuquerque, NM, 25–25 April 1997.

[103] F. Janabi-Sharifi and M. Marey. A Kalman-filter-based method for pose estimation in visual servoing. *IEEE Transactions on Robotics*, 26(5):939–947, 2010.

[104] B. Jia, J. Chen, and K. Zhang. Adaptive visual trajectory tracking of nonholonomic mobile robots based on trifocal tensor. In *Proceedings of IEEE/RSJ International Conference on Intelligent Robots and Systems*, pp. 3695–3700, Hamburg, Germany, 28 September–2 October 2015.

[105] B. Jia and S. Liu. Homography-based visual predictive control of tracked mobile robot with field-of-view constraints. *International Journal of Robotics and Automation*, 30(5), 2015.

[106] B. Jia, S. Liu, and Y. Liu. Visual trajectory tracking of industrial manipulator with iterative learning control. *Industrial Robot: An International Journal*, 42(1):54–63, 2015.

[107] B. Jia, S. Liu, K. Zhang, and J. Chen. Survey on robot visual servo control: Vision system and control strategies. *Acta Automatica Sinica*, 41(5):861–873, 2015.

[108] P. Jiang, L. C. Bamforth, Z. Feng, J. E. Baruch, and Y. Chen. Indirect iterative learning control for a discrete visual servo without a camera-robot model. *IEEE Transactions on Systems, Man, and Cybernetics, Part B (Cybernetics)*, 37(4):863–876, 2007.

[109] P. Jiang and R. Unbehauen. Robot visual servoing with iterative learning control. *IEEE Transactions on Systems, Man, and Cybernetics, Part A: Systems and Humans*, 32(2):281–287, 2002.

[110] Z. P. Jiang and H. Nijmeijer. Tracking control of mobile robots: A case study in backstepping. *Automatica*, 33(7):1393–1399, 1997.

[111] V. Kallem, M. Dewan, J. P. Swensen, G. D. Hager, and N. J. Cowan. Kernel-based visual servoing. In *IEEE/RSJ International Conference on Intelligent Robots and Systems*, pp. 1975–1980, 2007.

[112] D. Karagiannis and A. Astolfi. A new solution to the problem of range identification in perspective vision systems. *IEEE Transactions on Automatic Control*, 50(12):2074–2077, 2005.

[113] G. C. Karras, S. G. Loizou, and K. J. Kyriakopoulos. Towards semi-autonomous operation of under-actuated underwater vehicles: Sensor fusion, on-line identification and visual servo control. *Autonomous Robots*, 31(1):67–86, 2011.

[114] M. Kellner, U. Hofmann, M. E. Bouzouraa, and S. Neumaier. Multi-cue, model-based detection and mapping of road curb features using stereo vision. In *Proceedings of the IEEE International Conference on Intelligent Transportation Systems*, pp. 1221–1228, Las Palmas, Spain, 15–18 September 2015.

[115] H. K. Khalil. *Nonlinear Systems*. Upper Saddle River, NJ: Prentice-Hall, 2002.

[116] C. S. Kim, E. J. Mo, S. M. Han, M. S. Jie, and K. W. Lee. Robust visual servo control of robot manipulators with uncertain dynamics and camera parameters. *International Journal of Control, Automation, and Systems*, 8(2):308, 2010.

[117] D. E. Koditschek and E. Rimon. Robot navigation functions on manifolds with boundary. *Advances in Applied Mathematics*, 11(4):412–442, 1990.

[118] H. Kong, J. Y. Audibert, and J. Ponce. General road detection from a single image. *IEEE Transactions on Image Processing*, 19(8):2211–2220, 2010.

[119] H. Kong, S. E. Sarma, and F. Tang. Generalizing Laplacian of Gaussian filters for vanishing-point detection. *IEEE Transactions on Intelligent Transportation Systems*, 14(1):408–418, 2013.

[120] M. Krstić, I. Kanellakopoulos, and P. Kokotović. *Nonlinear and Adaptive Control Design*. New York: John Wiley and Sons, 1995.

[121] R. Labayrade, D. Aubert, and J. P. Tarel. Real time obstacle detection in stereovision on non flat road geometry through "v-disparity" representation. In *Proceedings of IEEE Intelligent Vehicle Symposium*, pp. 646–651, Versailles, France, 17–21 June 2002.

[122] J. F. Lalonde, A. A. Efros, and S. G. Narasimhan. Detecting ground shadows in outdoor consumer photographs. In *Proceedings of European Conference on Computer Vision*, pp. 322–335, Heraklion, Greece, 5–11 September 2010.

[123] S. Lee, G. Wolberg, and S. Y. Shin. Scattered data interpolation with multilevel B-splines. *IEEE Transactions on Visualization and Computer Graphics*, 3(3):228–244, 1997.

[124] T.-C. Lee, K.-T. Song, C.-H. Lee, and C.-C. Teng. Tracking control of unicycle-modeled mobile robots using a saturation feedback controller. *IEEE Transactions on Control Systems Technology*, 9(2):305–318, 2001.

[125] A. Levant. Chattering analysis. *IEEE Transactions on Automatic Control*, 55(6):1380–1389, 2010.

[126] D. Levi, N. Garnett, and E. Fetaya. Stixelnet: A deep convolutional network for obstacle detection and road segmentation. In *Proceedings of the British Machine Vision Conference (BMVC)*, pp. 109.1–109.12, 2015.

[127] B. Li, Y. Fang, G. Hu, and X. Zhang. Model-free unified tracking and regulation visual servoing of wheeled mobile robots. *IEEE Transactions on Control Systems Technology*, 24(4):1328–1339, 2016.

[128] L. Li, Y. H. Liu, K. Wang, and M. Fang. Estimating position of mobile robots from omnidirectional vision using an adaptive algorithm. *IEEE Transactions on Cybernetics*, 45(8):1633–1646, 2015.

[129] X. Li and C. C. Cheah. Global task-space adaptive control of robot. *Automatica*, 49(1):58–69, 2013.

[130] X. Li and C. C. Cheah. Human-guided robotic manipulation: Theory and experiments. In *Proceedings of 2014 IEEE International Conference on Robotics and Automation (ICRA)*, pp. 4594–4599, Hong Kong, China, 31 May–7 June 2014.

[131] X. Liang, X. Huang, and M. Wang. Improved stability results for visual tracking of robotic manipulators based on the depth-independent interaction matrix. *IEEE Transactions on Robotics*, 27(2):371–379, 2011.

[132] C. H. Lin, S. Y. Jiang, Y. J. Pu, and K. T. Song. Robust ground plane detection for obstacle avoidance of mobile robots using a monocular camera. In *Proceedings of IEEE/RSJ International Conference on Intelligent Robots and Systems*, pp. 3706–3711, Taipei, Taiwan, 18–22 October 2010.

[133] C. H. Lin and K. T. Song. Robust ground plane region detection using multiple visual cues for obstacle avoidance of a mobile robot. *Robotica*, 33(2):436–450, 2015.

[134] D. Liu, X. Wu, Y. Yang, and J. Xin. An improved self-calibration approach based on enhanced mutative scale chaos optimization algorithm for position-based visual servo. *Acta Automatica Sinica*, 34(6):623–631, 2008.

[135] X. Liu, X. Xu, and B. Dai. Vision-based long-distance lane perception and front vehicle location for full autonomous vehicles on highway roads. *Journal of Central South University*, 19:1454–1465, 2012.

[136] Y. Liu, H. Wang, and C. Wang. Uncalibrated visual servoing of robots using a depth-independent interaction matrix. *IEEE Transactions on Robotics*, 22(4):804–817, 2006.

[137] Y. H. Liu, H. Wang, W. Chen, and D. Zhou. Adaptive visual servoing using common image features with unknown geometric parameters. *Automatica*, 49(8):2453–2460, 2013.

[138] F. Lizarralde, A. C. Leite, and L. Hsu. Adaptive visual servoing scheme free of image velocity measurement for uncertain robot manipulators. *Automatica*, 49(5):1304–1309, 2013.

[139] M. S. Loffler, N. P. Costescu, and D. M. Dawson. QMotor 3.0 and the QMotor robotic toolkit: A PC-based control platform. *IEEE Control Systems*, 22(3):12–26, 2002.

[140] G. López-Nicolás, N. R. Gans, S. Bhattacharya, C. Sagüés, J. J. Guerrero, and S. Hutchinson. Homography-based control scheme for mobile robots with nonholonomic and field-of-view constraints. *IEEE Transactions on Systems, Man, and Cybernetics, Part B (Cybernetics)*, 40(4):1115–1127, 2010.

[141] G. López-Nicolás, J. J. Guerrero, and C. Sagüés. Visual control of vehicles using two-view geometry. *Mechatronics*, 20(2):315–325, 2010.

[142] G. López-Nicolás, J. J. Guerrero, and C. Sagüés. Visual control through the trifocal tensor for nonholonomic robots. *Robotics and Autonomous Systems*, 58(2):216–226, 2010.

[143] G. López-Nicolás, J. J. Guerrero, and C. Sagüés. Visual control of vehicles using two-view geometry. *Mechatronics*, 20(2):315–325, 2010.

[144] G. López-Nicolás, C. Sagüés, and J. J. Guerrero. Parking with the essential matrix without short baseline degeneracies. In *Proceedings of IEEE International Conference on Robotics and Automation*, pp. 1098–1103, Kobe, Japan, 12–17 May 2009.

[145] G. López-Nicolás, C. Sagüés, J. J. Guerrero, D. Kragic, and P. Jensfelt. Switching visual control based on epipoles for mobile robots. *Robotics and Autonomous Systems*, 56(7):592–603, 2008.

[146] H. Lu, L. Jiang, and A. Zell. Long range traversable region detection based on superpixels clustering for mobile robots. In *Proceedings of the IEEE/RSJ International Conference on Intelligent Robots and Systems*, pp. 546–552, Hamburg, Germany, 28 September–2 October 2015.

[147] X. Lv and X. Huang. Fuzzy adaptive Kalman filtering based estimation of image jacobian for uncalibrated visual servoing. In *Proceedings of IEEE/RSJ International Conference on Intelligent Robots and Systems*, pp. 2167–2172. IEEE, Beijing, China, 9–15 October 2006.

[148] W. MacKunis, N. Gans, A. Parikh, and W. E. Dixon. Unified tracking and regulation visual servo control for wheeled mobile robots. *Asian Journal of Control*, 16(3):669–678, 2014.

[149] W. Maddern, A. Stewart, C. McManus, B. Upcroft, W. Churchill, and P. Newman. Illumination invariant imaging: Applications in robust vision-based localisation, mapping and classification for autonomous vehicles. In *Proceedings of the Visual Place Recognition in Changing Environments*

Workshop, IEEE International Conference on Robotics and Automation, Hong Kong, China, 31 May–07 June 2014.

[150] E. Malis and F. Chaumette. 2 1/2 D visual servoing with respect to unknown objects through a new estimation scheme of camera displacement. *International Journal of Computer Vision*, 37(1):79–97, 2000.

[151] E. Malis and F. Chaumette. Theoretical improvements in the stability analysis of a new class of model-free visual servoing methods. *IEEE Transactions on Robotics and Automation*, 18(2):176–186, 2002.

[152] E. Malis, F. Chaumette, and S. Boudet. 2 1/2 D visual servoing. *IEEE Transactions on Robotics and Automation*, 15(2):238–250, 1999.

[153] J. A. Marchant and C. M. Onyango. Shadow-invariant classification for scenes illuminated by daylight. *JOSA A*, 17(11):1952–1961, 2000.

[154] G. L. Mariottini, G. Oriolo, and D. Prattichizzo. Image-based visual servoing for nonholonomic mobile robots using epipolar geometry. *IEEE Transactions on Robotics*, 23(1):87–100, 2007.

[155] S. S. Mehta and J. W. Curtis. A geometric approach to visual servo control in the absence of reference image. In *Proceedings of 2011 IEEE International Conference on Systems, Man, and Cybernetics*, pp. 3113–3118, Anchorage, AK, 9–12 October 2011.

[156] S. S. Mehta, V. Jayaraman, T. F. Burks, and W. E. Dixon. Teach by zooming: A unified approach to visual servo control. *Mechatronics*, 22(4):436–443, 2012. Visual Servoing SI.

[157] C. C. T. Mendes, V. Frémont, and D. F. Wolf. Exploiting fully convolutional neural networks for fast road detection. In *Proceedings of the IEEE International Conference on Robotics and Automation*, pp. 3174–3179, Stockholm, Sweden, 16–21 May 2016.

[158] Y. Mezouar and F. Chaumette. Path planning for robust image-based control. *IEEE Transactions on Robotics and Automation*, 18(4):534–549, 2002.

[159] P. Miraldo, H. Araujo, and N. Gonalves. Pose estimation for general cameras using lines. *IEEE Transactions on Cybernetics*, 45(10):2156–2164, 2015.

[160] P. Moghadam, J. A. Starzyk, and W. S. Wijesoma. Fast vanishing-point detection in unstructured environments. *IEEE Transactions on Image Processing*, 21(1):497–500, 2012.

[161] F. Morbidi and D. Prattichizzo. Range estimation from a moving camera: An immersion and invariance approach. In *Proceedings of IEEE International Conference on Robotics and Automation*, pp. 2810–2815, Kobe, Japan, 12–17 May 2009.

[162] D. Munoz, J. A. Bagnell, and M. Hebert. Stacked hierarchical labeling. In *Proceedings of the European Conference on Computer Vision*, pp. 376–389, Crete, Greece, 5–11 September 2010.

[163] F. Oniga and S. Nedevschi. Processing dense stereo data using elevation maps: Road surface, traffic isle, and obstacle detection. *IEEE Transactions on Vehicular Technology*, 59(3):1172–1182, 2010.

[164] K. E. Ozden, K. Schindler, and L. Van Gool. Multibody structure-from-motion in practice. *IEEE Transactions on Pattern Analysis and Machine Intelligence*, 32(6):1134–1141, 2010.

[165] B. E. Paden and S. Sastry. A calculus for computing Filippov's differential inclusion with application to the variable structure control of robot manipulators. *IEEE Transactions on Circuits and Systems*, 34(1):73–82, 1987.

[166] L. Pari, J. Sebastián, A. Traslosheros, and L. Angel. A comparative study between analytic and estimated image Jacobian by using a stereoscopic system of cameras. In *Proceedings of IEEE/RSJ International Conference on Intelligent Robots and Systems*, pp. 6208–6215. IEEE, Taipei, Taiwan, 18–22 October 2010.

[167] F. Pasteau, A. Krupa, and M. Babel. Vision-based assistance for wheelchair navigation along corridors. In *Proceedings of 2014 IEEE International Conference on Robotics and Automation (ICRA)*, pp. 4430–4435, Hong Kong, China, 31 May–7 June 2014.

[168] P. M. Patre, W. MacKunis, C. Makkar, and W. E. Dixon. Asymptotic tracking for systems with structured and unstructured uncertainties. *IEEE Transactions on Control Systems Technology*, 2(16):373–379, 2008.

[169] J. A. Piepmeier, G. V. McMurray, and H. Lipkin. Uncalibrated dynamic visual servoing. *IEEE Transactions on Robotics and Automation*, 20(1):143–147, 2004.

[170] S. Qu and C. Meng. Statistical classification based fast drivable region detection for indoor mobile robot. *International Journal of Humanoid Robotics*, 11(1), 2014.

[171] M. S. De Queiroz, D. M. Dawson, S. P. Nagarkatti, and F. Zhang. *Lyapunov-Based Control of Mechanical Systems*. New York: Springer Science & Business Media, 2012.

[172] S. Ramalingam and P. Sturm. A unifying model for camera calibration. *IEEE Transactions on Pattern Analysis and Machine Intelligence*, 39:1309–1319, 2017.

[173] F. Remondino and C. Fraser. Digital camera calibration methods: Considerations and comparisons. *International Archives of Photogrammetry, Remote Sensing and Spatial Information Sciences*, 36(5):266–272, 2006.

[174] E. Rimon and D. E. Koditschek. Exact robot navigation using artificial potential functions. *IEEE Transactions on Robotics and Automation*, 8(5):501–518, 1992.

[175] E. Rohmer, S. P. Singh, and M. Freese. V-REP: A versatile and scalable robot simulation framework. In *Proceedings of IEEE/RSJ International Conference on Intelligent Robots and Systems*, pp. 1321–1326. Tokyo, Japan, 3–7 November 2013.

[176] G. Ros and J. M. Alvarez. Unsupervised image transformation for outdoor semantic labelling. In *Proceedings of IEEE Intelligent Vehicles Symposium*, pp. 537–542, Seoul, South Korea, 28 June–1 July 2015.

[177] D. Sabatta and R. Siegwart. Vision-based path following using the 1D trifocal tensor. In *Proceedings of IEEE International Conference on Robotics and Automation*, pp. 3095–3102. Karlsruhe, Germany, 6–10 May 2013.

[178] P. Salaris, A. Cristofaro, L. Pallottino, and A. Bicchi. Epsilon-optimal synthesis for vehicles with vertically bounded field-of-view. *IEEE Transactions on Automatic Control*, 60(5):1204–1218, 2015.

[179] P. Salaris, D. Fontanelli, L. Pallottino, and A. Bicchi. Shortest paths for a robot with nonholonomic and field-of-view constraints. *IEEE Transactions on Robotics*, 26(2):269–281, 2010.

[180] S. Sastry and M. Bodson. *Adaptive Control: Stability, Convergence and Robustness*. Upper Saddle River, NJ: Prentice-Hall, 1989.

[181] D. Scaramuzza and F. Fraundorfer. Visual odometry [Tutorial]. *IEEE Robotics and Automation Magazine*, 18(4):80–92, 2011.

[182] S. Šegvić, K. Brkić, Z. Kalafatić, and A. Pinz. Exploiting temporal and spatial constraints in traffic sign detection from a moving vehicle. *Machine Vision and Applications*, 25(3):649–665, 2011.

[183] J. Shen, X. Yang, Y. Jia, and X. Li. Intrinsic images using optimization. In *Proceedings of IEEE Conference on Computer Vision and Pattern Recognition*, pp. 3481–3487. Colorado Springs, CO, 20–25 June 2011.

[184] B. Shin, Z. Xu, and R. Klette. Visual lane analysis and higher-order tasks: A concise review. *Machine Vision and Applications*, 25(6):1519–1547, 2014.

[185] J. Shin, H. J. Kim, Y. Kim, and W. E. Dixon. Autonomous flight of the rotorcraft-based UAV using RISE feedback and NN feedforward terms. *IEEE Transactions on Control Systems Technology*, 20(5):1392–1399, 2012.

[186] P. Y. Shinzato, D. Gomes, and D. F. Wolf. Road estimation with sparse 3D points from stereo data. In *Proceedings of IEEE International Conference on Intelligent Transportation Systems*, pp. 1688–1693. Qingdao, China, 8–11 October 2014.

[187] P. Y. Shinzato, V. Grassi, F. S. Osorio, and D. F. Wolf. Fast visual road recognition and horizon detection using multiple artificial neural networks. In *Proceedings of the IEEE Intelligent Vehicles Symposium*, pp. 1090–1095, Madrid, Spain, 3–7 June 2012.

[188] G. Silveira. On intensity-based nonmetric visual servoing. *IEEE Transactions on Robotics*, 30(4):1019–1026, 2014.

[189] G. Silveira and E. Malis. Direct visual servoing: Vision-based estimation and control using only nonmetric information. *IEEE Transactions on Robotics*, 28(4):974–980, 2012.

[190] I. Siradjuddin, L. Behera, T. M. McGinnity, and S. Coleman. Image-based visual servoing of a 7-DOF robot manipulator using an adaptive distributed fuzzy PD controller. *IEEE/ASME Transactions on Mechatronics*, 19(2):512–523, 2014.

[191] S. Sivaraman and M. M. Trivedi. Active learning for on-road vehicle detection: A comparative study. *Machine Vision and Applications*, 25(3):599–611, 2011.

[192] J. J. E. Slotine and W. Li. *Applied Nonlinear Control*. Englewood Cliffs, NJ: Prentice Hall, 1991.

[193] P. Soille. *Morphological Image Analysis: Principles and Applications*. Secaucus, NY: Springer, 2 edition, 2003.

[194] K. T. Song, C. H. Chang, and C. H. Lin. Robust feature extraction and control design for autonomous grasping and mobile manipulation. In *Proceedings of International Conference on System Science and Engineering*, pp. 445–450, Taipei, Taiwan, 1–3 July 2010.

[195] R. Spica, P. R. Giordano, and F. Chaumette. Coupling image-based visual servoing with active structure from motion. In *Proceedings of IEEE International Conference on Robotics and Automation*. Hong Kong, China, 31 May–7 June 2014.

[196] R. Spica and P. Robuffo Giordano. A framework for active estimation: Application to structure from motion. In *Proceedings of the IEEE Conference on Decision and Control*, pp. 7647–7653, Florence, Italy, 10–13 December 2013.

[197] M. W. Spong and M. Vidyasagar. *Robot Dynamics and Control*. New York: John Wiley & Sons, 2008.

[198] C. Tan, T. Hong, T. Chang, and M. Shneier. Color model-based real-time learning for road following. In *Proceedings of IEEE Conference on Intelligent Transportation Systems*, pp. 939–944. Toronto, Canada, 17–20 September 2006.

[199] S. Tarbouriech and P. Souères. Image-based visual servo control design with multi-constraint satisfaction. In: Chesi G., Hashimoto K. (eds) *Visual Servoing via Advanced Numerical Methods*, Lecture Notes in Control and Information Sciences, volume 401, pp. 275–294, London: Springer, 2010.

[200] J. Tian, X. Qi, L. Qu, and Y. Tang. New spectrum ratio properties and features for shadow detection. *Pattern Recognition*, 51:85–96, 2015.

[201] D. Tick, A. C. Satici, J. Shen, and N. Gans. Tracking control of mobile robots localized via chained fusion of discrete and continuous epipolar geometry, IMU and odometry. *IEEE Transactions on Cybernetics*, 43(4):1237–1250, 2013.

[202] M. A. Treiber. *Optimization for Computer Vision*. London: Springer, 2013.

[203] C. Y. Tsai and K. T. Song. Visual tracking control of a wheeled mobile robot with system model and velocity quantization robustness. *IEEE Transactions on Control Systems Technology*, 17(3):520–527, 2009.

[204] C. Y. Tsai, K. T. Song, X. Dutoit, H. Van Brussel, and M. Nuttin. Robust visual tracking control system of a mobile robot based on a dual-Jacobian visual interaction model. *Robotics and Autonomous Systems*, 57(6): 652–664, 2009.

[205] R. Tsai. A versatile camera calibration technique for high-accuracy 3D machine vision metrology using off-the-shelf TV cameras and lenses. *IEEE Journal on Robotics and Automation*, 3(4):323–344, 1987.

[206] Y. W. Tu and M. T. Ho. Design and implementation of robust visual servoing control of an inverted pendulum with an FPGA-based image co-processor. *Mechatronics*, 21(7):1170–1182, 2011.

[207] F. Vasconcelos, J. P. Barreto, and E. Boyer. Automatic camera calibration using multiple sets of pairwise correspondences. *IEEE Transactions on Pattern Analysis and Machine Intelligence*, 40(4): 791–803, 2017.

[208] R. Vidal and R. Hartley. Three-view multibody structure from motion. *IEEE Transactions on Pattern Analysis and Machine Intelligence*, 30(2):214–227, 2008.

[209] H. Wang. Adaptive visual tracking for robotic systems without visual velocity measurement. *Automatica*, 55:294–301, 2015.

[210] H. Wang, M. Jiang, and W. Chen. Visual servoing of robots with uncalibrated robot and camera parameters. *Mechatronics*, 22(6):661–668, 2012.

[211] H. Wang and M. Liu. Design of robotic visual servo control based on neural network and genetic algorithm. *International Journal of Automation and Computing*, 9(1):24–29, 2012.

[212] H. Wang, Y. H. Liu, and W. Chen. Uncalibrated visual tracking control without visual velocity. *IEEE Transactions on Control Systems Technology*, 18(6):1359–1370, 2010.

[213] K. Wang, Y. Liu, and L. Li. Visual servoing trajectory tracking of nonholonomic mobile robots without direct position measurement. *IEEE Transactions on Robotics*, 30(4):1026–1035, 2014.

[214] Y. Wang, Z. Miao, H. Zhong, and Q. Pan. Simultaneous stabilization and tracking of nonholonomic mobile robots: A Lyapunov-based approach. *IEEE Transactions on Control Systems Technology*, 23(4):1440–1450, 2015.

[215] M. Wu, S. Lam, and T. Srikanthan. Nonparametric technique based high-speed road surface detection. *IEEE Transactions on Intelligent Transportation Systems*, 16(2):874–884, 2015.

[216] P. Wu, C. Chang, and C. Lin. Lane-mark extraction for automobiles under complex conditions. *Pattern Recognition*, 47(8):2756–2767, 2014.

[217] B. Xian, D. M. Dawson, M. S. Queiroz, and J. Chen. A continuous asymptotic tracking control strategy for uncertain nonlinear systems. *IEEE Transactions on Automatic Control*, 49(7):1206–1211, 2004.

[218] B. Xian and Y. Zhang. A new smooth robust control design for uncertain nonlinear systems with non-vanishing disturbances. *International Journal of Control*, 89(6):1285–1302, 2016.

[219] L. Tian Y. Li, Z. Mao. Visual servoing of 4DOF using image moments and neural network. *Control Theory and Applications*, 26(9):1162–1166, 2009.

[220] Q. Yang, S. Jagannathan, and Y. Sun. Robust integral of neural network and error sign control of MIMO nonlinear systems. *IEEE Transactions on Neural Networks and Learning Systems*, 26(12):3278–3286, 2015.

[221] Q. Yang, K. Tan, and N. Ahuja. Real-time O(1) bilateral filtering. In *Proceedings of IEEE Conference on Computer Vision and Pattern Recognition*, pp. 557–564, Miami, FL, 20–25 June 2009.

[222] E. Zergeroglu, D. M. Dawson, M. S. de Queiroz, and P. Setlur. Robust visual-servo control of robot manipulators in the presence of uncertainty. *Journal of Field Robotics*, 20(2):93–106, 2003.

[223] X. Zhang, Y. Fang, and N. Sun. Visual servoing of mobile robots for posture stabilization: From theory to experiments. *International Journal of Robust and Nonlinear Control*, 25(1):1–15, 2015.

[224] Z. Zhang. A flexible new technique for camera calibration. *IEEE Transactions on Pattern Analysis and Machine Intelligence*, 22(11):1330–1334, 2000.

[225] Z. Zhang and A. R. Hanson. Scaled euclidean 3D reconstruction based on externally uncalibrated cameras. In *Proceedings of International Symposium on Computer Vision*, pp. 37–42. Coral Gables, FL, 21–23 November 1995.

[226] Z. Zhang and A. R. Hanson. 3D reconstruction based on homography mapping. In *ARPA Image Understanding Workshop*, pp. 249–6399, 1996.

Index